Functional Chitosan-Based Composites

Functional Chitosan-Based Composites

Editors

Luminita Marin
Maria Bardosova

MDPI • Basel • Beijing • Wuhan • Barcelona • Belgrade • Manchester • Tokyo • Cluj • Tianjin

Editors
Luminita Marin
"Petru Poni" Institute of
Macromolecular Chemistry
Iasi
Romania

Maria Bardosova
Institute of Informatics SAS
Bratislava
Slovakia

Editorial Office
MDPI
St. Alban-Anlage 66
4052 Basel, Switzerland

This is a reprint of articles from the Special Issue published online in the open access journal *Polymers* (ISSN 2073-4360) (available at: www.mdpi.com/journal/polymers/special_issues/Func_Chitosan_Compos).

For citation purposes, cite each article independently as indicated on the article page online and as indicated below:

LastName, A.A.; LastName, B.B.; LastName, C.C. Article Title. *Journal Name* **Year**, *Volume Number*, Page Range.

ISBN 978-3-0365-4650-6 (Hbk)
ISBN 978-3-0365-4649-0 (PDF)

© 2022 by the authors. Articles in this book are Open Access and distributed under the Creative Commons Attribution (CC BY) license, which allows users to download, copy and build upon published articles, as long as the author and publisher are properly credited, which ensures maximum dissemination and a wider impact of our publications.

The book as a whole is distributed by MDPI under the terms and conditions of the Creative Commons license CC BY-NC-ND.

Contents

About the Editors vii

Preface to "Functional Chitosan-Based Composites" ix

Yuanbing Wu, Ania Rashidpour, María Pilar Almajano and Isidoro Metón
Chitosan-Based Drug Delivery System: Applications in Fish Biotechnology
Reprinted from: *Polymers* 2020, 12, 1177, doi:10.3390/polym12051177 1

Ioana A. Duceac, Liliana Verestiuc, Cristina D. Dimitriu, Vasilica Maier and Sergiu Coseri
Design and Preparation of New Multifunctional Hydrogels Based on Chitosan/Acrylic Polymers for Drug Delivery and Wound Dressing Applications
Reprinted from: *Polymers* 2020, 12, 1473, doi:10.3390/polym12071473 25

Shunli Chen, Min Wu, Caixia Wang, Shun Yan, Peng Lu and Shuangfei Wang
Developed Chitosan/Oregano Essential Oil Biocomposite Packaging Film Enhanced by Cellulose Nanofibril
Reprinted from: *Polymers* 2020, 12, 1780, doi:10.3390/polym12081780 45

Cha Yee Kuen, Tieo Galen, Sharida Fakurazi, Siti Sarah Othman and Mas Jaffri Masarudin
Increased Cytotoxic Efficacy of Protocatechuic Acid in A549 Human Lung Cancer Delivered via Hydrophobically Modified-Chitosan Nanoparticles As an Anticancer Modality
Reprinted from: *Polymers* 2020, 12, 1951, doi:10.3390/polym12091951 59

Evgen Prokhorov, Gabriel Luna-Bárcenas, José Martín Yáñez Limón, Alejandro Gómez Sánchez and Yuriy Kovalenko
Chitosan-ZnO Nanocomposites Assessed by Dielectric, Mechanical, and Piezoelectric Properties
Reprinted from: *Polymers* 2020, 12, 1991, doi:10.3390/polym12091991 83

Andra-Cristina Humelnicu, Petrisor Samoila, Mihai Asandulesa, Corneliu Cojocaru, Adrian Bele and Adriana T. Marinoiu et al.
Chitosan-Sulfated Titania Composite Membranes with Potential Applications in Fuel Cell: Influence of Cross-Linker Nature
Reprinted from: *Polymers* 2020, 12, 1125, doi:10.3390/polym12051125 97

Nurul Illya Muhamad Fauzi, Yap Wing Fen, Nur Alia Sheh Omar, Silvan Saleviter, Wan Mohd Ebtisyam Mustaqim Mohd Daniyal and Hazwani Suhaila Hashim et al.
Nanostructured Chitosan/Maghemite Composites Thin Film for Potential Optical Detection of Mercury Ion by Surface Plasmon Resonance Investigation
Reprinted from: *Polymers* 2020, 12, 1497, doi:10.3390/polym12071497 115

Mengjie Wang, Yonggang Shangguan and Qiang Zheng
Dynamics and Rheological Behavior of Chitosan-Grafted-Polyacrylamide in Aqueous Solution upon Heating
Reprinted from: *Polymers* 2020, 12, 916, doi:10.3390/polym12040916 129

Nareekan Chaiwong, Pimporn Leelapornpisid, Kittisak Jantanasakulwong, Pornchai Rachtanapun, Phisit Seesuriyachan and Vinyoo Sakdatorn et al.
Antioxidant and Moisturizing Properties of Carboxymethyl Chitosan with Different Molecular Weights
Reprinted from: *Polymers* 2020, 12, 1445, doi:10.3390/polym12071445 143

Manuela Maria Iftime, Stefan Andrei Irimiciuc, Maricel Agop, Marian Angheloiu, Lacramioara Ochiuz and Decebal Vasincu
A Theoretical Multifractal Model for Assessing Urea Release from Chitosan Based Formulations
Reprinted from: *Polymers* **2020**, *12*, 1264, doi:10.3390/polym12061264 . . . . **157**

Iuliana Spiridon, Narcis Anghel, Maria Valentina Dinu, Stelian Vlad, Adrian Bele and Bianca Iulia Ciubotaru et al.
Development and Performance of Bioactive Compounds-Loaded Cellulose/Collagen/ Polyurethane Materials
Reprinted from: *Polymers* **2020**, *12*, 1191, doi:10.3390/polym12051191 . . . . **171**

About the Editors

Luminita Marin

Luminita Marin is a Senior Researcher at the "Petru Poni"Institute of Macromolecular Chemistry Iasi, Romania. She defended her PhD thesis in 2007 in the field of liquid crystals based on imine bonds. In 2006, she performed a doctoral stage at Instituto per lo Studio delle Macromolecole, Milan, Italy, in the field of highly conjugated compounds for optoelectronics, and two postdoctoral stages at Institute Europeen des Membranes, Montpellier, France in the field of dynamic materials based on the imine linkage. Currently, she is the Leader of the "Polycondensation and Thermostable Polymers"Department, and the team leader of a young research group with a center of interest on the development of imino-chitosan biomaterials for contemporary applications. Scientometric profile: 102 ISI papers, 1889 citations in Web of Science, Hirsh index 27, ID Researcher: F-7588-2011; orcid.org/0000-0003-3987-4912.

Maria Bardosova

Maria Bardosova obtained her first degree in Physics from the Odessa State University (1976) and PhD from the Slovak Academy of Sciences (1983). She is also an alumna of the University of Manchester (2001), where she studied Science and Technology Policy. She received research fellowships from the Royal Society, the Canon Foundation, JSPS and Science Foundation Ireland. She carried out research at different institutions worldwide, including the Institute of Electrical Engineering SAS, Bratislava, the University of Manchester, the National Institute of Materials Chemistry, Tsukuba and the Institute of Physics, University of Sao Carlos. Her research interests include nanomaterials and nanoarchitectures generally, and in particular thin organic films, colloids and photonic crystals, and smart materials for applications in medicine. She is an expert in the Langmuir-Blodgett technique. Her other activities include studies of complex networks as a tool for analyzing future trends in research. For over 15 years, she worked as a senior research fellow and a staff researcher at the Tyndall National Institute, UCC Cork. She coordinated two Framework 7 EU research projects, which were completed successfully. Currently, she is with the Institute of Informatics SAS, Bratislava, and concentrates her efforts onto a Horizon 2020 project "Smart Wound monitoring Restorative Dressings"(acronym SWORD).

Preface to "Functional Chitosan-Based Composites"

Chitosan is a nature-originated biopolymer prepared from chitin, the dominant renewable polysaccharide found in the marine environment and the second most abundant on Earth, after cellulose. Its history starts in 1859, when boiling chitin in concentrated KOH solution under reflux, Charles Roguet obtained a product soluble in dilute solutions of organic acids. He was thus the first to describe the deacetylation of chitin, an important step forward that opened new horizons for its use in applications [1]. In 1894, Hoppe-Seyler named this modified chitin "chitosan", and in 1971, chitosan was produced industrially for the first time in Japan by Kyowa Yushi Co., Ltd. Nowadays, chitin, chitosan, and their numerous derivatives are used industrially in more than 2000 applications. Presently, over 10000 papers/year have been published (WoS data) since 2019, the number steadily growing by an average of 10.6% over the last 10 years. The great expansion of studies focused on chitosan is related to the large variety of material types that can be prepared based on it, such as hydrogels, fibers, nanostructures, films and coatings, which can then be further applied in diverse fields such as medicine, food packaging, environmental protection, cosmetics, agriculture, textiles, the paper industry and so on. The aim of this book is to present some of the latest developments in the field of chitosan biomaterials and their potential applications in, but not limited to, the areas mentioned above. The individual chapters of this reprint have been previously published in a Special Issue of the MDPI journal, Polymers.

The first chapter, co-authored by Yuanbing Wu et al., comprises a review focusing on all aspects of the use of chitosan for the fish farming industry, including drug delivery, fish immunization and chitosan-mediated gene delivery into fish. The authors of Chapter 2, Ioana A. Duceac et al., investigated the properties of chitosan-based hydrogels conjugated with arginine using FTIR spectroscopy, elemental analysis and SEM. The materials are superabsorbent and their tunable properties make them promising candidates for wound dressings and drug delivery applications. In Chapter 3, the authors (Shunli Chen et al.) discuss a novel biocomposite packaging film featuring antibacterial activity combined with good mechanical and barrier properties. The material consists of a chitosan matrix, cellulose nanofibrils as a reinforcing filler and oregano essential oil as an antibacterial agent. Chapter 4 (Cha Yee Kuen et al.) describes a modified chitosan nanoparticle system for the encapsulation of natural phenolic compounds to be used for the delivery of cancer therapeutics. Prokhorov et al. describe the dielectric, conductivity, mechanical, and piezoelectric properties of chitosan-ZnO nanocomposites for applications in flexible electronics, and in biomedicine as biocompatible sensors, actuators, and nanogenerators (Chapter 5). The authors of Chapter 6 (Andra-Cristina Humelnicu et al.) studied biopolymer-based membranes with the aim of developing cheap and environmentally friendly polymer electrolyte membranes for fuel cells. To this end, novel chitosan-based composite materials were prepared, their properties analyses and three different crosslinkers were studied. Nurul Illya Muhamad Fauzi et al. in Chapter 7 investigated the properties of chitosan/Fe_2O_3 films to determine their suitability as active layers for the Surface Plasmon Resonance technique. The aim of the study was to develop a system capable of detecting concentrations of mercury ions Hg^{2+} as low as 0.01 ppm in solutions of different concentrations. Chapter 8 (Mengjie Wang et al.) discusses the thermo-thickening behavior of a chitosan derivative, chitosan-grafted-polyacrylamide. The authors argue that the formation of larger aggregates upon heating is responsible for the high viscosity of the composite and postulate that such materials have potential applications in oil recovery. Authors Nareekan Chaiwong et al. (Chapter 9) synthesized and studied the properties of carboxymethyl chitosan prepared from chitosans having different molecular weights, concentrating, in particular, on the antioxidant and moisturising properties. In Chapter 10, the authors (Manuela Maria Iftime et al.) studied urea release from a series of soil conditioner systems containing chitosan, salicylaldehyde and different amounts of urea. The empirical in vitro urea release data were then used in a theoretical multifractal model that was found to be in good agreement with the release profile after calibration. It is proposed that this methodology can be adapted to also describe drug release mechanisms. Chapter 11 is a feature article co-authored by Iuliana Spiridon et al. dealing with a cellulose-based material that is a polysaccharide closely related structurally to chitosan. A matrix based on cellulose, collagen and polyurethane was prepared, into which several bioactive substances having antioxidant properties were incorporated. The biomaterials studied were characterized by FTIR and SEM, and their mechanical and biological properties were tested, confirming their potential for prospective medical and cosmetic applications.

References

[1] Crini G. Historical review on chitin and chitosan biopolymers. Environmental Chemistry Letters 2019, 17, 1623–1643. [https://doi.org/10.1007/s10311-019-00901-0]

Luminita Marin and Maria Bardosova
Editors

Review

Chitosan-Based Drug Delivery System: Applications in Fish Biotechnology

Yuanbing Wu [1,†], **Ania Rashidpour** [1,†], **María Pilar Almajano** [2] **and Isidoro Metón** [1,*]

1. Secció de Bioquímica i Biologia Molecular, Departament de Bioquímica i Fisiologia, Facultat de Farmàcia i Ciències de l'Alimentació, Universitat de Barcelona, Joan XXIII 27–31, 08028 Barcelona, Spain; wuuanbing@gmail.com (Y.W.); aniyarashidpoor2017@gmail.com (A.R.)
2. Departament d'Enginyeria Química, Universitat Politècnica de Catalunya, Diagonal 647, 08028 Barcelona, Spain; m.pilar.almajano@upc.edu
* Correspondence: imeton@ub.edu; Tel.: +34-93-4024521
† Both authors contributed equally to this work.

Received: 29 April 2020; Accepted: 19 May 2020; Published: 21 May 2020

Abstract: Chitosan is increasingly used for safe nucleic acid delivery in gene therapy studies, due to well-known properties such as bioadhesion, low toxicity, biodegradability and biocompatibility. Furthermore, chitosan derivatization can be easily performed to improve the solubility and stability of chitosan–nucleic acid polyplexes, and enhance efficient target cell drug delivery, cell uptake, intracellular endosomal escape, unpacking and nuclear import of expression plasmids. As in other fields, chitosan is a promising drug delivery vector with great potential for the fish farming industry. This review highlights state-of-the-art assays using chitosan-based methodologies for delivering nucleic acids into cells, and focuses attention on recent advances in chitosan-mediated gene delivery for fish biotechnology applications. The efficiency of chitosan for gene therapy studies in fish biotechnology is discussed in fields such as fish vaccination against bacterial and viral infection, control of gonadal development and gene overexpression and silencing for overcoming metabolic limitations, such as dependence on protein-rich diets and the low glucose tolerance of farmed fish. Finally, challenges and perspectives on the future developments of chitosan-based gene delivery in fish are also discussed.

Keywords: chitosan; gene delivery; gene overexpression; gene silencing; fish biotechnology

1. Introduction

Chitosan is a cationic polymer of β (1-4)-linked 2-amino-2-deoxy-D-glucose interspersed by residual 2-acetamido-2-deoxy-β-D-glucose, derived from chitin by deacetylation under alkaline conditions. Chitin is the second most abundant polysaccharide in nature, after cellulose, and it is obtained from the external skeleton and skin of arthropods and insects. Chitin is also found in some microorganisms, yeast and fungi. Mucoadhesion, low toxicity, biodegradability and biocompatibility, as well as antioxidant, antibacterial, antifungal, antitumor and anti-inflammatory properties led, in recent years, to the increasing use of chitosan in a wide variety of pharmaceutical, biomedical and biotechnological fields, including wound healing, tissue engineering, bone regeneration, gene therapy, food industry and agriculture [1–6].

Chitosan has many desirable biological properties that make it a highly suitable carrier to deliver nucleic acids for the development of gene therapy assays. The goal of gene therapy is to introduce exogenous genetic material into target cells, with the aim of modifying the expression of specific genes. The efficient delivery of plasmid DNA to express exogenous genes or siRNA to knockdown the expression of target genes must overcome systemic and cell barriers, depending on the target tissue and nature of the molecular mechanism triggered by the gene therapy. Ideally, for safe nucleic acid delivery, the vector must establish a stable interaction with the cargo, protect it from the action of nucleases, reach target cells, enable crossing the cell membrane and, once inside the cell, facilitate escape from endosomes and lysosomes. Decomplexation from the carrier must allow plasmid DNA to cross the nuclear membrane and become transcribed, or in the case of siRNA, render the cargo in the cytosol [7–9].

Nucleic acid delivery into cells is facilitated by viral and non-viral vectors. The choice of the vector for gene therapy is a key step to properly reach target cells, confer protection from nucleases, cross the cell membrane, nucleic acid escape from endosomal vesicles, determine transient or permanent effects, allow transcription of delivered plasmid DNA and knockdown the expression of target genes by RNA interference (RNAi) [7,10].

Due to its high transfection efficiency, viral vectors are still used in most gene therapy assays. However, immunogenicity, acute inflammation and other unwanted effects, such as reversal of the wild-type phenotype associated with the use of viral vectors, have focused attention on the development of safer alternative gene delivery systems [9,11,12]. Non-viral vectors include lipid-based vectors and cationic polymers. Low transfection efficiency in vivo, reduced half-life of lipoplex circulation, cytotoxicity and other non-desired effects, such as complement activation, limit in vivo use of cationic lipids and lipid-based vectors [10,13–16]. Unlike viral vectors, cationic polymers, such as chitosan and its derivatives, exhibit increased ability to select target tissues, easy large-scale production, low toxicity and immunogenicity in vivo and biocompatibility [4,9,10]. In this review, we will summarize recent advances in chitosan-based formulations for delivering nucleic acids, and address current progress of the use of chitosan for fish biotechnology applications and gene therapy.

2. Chitosan as a Nucleic Acid Delivery Vector

The use of chitosan as a vector for nucleic acid delivery was proposed in 1995 [17]. A few years later, in 1998, in vivo administration of chitosan complexed with plasmid DNA to express a reporter gene in the upper small intestine and colon of rabbits was published [18]. It was in 2006 when chitosan nanoparticles encapsulating small interfering RNA (siRNA) were shown to be also effective for silencing the expression of target genes [19]. Since pioneering studies, much progress has been made in this area, and chitosan is considered, at present, one of the most effective non-viral gene delivery systems. Figure 1 shows Web of Science (Clarivate Analytics) citations, with the topics chitosan, fish and gene delivery until 2019.

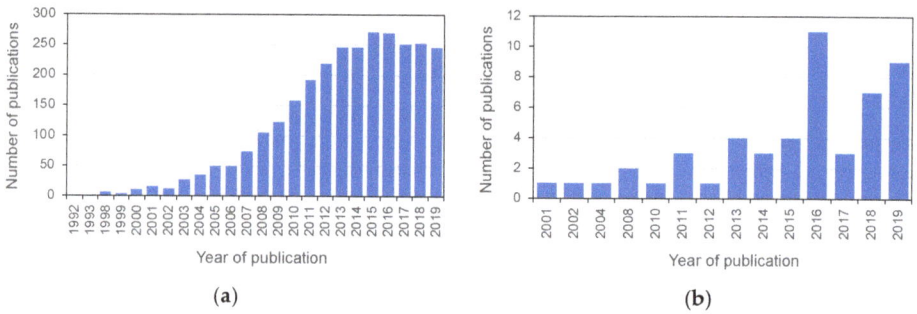

Figure 1. Web of Science (Clarivate Analytics) citations published until 2019 with the topics: (**a**) chitosan and gene therapy; (**b**) chitosan, fish and gene therapy.

The presence of numerous primary amine groups that are protonated at slightly acidic pH in chitosan allows electrostatic interaction with negatively charged nucleic acids. The stability of the complex formed between chitosan and nucleic acids allows oral, nasal, intravenous and intraperitoneal administration of chitosan–DNA complexes, and prevents dissociation before reaching the intracellular compartment [20–22]. Oral delivery would mainly result in intestinal absorption of the product [22]. Biodistribution of radioiodinated chitosan fractions with different molecular mass, intravenously injected to rats, showed rapid plasma clearance (<15% in the blood 5 min following treatment) and localization in the liver of most of the chitosan with diameter size >10 kDa (>50% at 5 min following intravenous administration and >80% at 60 min post-treatment). However, low molecular weight chitosan (<5 kDa) was cleared more slowly from the circulation and significantly less retained in the liver at the short- and long-term [20].

2.1. Chitosan Derivatization

Derivatization can greatly influence biodistribution of chitosan complexes. An illustrative example was developed by Kang et al. to down-regulate Akt2 expression for treatment of colorectal liver metastases in mice [23]. To protect siRNA from gastrointestinal degradation, facilitate active transport into enterocytes and enhance transportation to the liver through the enterohepatic circulation, the authors first obtained gold nanoparticles conjugated with thiolated siRNA (AR). The resulting complex was subsequently complexed with glycol chitosan–taurocholic acid (GT) through electrostatic interaction to generate AR-GT nanoparticles. Derivatization with taurocholic acid successfully protected Akt2-siRNA from gastrointestinal degradation and favored targeting to the liver through the enterohepatic circulation. Chitosan derivatization with hydrophilic ethylene glycol (glycol chitosan) increases solubility in water at a neutral/acidic pH. In addition, the reactive functional groups of glycol chitosan facilitate chemical modifications and formation of different derivatives useful for targeting gene delivery [24]. In addition to the properties of chitosan derivatives, the efficient delivery of the cargo greatly depends on chitosan polyplex properties, such as pH, molecular weight, deacetylation degree and N/P ratio [7,9].

The molecular weight of chitosan is a major factor affecting polyplex formation, the stability of the chitosan/DNA complex, cell entry, DNA unpacking after endosomal escape and transfection efficiency. Furthermore, the average particle size is highly dependent on the molecular weight of chitosan [7,9,25]. Chitosan between ~20–150 kDa forms chitosan–plasmid DNA complexes with diameter size of ~155–200 nm. High molecular weight chitosan >150 kDa losses solubility and favors aggregate formation, whereas chitosan of molecular weight <20 kDa tends to form polyplexes with diameter size >200 nm [26]. The optimal molecular weight range for stable chitosan–siRNA nanoparticle formation and efficient transfection and silencing effect is considered to be ~65–170 kDa [27].

Chemical modification of chitosan can greatly improve desirable properties for gene delivery. Functional groups of chitosan include C_3-OH, C_6-OH, C_2-NH_2, acetyl amino and glycoside bonds [6,28]. Two of the functional groups, C_6-OH and C_2-NH_2, have chemical properties that make them of particular interest for derivatization (Figure 2).

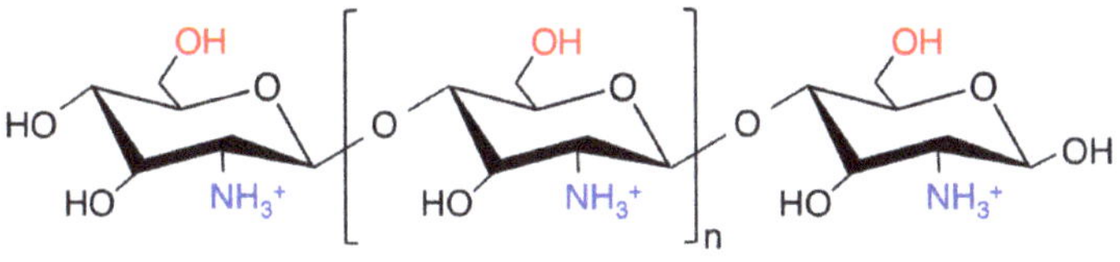

Figure 2. Schematic representation of chitosan. Functional groups C_2-NH_2 and C_6-OH and are represented in blue and red color, respectively.

2.2. Chitosan Solubility

The water solubility of chitosan is low due to the presence of highly crystalline intermolecular and intramolecular hydrogen bonds, and can be greatly influenced by the pH, molecular weight and deacetylation degree [6,9,29]. The solubility of chitosan has been improved by introducing a hydrophilic group on amino or hydroxyl groups. Examples include: N-acylated chitosan derivatives, which exhibit enhanced biocompatibility, anticoagulability, blood compatibility and sustained drug release [6,30]; chitosan conjugation with saccharides through N-alkylation, such as glycosylation [3,31,32]; and the introduction of a quaternary ammonium salt group, which increases chargeability, mucoadhesion, crossing of mucus layers and binding to epithelial surfaces [6,33,34].

2.3. Stability of Chitosan Polyplexes

To increase the stability of chitosan-based formulations, a number of chitosan derivatives have been developed. Among them, PEGylation [35–37], glycosylation [3,38,39] and quaternization [39–42]. The choice of the method for preparing chitosan–nucleic acid complexes can also significantly affect stability of the complex and transfection efficiency. Katas and Alpar showed that for efficient siRNA-mediated silencing of the expression of target genes in CHO K1 and HEK 293 cells, nanoparticles produced by ionic gelation of tripolyphosphate (TPP) with chitosan were more efficient in delivering siRNA than chitosan–siRNA complexes and siRNA adsorbed onto chitosan–TPP nanoparticles. Chitosan–TPP-siRNA nanoparticles generated by ionic gelation presented higher binding capacity and loading efficiency [19]. During ionic gelation, TPP is a polyanion that crosslinks with positively charged chitosan through electrostatic interaction, avoiding the use of toxic reagents for chemical crosslinking, and allowing for the easy modulation of size and surface charge of the nanoparticles (Figure 3). The addition of TPP was shown to reduce the particle size and increase the stability of complexes in biological fluids [19,43–47]. The inclusion of hyaluronic acid in chitosan–siRNA polyplexes can be also a promising strategy to increase stability and targeting capacity, while lowering aggregation in the presence of serum proteins [48].

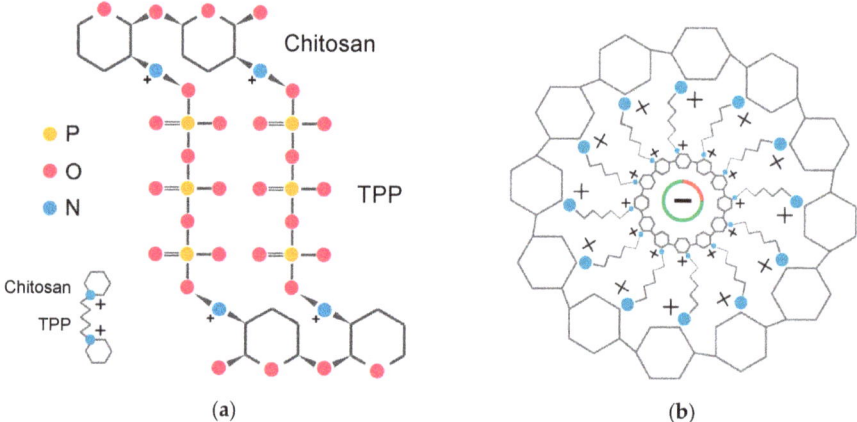

Figure 3. Molecular structure and electrostatic interactions of chitosan–tripolyphosphate (TPP) (**a**), and chitosan–TPP–plasmid DNA nanoparticles (**b**).

One major advantage of chitosan is that chitosan–DNA complexation protects DNA from DNase-mediated degradation, possibly as a result of modification of the DNA tertiary structure [20,49]. Cell penetration of chitosan-based gene delivery systems involves interaction between positively charged chitosan–nucleic acid polyplexes and negatively charged cell membrane components, such as heparan sulfate proteoglycans, enabling ATP-driven crossing of the cell membrane, or receptor-mediated endocytosis. In any case, chitosan polyplexes are internalized following the endocytic-lysosomal pathway [7].

2.4. Targeting Drug Delivery, Cellular Uptake and Intracellular Trafficking

Safe and effective therapies can be performed by using chitosan derivatives to improve target drug delivery. To this end, a variety of molecules can be conjugated to chitosan, such as proteins and peptides, polysaccharides, oligonucleotides and other molecules [4].

2.4.1. Targeting Drug Delivery with Chitosan Derivatives

A common strategy to target drug delivery is based on ligand-receptor specificity. Cell-target delivery drugs can be thus enhanced by conjugation of chitosan–nucleic acid complexes with ligands that enable binding to receptors specifically found in the target cell membrane. Examples of ligands conjugated to chitosan formulations include transferrin, galactose and mannose. For instance, transferrin can be used as a targeting ligand for delivery into tumor cells through binding to the transferrin receptor, whose expression is enhanced in tumor cells to provide iron as a necessary cofactor for DNA synthesis and rapid cell proliferation [50–52]. The presence of asialoglycoprotein receptors on the hepatocyte surface and selective binding of asialoglycoprotein receptors to galactose allow galactosylated chitosan to target hepatocytes [53,54]. Mannosylated chitosan takes advantage of mannose recognition by mannose receptors to target dendritic cells [55].

Chitosan derivatives generally achieve mucosal adhesion through hydrogen bonding or non-specific, non-covalent, electrostatic interactions. Thiolated chitosan increases mucoadhesion and enhances crossing capability trough the cell membrane and ophthalmic drug delivery [56–60]. The mucoadhesive properties of chitosan derivatives allow oral administration and nasal immunization to treat respiratory diseases [61]. Other examples include O-carboxymethyl chitosan, which can be used for intestine-targeted drug delivery [62], and acetylated low molecular weight chitosan, for targeting the kidneys [63].

2.4.2. Endosomal Escape, Unpacking and Nuclear Import of DNA

The proton sponge effect of chitosan gene delivery formulations allows endosomal escape before the maturation of early endosomes into late endosomes, and the ultimate fusion with lysosomes. The increasing acidification in early endosomes generated by the V-type ATPase proton pump results in progressive protonation of the amine groups of chitosan (pKa value of ~6.5), leading to the influx of water and chloride ions into the endosomes, increased osmotic swelling, endosome lysis and cytosolic release of the endosomal content [9,64]. The endosomal release of chitosan polyplexes can be enhanced by fusogenic peptides [65,66] and pH-sensitive neutral lipids [67]. Efficient transfection and endosomal escape of chitosan polyplexes can be also enhanced by chitosan–polyethylenimine (PEI) copolymeric delivery systems. PEI is a cationic polymer non-viral vector with high transfection efficiency and a strong buffering capacity, which may enhance the influx of chloride anions, osmotic swelling and endosomal lysis. However, PEI-dependent cytotoxic effects constitute a major concern when using PEI for gene delivery [7,68–70]. In contrast, chitosan–PEI complexes exhibit efficient uptake by target cells, high transfection efficiency and negligible toxicity [36,71–75].

Following endosomal escape into the cytosol, chitosan polyplexes carrying DNA must be unpacked, and the entrance of loaded DNA into the nucleus is needed for transfection. The molecular events that mediate DNA unpacking after endosomal release and translocation to the nucleus remain not fully understood. It is generally accepted that, in non-dividing cells, molecules smaller than ~40 kDa can passively diffuse through the nuclear pores, while larger molecules must carry nuclear localization signals for active transportation [68]. Sun et al. largely improved DNA unpacking from chitosan and transfection efficiency upon the conjugation of chitosan with small peptides that can be phosphorylated [76]. The phosphorylation of conjugated peptides mimics the process leading to genomic DNA release and the activation of transcription, mediated by histone phosphorylation. In addition, the introduction of negatively charged phosphate groups may result in electric repulsion between DNA and chitosan conjugated with phosphorylated peptides. Hence, further enhancement of transfection was obtained by conjugating chitosan with small peptides carrying a nuclear localization signal, in addition to a potentially phosphorylatable serine residue [77]. Exogenous gene expression was improved through a mechanism that enabled DNA import into the nucleus, and enhanced unpacking by the action of nuclear histone kinases. Miao et al. improved endosomal escape and intracellular drug release in HepG2.2.15 cells by loading DNA into a redox-responsive chitosan oligosaccharide-SS-octadecylamine (CSSO) polymer. Intracellular reduction and cleavage of CSSO disulfide bonds '–SS-' by gluthation allowed rapid DNA release [78].

For strategies aiming RNAi on target genes, chitosan has been mostly complexed with siRNA, microRNA (miRNA) and plasmids expressing short hairpin RNA (shRNA). After unpacking, siRNA/miRNA associates with RNA-induced silencing complex (RISC) in the cytosol. The RNAi-guided complex hybridizes with target mRNA, leading to mRNA cleavage and/or translation repression, and subsequent inhibition of protein synthesis [9,10,48,79]. The use of shRNA expression plasmids allowing long lasting expression of siRNA may improve RNAi in vivo. Following plasmid DNA transcription in the nucleus, the transcribed shRNA is processed by Drosha, exported to the cytosol and processed by Dicer, leading to cleavage of double-stranded shRNA and the formation of specific siRNA [75,80–85].

Sequential events associated with three illustrative examples using chitosan to deliver nucleic acids are represented in Figure 4 (chitosan–TPP complexed with a plasmid construct, to express an exogenous protein), Figure 5 (chitosan–TPP complexed with a plasmid construct, to express a shRNA designed for target gene silencing) and Figure 6 (chitosan loading siRNA for target gene silencing).

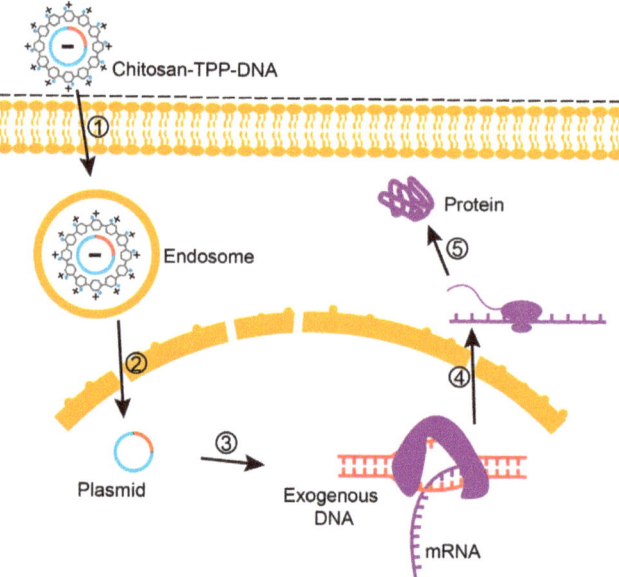

Figure 4. Cellular events associated with chitosan-based plasmid delivery for exogenous gene expression. 1, Cellular uptake of chitosan–DNA by endocytosis. 2, Endosomal escape of the chitosan–DNA complex, plasmid dissociation from chitosan and translocation to the nucleus. 3, Transcription of plasmid (exogenous DNA) in the nucleus and mRNA generation. 4, Translation of newly transcribed mRNA in the cytosol. 5, Exogenous protein assembly.

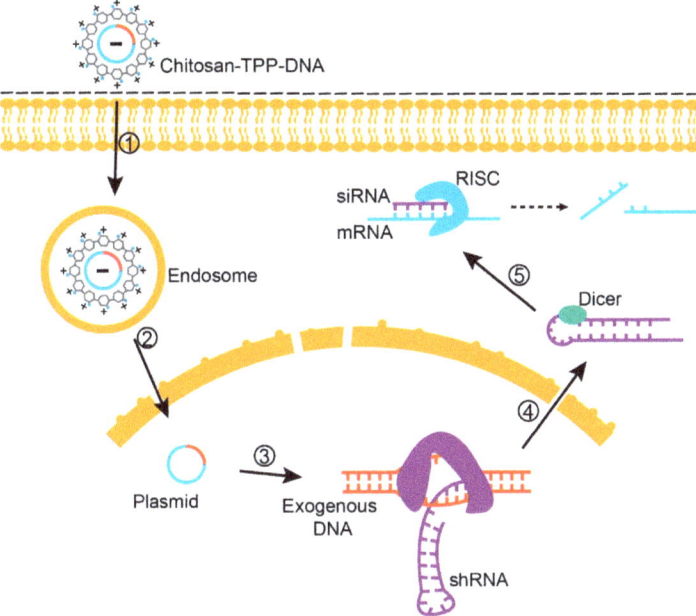

Figure 5. Cellular events associated with chitosan-based plasmid delivery for short hairpin RNA (shRNA) expression, siRNA formation and target gene silencing. 1, Cellular uptake of chitosan–DNA by endocytosis. 2, Endosomal escape of chitosan–DNA complex, plasmid dissociation from chitosan and translocation to the nucleus. 3, Transcription of plasmid (exogenous DNA) in the nucleus and generation of shRNA. 4, Transportation of shRNA to the cytosol and association with Dicer to generate siRNA. 5, siRNA association with RNA-induced silencing complex (RISC) and target mRNA by base pairing, resulting in mRNA cleavage and/or translation repression, and subsequent inhibition of protein synthesis.

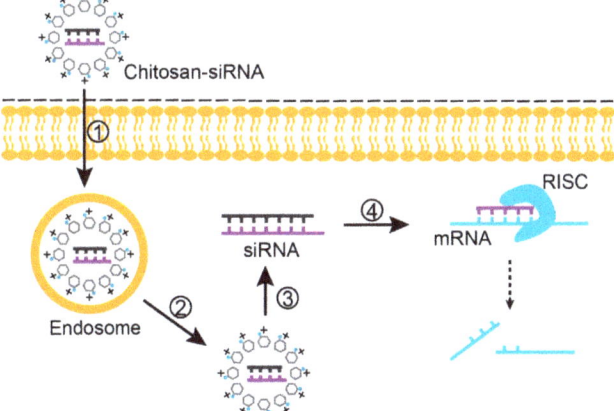

Figure 6. Cellular events associated with chitosan-based siRNA delivery for target gene silencing. 1, Cellular uptake of chitosan–siRNA by endocytosis. 2, Endosomal escape of chitosan–siRNA. 3, Dissociation of siRNA from chitosan. 4, siRNA association with RISC and target mRNA by base pairing, resulting in target mRNA cleavage and/or translation repression, and subsequent inhibition of protein synthesis.

3. Use of Chitosan in Fish Biotechnology

Chitosan and its derivatives are widely used in aquaculture. Low toxicity, biodegradability, biocompatibility, bioadhesion and immunomodulatory properties make chitosan and its derivatives of increasing interest for the fish farming industry as dietary additives, non-viral vectors enabling fish vaccination and protection against diseases, control of gonadal development and for the gene therapy-based modulation of fish metabolism.

3.1. Chitosan and Its Derivatives as Dietary Additives

Dietary supplementation with chitosan and its derivatives has been shown to improve fish growth performance, non-specific immunity and antioxidant effects [86,87]. However, the strategy for chitosan dietary supplementation in fish requires extensive investigation, according to the species and the growth stage of fish.

3.1.1. Dietary Supplementation with Chitosan

The inclusion of chitosan as feed additive for fish has been receiving attention since the 1980s [88]. Shiau et al. reported that inclusion of dietary levels of chitosan from 2% to 10% for 28 days decreases the weight gain and increases the feed conversion ratio (FCR) in hybrid tilapia (*Oreochromis niloticus* × *Oreochromis aureus*) [89]. However, other studies performed in *Oreochromis niloticus* showed positive effects of chitosan on fish growth. Feed supplementation of tilapia with chitosan (0–8 g/kg dry diet) for 56 days led to the conclusion that 4 g/kg of chitosan was the optimal dose to promote the highest body weight gain (BWG) rate and specific growth rate (SGR) [90]. Similarly, chitosan supplementation at 5 g/kg diet for 60 days improved growth performance, BWG, SGR and FCR in tilapia [91]. The contradictory effects reported for chitosan on tilapia growth could be attributed to the fact that the studies were performed using different fish growth stages. Indeed, the initial weight of fish in the study by Shiau et al. was of 0.99 ± 0.01 g, while the latter two reports used a significantly higher initial body weight (50.1 ± 4.1 g and 39.3 ± 0.3 g, respectively).

In addition to the developmental stage and amount of dietary chitosan supplied, chitosan effects exerted on fish growth performance also seem to depend on the species [87]. According to the effect observed on SGR, the apparent digestibility coefficient of dry matter and the apparent digestibility coefficient of protein, 75 days of feeding on diets supplemented with 10–20 g chitosan/kg significantly reduced the growth performance of gibel carp (*Carassius gibelio*) (initial body weight, 4.80 ± 0.01 g) [92]. However, the supply of 0–0.2 g chitosan/kg diet caused a dose dependent increase of the average daily weight and SGR in post-larvae sea bass (*Dicentrarchus labrax*) [93]. Yan et al. also reported that dietary supplementation of 0%–5% chitosan improved growth performance by inducing dose dependent increases of BWG and SGR, while FCR decreased [94]. Similarly, 70 days of supplementation with 1–5 g chitosan/kg diet of loach fish (*Misgurnus anguillicadatus*) with an average body weight of 3.14 ± 0.05 g, significantly increased BWG, SGR and condition factor (CF), whereas it decreased FCR [95]. In contrast, Najafabad et al. found that Caspian kutum (*Rutilus kutum*) fingerlings (1.7 ± 0.15 g) supplied with 0–2 g chitosan/kg diet for 60 days showed no effect of final weight, SGR and condition factor [96].

The positive effect of chitosan on the growth performance of some fish species might result from its role in nonspecific immunity. Chitosan acts as an immunostimulary drug through induction of nonspecific immunity in fish. In loach fish, the dietary supplement of chitosan increased the serum levels of factors considered as immune boosters, such as the content of immunoglobulin M (IgM), complement component 3 (C3) levels, the activity of lysozyme, acid phosphatase and alkaline phosphatase, as well as increased the survival rate after being challenged by *Aeromonas hydrophila* [95]. In accordance with the immune boost, other investigations also showed immune reinforcement by chitosan, when fish were challenged by bacteria in regard to immunoglobulin content, serum lysozyme, bactericidal activity, immune-related gene expression, phagocytosis and respiratory burst activity [90,92,94,97]. Consistently, chitosan was shown to modify hematological parameters of fish, which are also considered important indicators of immunostimulation. In Asian seabass (*Lates calcarifer*), chitosan supplement during 60 days at 5–20 g/kg diet increased red blood cells (RBC), white blood cells (WBC), total serum protein, albumin and globulin [98]. Supplementation with chitosan was reported also to increase RBC, WBC, haemoglobin, lymphocytes, monocytes, neutrophils and thrombocytes in mrigal carp (*Cirrhinus mrigala*) and kelp grouper (*Epinephelus bruneus*) [99–101].

Concomitant to the effects on immunity, chitosan also elevates antioxidant responses in fish. In loach fish, the activity of phenoloxidase, superoxide dismutase (SOD) and glutathione peroxidase (GPx) increased after 12 weeks of chitosan supplementation [95]. Similarly, chitosan induced the activity of SOD and catalase (CAT) after 56 days of dietary supplementation in tilapia [90], and the mRNA levels of SOD, CAT, GPx and nuclear factor erythroid 2-related factor 2 [94]. The protective effect of chitosan from oxidative stress was also reported in olive flounder (*Paralichthys olivaceus*) challenged with H_2O_2 [97]. The authors observed that chitosan-coated diets significantly narrowed the increase of protein carbonyl formation and DNA damage in the plasma.

3.1.2. Dietary Supplementation with Chitosan Nanoparticles

Wang et al. reported that BWG significantly increased in tilapia (initial body weight, 23.6 ± 1.2 g) fed with chitosan nanoparticles (5 g/kg dry diet) [102]. Similar results were described by other authors. Chitosan nanoparticle intake increased final weight, weight gain, SGR and FCR in tilapia supplied for 45 days with 0–2 g/kg (initial body weight, 19.8 ± 0.6 g) and 70 days for 1–5 g/kg (initial body weight, 5.66 ± 0.02 g). In these reports, innate immunity was also enhanced and fish exhibited increased respiratory burst activity, lysozyme malondialdehyde, CAT and SOD activity, and hematological parameters such as RBC, hematocrit, hemoglobin, mean corpuscular volume, WBC and platelets [103,104]. Remarkably, optimal supplement of dietary chitosan nanoparticles to improve growth and immunity against pathogens may vary, according to parameters such as developmental growth stage and species.

Dietary supplementation of chitosan nanoparticles complexed with vitamin C and thymol is more effective in enhancing immunity than supplementation with the single additives. Dietary chitosan–vitamin C nanoparticles slightly improved growth performance of tilapia, while inducing the viscerosomatic index, therefore decreasing economic performance. However, when fish fed chitosan–vitamin C nanoparticles were challenged by imidacloprid-polluted water, chitosan–vitamin C supplementation significantly strengthened immunity and antioxidant activity, including the activity of lysozyme, glutathione reductase and CAT, C3 and immunoglobulins [105]. Growth effects of dietary supplementation with chitosan nanoparticles mixed with thymol, the most important phenolic compound in *Thymus vulgaris* essential oil, were evaluated on hematological parameters, and the liver and kidney function in tilapia [106]. The results showed that chitosan–thymol nanoparticle supplementation increased feed efficiency and protein efficiency ratio, while it had moderated effects on final weight, weight gain and SGR. Nevertheless, chitosan–thymol produced a synergistic effect on lymphocytes and monocyte leukocytes. The use of chitosan nanoparticles as feed additive is limited by the fact that it can exhibit toxic effects at high levels. In this regard, chitosan nanoparticles significantly decreased hatching rate and survival rate of zebrafish (*Danio rerio*) when the immersion concentration reached 20 and 30 µg/mL or higher [107,108].

3.1.3. Dietary Supplementation with Chitin and Chitooligosaccharide

Meanwhile the inclusion of chitin in the diet has no significant effects on fish growth performance [109–111], chitooligosaccharide (COS) enhances growth performance parameters such as BWG, hepatosomatic and intestosomatic index, SGR and FCR in a number of fish species, including juvenile largemouth bass (*Micropterus salmoides*) [112], striped catfish (*Pangasianodon hypophthalmus*) [113], Nile tilapia (*Oreochromis niloticus*) [114], tiger puffer (*Takifugu rubripes*) [115], koi (*Cyprinus carpio koi*) [116], and silverfish (*Trachinotus ovatus*) [117]. Similarly as in most fish species, dietary supplementation with low molecular weight and highly deacetylated COS enhances growth performance, innate immunity and digestive enzyme activity in Pacific white shrimp (*Litopenaeus vannamei*) [118]. However, the effect of dietary COS may depend on the species. In this regard, dietary COS supplementation was reported to cause not significant effects on weight gain, FCR and the survival rate in hybrid tilapia (*Oreochromis niloticus×O. aureus*) [109]. Similar results were reported for rainbow trout (*Oncorhynchus mykiss*) [119]. Incomplete intestinal development in early developmental stages may contribute to the lack of COS effect on growth performance observed in several fish species.

A number of studies showed that both chitin and COS can be potentially utilized as immunostimulants in fish. Respiratory burst activity, phagocytic activity and lysozyme activity, which are considered indicators of non-specific immunity, have been shown to be significantly stimulated by chitin and COS in a number of fish species, including juvenile largemouth bass (*Micropterus salmoides*) [112], Nile tilapia (*Oreochromis niloticus*) [114], striped catfish (*Pangasianodon hypophthalmus*) [113] and mrigal carp (*Cirrhina mrigala*) [99]. Chitin and COS also induce other immunity parameters, such as nitric oxide production, inducible nitric oxide synthase (iNOS) activity and gene expression [112,120], leukocyte count [99,112,116] and complement activity [99,100].

3.2. Chitosan as a Carrier for Drug Delivery in Fish

Chitosan is nanoscale, biodegradable, biocompatible, hemocompatible, simple and mild for preparation conditions, and is highly efficient for drug loading. Therefore, chitosan has been used for loading a variety of bioactive compounds, such as vitamins, metal ions, inactivated pathogens for vaccines, proteins and nucleic acids in a variety of applications in fish farming. In addition, loading into chitosan can significantly boost the bioeffects of these compounds.

3.2.1. Chitosan Loading Chemical Compounds

The sustained release of compounds complexed with chitosan nanoparticles fulfills the requirements of artificial breeding in fish farming and enable delivery and cell uptake of compounds with low toxicity [121,122]. Chitosan nanoparticles loaded with vitamin C, an important but labile antioxidant, were proven to enhance sustained vitamin C release in the stomach, the intestine and in serum after oral administration in rainbow trout (*Oncorhynchus mykiss*) [123]. Chitosan–vitamin C nanoparticles exhibited a markedly high antioxidant activity and no toxicity up to 2.5 mg/mL in the culture medium of ZFL cells, a zebrafish liver-derived cell line. In addition, chitosan–vitamin C nanoparticles showed the capability to penetrate the intestinal epithelium of *Solea senegalensis* [124]. Several studies evaluated chitosan nanoparticles loading aromatase inhibitors and eurycomanone, compounds that promote gonadal development. Chitosan-mediated delivery of aromatase inhibitors and eurycomanone prolonged serum presence, improved testicular development with lack of testicular toxicity, and led to higher serum concentrations of reproductive hormones [125–128].

3.2.2. Chitosan Loading Metal Ions

Loading with chitosan facilitates delivery of metal ions that are micronutrients and antibacterial factors, such as selenium and silver, to fish in culture. Barakat et al. showed that chitosan–silver nanoparticles successfully treated European sea bass larvae infected with *Vibrio anguillarum*. Chitosan–silver nanoparticles significantly decreased the bacterial number and improved fish survival [129]. In addition, dietary supplementation with chitosan–silver nanoparticles were shown to altering gut morphometry and microbiota in zebrafish. Feeding with chitosan–silver nanoparticles increased *Fusobacteria* and *Bacteroidetes* phyla, goblet cell density and villi height, while upregulated the expression of immune-related genes [130]. Similarly, chitosan–selenium nanoparticles had immunostimulary roles and increased disease resistance in zebrafish and *Paramisgurnus dabryanus* by improving the activity of lysozyme, acid phosphatase and alkaline phosphatase, phagocytic respiratory burst and splenocyte-responses towards concanavalin A [131,132].

3.2.3. Chitosan Loading Inactivated Pathogens

Vaccines against pathogens is a major challenge in aquaculture. In this regard, chitosan can be used as proper carrier and adjuvant to enhance effectiveness of vaccination. A number of inactivated bacteria and virus have been evaluated with chitosan or its derivatives as adjuvant against infections in fish. Vaccines, such as inactivated *Edwardsiella ictaluri* and infectious spleen and kidney necrosis virus, have been tested with chitosan in yellow catfish (*Pelteobagrus fulvidraco*) and Chinese perch (*Siniperca chuasi*), respectively. Chitosan enhanced incorporation into the host cells and improved fish survival rate and immune response, increasing IgM content, lysozyme activity and mRNA levels of interleukin (IL)-1β, IL-2 and interferon (IFN)-γ2 [133,134]. A mixture of COS and inactivated *Vibrio anguillarum* vaccine significantly reduced zebrafish mortality against *Vibro anguillarum* [135], while COS combined with inactivated *Vibrio harveyi* also markedly increased survival rate, IgM and the expression of immune-related genes, such as IL-1β, IL-16, tumor necrosis factor-alpha (TNF-α) and major histocompatibility complex class I alpha (MHC-Iα), in the grouper ♀*Epinephelus fuscoguttatus*×♂*Epinephelus lanceolatus* [136]. Similarly, rainbow trout (*Oncorhynchus mykiss*) immunized against bacterial infection (*Lactococcus garvieae* and *Streptococcus iniae*) through chitosan–alginate coated vaccination exhibited a higher survival rate, immune-related gene expression, and antibody titer than fish submitted to non-coated vaccination [137].

Olive flounder (*Paralichthys olivaceus*) vaccinated against inactivated viral haemorrhagic septicaemia virus encapsulated with chitosan through oral and immersion routes showed effective immunization in the head kidney, which is considered as the primary organ responsible for the initiation of adaptive immunity in fish, skin and intestine, which are regarded as the main sites for antigen uptake and mucosal immunity. Additionally to upregulation of IgM, immunoglobulin T (IgT), polymeric Ig receptor (pIgR), MHC-I, major histocompatibility complex class II (MHC-II) and IFN-γ in the three tissues, caspase 3 was also highly induced 48 h post-challenge, suggesting cytotoxicity due to rapid T-cell response and impairment of viral proliferation [138].

Coating chitosan with membrane vesicles from pathogens such as *Piscirickettsia salmonis* was also shown to be an effective strategy to induce immune response in zebrafish (*Danio rerio*) and upregulation of CD 4, CD 8, MHC-I, macrophage-expressed 1, tandem duplicate 1 (Mpeg1.1), TNFα, IL-1β, IL-10, and IL-6 [139].

3.2.4. Chitosan Loading Proteins

Effectiveness of fish vaccination against infections can be also improved with antigenic proteins derived from bacteria and virus. For example, chitosan nanoparticles encapsulated with the recombinant outer membrane protein A of *Edwardsiella tarda* was used for oral vaccination of fringed-lipped peninsula carp (*Labeo fimbriatus*). Treated fish showed significant higher levels of post-vaccination antibody in circulation and survival rate against *Edwardsiella tarda* [140]. In another study, oral vaccination with alginate-chitosan microspheres encapsulating the recombinant protein serine-rich repeat (rSrr) of *Streptococcus iniae* were evaluated and the results showed that lysozyme activity and immune-related genes were induced, leading to a 60% increased survival rate of channel catfish (*Ictalurus punctatus*) against *Streptococcus iniae* infection [141]. In grass carp (*Ctenopharyngodon idella*), chitosan was also used for carrying the immunomodulatory factor IFN-γ2. Treatment with chitosan–*Ctenopharyngodon idella* IFN-γ2 highly upregulated inflammatory factors, leading to severe inflammatory damage in the intestine, hepatopancreas and decreased survival rate [142].

3.2.5. Chitosan Loading Nucleic Acids

Compared to chitosan-based gene delivery in other organisms, gene therapy methodologies using chitosan for improving desirable traits in farmed fish have great potential for development (Figure 1b). A number of studies addressed the characterization of factors that can influence the efficiency of chitosan loading and nucleic acid release, such as the average diameter, zeta potential and encapsulation efficiency of chitosan–DNA microspheres or nanospheres. Table 1 summarizes chitosan–plasmid DNA encapsulation efficiency and changes in particle diameter and zeta potential before and after encapsulation for fish biotechnology studies. Existing data show that the diameter of chitosan nanospheres before loading DNA mostly ranged from ~30 to ~230 nm, while encapsulation with plasmid DNA led to ~40–190 nm diameter increase. The zeta potential indicates the surface charge on the particles. A higher positive zeta potential suggests higher stability of nanoparticles in the suspension [143]. The zeta potential before loading plasmid DNA were ~25–33 mV, which mostly tended to decrease to ~14–18 mV. The exception was reported by Rather et al., who found that zeta potential of chitosan nanospheres increased ~6 mV following DNA encapsulation [144]. DNA encapsulation efficiency was generally higher than 80%, which indicates that chitosan is capable to load a high mass of DNA, which in turn may benefit many applications in aquaculture.

Table 1. Characteristics of chitosan–plasmid DNA polyplexes for studies performed in fish.

Preloading Diameter (nm)	Postloading Diameter (nm)	Preloading Zeta Potential (mV)	Postloading Zeta Potential (mV)	Encapsulation Efficiency	References
-	<10,000	-	-	94.5%	[145]
30–60	-	-	-	-	[146]
-	200	-	-	91.5%	[147]
-	-	-	-	83.6%	[148]
193 ± 53 [1]	246 ± 74 [1]	32.0 ± 1.0 [1]	14.4 ± 1.3 [1]	-	[80]
-	146 ± 2 [2]	-	24.3 ± 0.5 [2]	92.8% ± 1.4% [2]	[149]
-	133	-	34.3	63%	[150]
-	50-200	-	-	97.5%	[151]
87	156	30.3	36.5	60%	[144]
-	743	-	-	98.6%	[152]
135	-	26.7	-	86%	[153]
-	-	-	-	84.2%	[154]
224 ± 62 [1]	Similar to preloading diameter	33.0 ± 1.2 [1]	14.4 ± 1.3 [1]	-	[81]
-	750–950	-	-	98.6%	[155]
116	306	24.7	18.0	-	[156]
231 ± 18 [2]	272 ± 36 [2]	31.2 ± 1.5 [2]	14.1 ± 2.3 [2]	-	[157]
-	267	-	27.1	87.4%	[158]

[1] Mean ± SD; [2] mean ± SEM.

Chitosan-encapsulated DNA is more stable in vivo, exhibit sustained-release and increased cell uptake than naked DNA. Taken together, these factors confer chitosan-delivered DNA a particular expression profile regarding tissue distribution, persistence of expression and abundance in fish. Sáez et al. found that intramuscular injection led to a restricted expression to adjacent tissues of both chitosan-encapsulated DNA and naked DNA, while the oral administration of chitosan-encapsulated DNA, largely used for fish vaccination studies, showed enhanced expression not only in the intestine, but also in the liver of gilthead sea bream (*Sparus aurata*) [152,155]. Furthermore, oral administration of chitosan nanoparticles loaded with pCMVβ, a plasmid encoding for *Escherichia coli* β-galactosidase, enabled sustained detection of the exogenous plasmid and bacterial β-galactosidase activity in the liver and the intestine of *Sparus aurata* juveniles up to 60 days posttreatment [152].

Through the immersion route, Rao et al. showed that chitosan-coated DNA was confined to the surface area of rohu (*Labeo rohita*), i.e., gill, intestine and skin-muscle, while no detection was observed in the kidney and the liver. Naked DNA was undetectable due to degradation [158]. Oral delivery seems to have a wider distribution of chitosan-encapsulated DNA, being found in the stomach, spleen, intestine, gill, muscle, liver, heart and kidney [148,154,159]. Chitosan-encapsulated DNA has longer and more abundant presence than naked DNA after administration. For example, Rajesh Kumar et al. showed that antibody in serum from fish immunized with a chitosan–DNA vaccine was 30% higher than naked DNA after 21 days of oral immunity [160]. The presence of DNA vaccine was reported more than 90 days after oral administration of chitosan–DNA [145]. Additionally, Rather et al. reported that chitosan–DNA induced 2-fold longer and higher peak abundant expression of downstream genes than naked DNA [144].

3.3. Chitosan-Based Applications in Fish Biotechnology and Gene Therapy

In recent years, chitosan has been increasingly used for drug and gene delivery in fish biotechnology. Most of the studies used chitosan-based systems to improve oral vaccination, control of gonadal development, and the modification of fish intermediary metabolism.

3.3.1. Fish Vaccination

DNA vaccines delivered by chitosan significantly increase relative percent survival of fish at a range of 45%–82% against bacterial and viral infection [151,156]. Higher doses of chitosan–DNA vaccines resulted in concomitant increase of fish relative percent survival from ~47% to ~70% [154]. In addition, DNA vaccination with chitosan stimulated expression of immune-related genes. Zheng et al. reported upregulation of the expression of immune-related genes, such as interferon-induced GTP-binding protein Mx2 (MX2), IFN, chemokine receptor (CXCR), T-cell receptor (TCR), MHC-Iα and MHC-IIα, 7 days after oral vaccination against reddish body iridovirus in turbot (*Scophthalmus maximus*). A 10-fold higher expression of TNF-α gene expression was found in the hindgut [149].

In addition to the short-term modification of the expression levels of immune-related genes, the administration of chitosan–DNA vaccines also promote a sustained effect after treatment. Valero et al. found that European sea bass (*Dicentrarchus labrax*) orally vaccinated with chitosan-encapsulated DNA against nodavirus failed to induce circulating IgM. However, the expression of genes involved in cell-mediated cytotoxicity (TCRβ and CD8α) and the interferon pathway (IFN, MX and IFN-γ) were upregulated. Three months following vaccination, challenged fish exhibited partial protection with retarded onset of fish death and lower cumulative mortality [151]. Kole et al. immunized rohu (*Labeo rohita*) with chitosan nanoparticles complexed with a bicistronic DNA plasmid encoding the antigen *Edwardsiella tarda* glyceraldehyde 3-phosphate dehydrogenase and the immune adjuvant gene *Labeo rohita* IFN-γ [156]. Follow-up of the expression of immune-related genes in the the kidney, liver and spleen showed maximal upregulation of IgHC (IgM heavy chain), iNOS, toll like receptor 22 (TLR22), nucleotide binding and oligomerization domain-1 (NOD1) and IL-1β at 14 days post immunization. The authors also confirmed that oral and immersion vaccination with chitosan–DNA nanoparticles enhances the fish immune response to a greater extent than intramuscular injection of naked DNA. In another study, the oral vaccination of rainbow trout fry with chitosan–TPP nanoparticles complexed with pcDNA3.1-VP2, showed that the expression of genes related with innate immune response, IFN-1 and MX, reached maximal values at 3 days postvaccination and 7 days after boosting (22 days postvaccination), while, with regard to genes involved in the adaptive immune response, CD4 peaked at 15 days postvaccination and IgM and IgT at 30 days postvaccination [154].

3.3.2. Control of Gonadal Development

Chitosan nanoparticles have been used for drug delivery in studies aiming proper gonadal development in fish farming. Bhat et al. administered chitosan conjugated with salmon luteinizing hormone-releasing hormone (sLHRH) into walking catfish (*Clarias batrachus*) to promote gonadal development. Chitosan-conjugated sLHRH and naked sLHRH exerted similar effects: upregulation of Sox9 expression in the gonads and increase of circulating steroid hormonal levels, testosterone and 11-ketotestosterone in males and testosterone and 17β-estradiol in females. However, sLHRH conjugation with chitosan induced sustained and controlled release of the hormones with maximal levels observed in the last sampling point of the experiment (36 h posttreatment), while naked sLHRH peaked circulating steroid hormones at 12 h posttreatment [150]. Similarly, compared to the administration of naked kisspeptin-10, intramuscular injection of chitosan-encapsulated kisspeptin-10 in immature female *Catla catla* caused a delayed but greater increase of gonadotropin-releasing hormone, luteinizing hormone and follicle-stimulating hormone expression, as well as circulating levels of 11-ketotestosterone and 17β-estradiol [144].

With the ultimate goal of controlling gonadal development in fish, chitosan was also assayed for gene delivery. In walking catfish (*Clarias batrachus*), intramuscular administration of chitosan nanoparticles conjugated with an expression plasmid encoding steroidogenic acute regulatory protein (StAR), a major regulator of steroidogenesis, also resulted in long-lasting stimulatory effects than administration of the naked plasmid construct on the expression of key genes in reproduction, cytochrome P450 (CYP) 11A1, CYP17A1, CYP19A1, 3β-hydroxysteroid dehydrogenase and 173β-hydroxysteroid dehydrogenase [153].

3.3.3. Control of Fish Metabolism

Chitosan has been used for enhancing fish digestibility, the absorption of food constituents and increasing the utilization of dietary carbohydrate in carnivorous fish. To supplement exogenous proteolytic enzymes and thus facilitate protein digestion and amino acid absorption, Kumari et al. orally administered chitosan–TPP nanoparticles encapsulating trypsin to *Labeo rohita* over 45 days. Treatment with chitosan–TPP–trypsin enhanced nutrient digestibility, intestinal protease activity and transamination activity, alanine aminotransferase (ALT) and aspartate aminotransferase (AST), in the liver and the muscle [161].

The substitution of dietary protein by cheaper nutrients with reduced environmental impact in farmed fish is a challenging trend for sustainable aquaculture [162]. However, the metabolic features of fish, particularly carnivorous fish, constrain the replacement of dietary protein by other nutrients in aquafeeds. Carnivorous fish exhibit a preferential use of amino acids as fuel and gluconeogenic substrates, and thus require high levels of dietary protein for optimal growth. Instead, carbohydrates are metabolized markedly slower than in mammals, and give rise to prolonged hyperglycemia [163,164]. The essential role of the liver in controlling the intermediary metabolism makes this organ an ideal target for investigating and modifying the glucose tolerance of farmed fish.

To overcome metabolic limitations of carnivorous fish, in recent years we synthesized chitosan–TPP nanoparticles, complexed with plasmid DNA, to induce in vivo transient overexpression and the silencing of target genes in the liver of gilthead sea bream (*Sparus aurata*). With the aim of decreasing the use of amino acids for gluconeogenic purposes and improving carbohydrate metabolism in the liver, chitosan–TPP nanoparticles complexed with a plasmid overexpressing a shRNA designed to silence the expression of cytosolic ALT (cALT) were intraperitoneally administered to *Sparus aurata* juveniles. Seventy-two hours posttreatment, a significant decrease in cALT1 mRNA levels, immunodetectable ALT and ALT activity was observed in the liver of treated fish. Knockdown of cALT expression to ~63%–70% of the values observed in control fish significantly increased the hepatic activity of key enzymes in glycolysis, 6-phosphofructo 1-kinase (PFK1) and pyruvate kinase, and protein metabolism, glutamate dehydrogenase (GDH). In addition to showing efficient gene silencing after administration of chitosan–TPP–DNA nanoparticles, the findings supported evidence that the downregulation of liver transamination increased the use of dietary carbohydrates to obtain energy, and thus made it possible to spare protein in carnivorous fish [80].

Following the same methodology, we showed that the shRNA-mediated knockdown of GDH significantly decreased GDH mRNA and immunodetectable levels in the liver, which, in turn, reduced GDH activity to ~53%. Downregulation of GDH decreased liver glutamate, glutamine and 2-oxoglutarate, as well as the hepatic activity of AST, while it increased 2-oxoglutarate dehydrogenase activity and the PFK1/fructose-1,6-bisphosphatase (FBP1) activity ratio. Therefore, by reducing hepatic transdeamination and gluconeogenesis, the knockdown of GDH could impair the use of amino acids as gluconeogenic substrates and facilitate the metabolic use of dietary carbohydrates [81].

With the aim of inducing a multigenic action leading to a stronger protein-sparing effect, *Sparus aurata* were intraperitoneally injected with chitosan–TPP nanoparticles complexed with a plasmid expressing the N-terminal nuclear fragment of hamster SREBP1a, a transcription factor that—in addition to exhibiting strong transactivating capacity of genes required for fatty acid, triglycerides and cholesterol synthesis—previous reports showed can also transactivate the promoter of genes encoding key enzymes in hepatic glycolysis, glucokinase (GK) and 6-phosphofructo 2-kinase/fructose 2,6-bisphosphatase (PFKFB1) in fish [165,166]. Overexpression of exogenous SREBP1a in the liver of *Sparus aurata* enhanced the expression of glycolytic enzymes GK and PFKFB1, decreased the activity of the gluconeogenic enzyme FBP1 and increased the mRNA levels of key enzymes in fatty acid synthesis, elongation and desaturation (acetyl-CoA carboxylase 1, acetyl-CoA carboxylase 2, elongation of very long chain fatty acids protein 5, fatty acid desaturase 2), as well as induced NADPH formation (glucose 6-phophate dehydrogenase) and cholesterol synthesis (3-hydroxy-3-methylglutaryl-coenzyme A reductase). As a result, chitosan-mediated SREBP1a overexpression caused a multigenic action that enabled the conversion of dietary carbohydrates into lipids (Figure 7), leading to increased circulating levels of triglycerides and cholesterol in carnivorous fish [157].

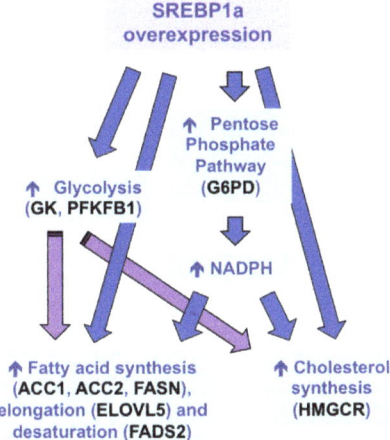

Figure 7. Multigenic action and metabolic effects in the liver of *Sparus aurata* after intraperitoneal administration of chitosan–TPP–DNA nanoparticles to overexpress exogenous SREBP1a [157]. ACC1, acetyl-CoA carboxylase 1; ACC2, acetyl-CoA carboxylase 2; ELOVL5, elongation of very long chain fatty acids protein 5; FADS2, fatty acid desaturase 2; G6PD, glucose 6-phophate dehydrogenase; GK, glucokinase; HMGCR, 3-hydroxy-3-methylglutaryl-coenzyme A reductase; PFKFB1, 6-phosphofructo 2-kinase/fructose 2,6-bisphosphatase.

4. Conclusions

Characteristics such as nanoscale, low-toxicity, biodegradability, biocompatibility, derivatization, immunomodulatory effects, and easily affordable preparation conditions make chitosan a strong candidate for drug delivery into fish. Therefore, the use of chitosan in fish biotechnology has received growing attention in recent years. However, applications based on novel chitosan-based gene therapy methodologies to improve desirable traits in farmed fish have enormous potential for development. Most remarkable advances in the field addressed fish immunization, the control of reproduction for broodstock management and the modulation of gene expression to spare protein and overcome metabolic limitations of farmed fish. Further studies are needed for a better understanding of the extracellular and intracellular process, following chitosan-mediated gene delivery into fish. In addition, future trends in fish farming may greatly benefit from improved and more efficient chitosan formulations for enhancing gene delivery targeting and intracellular traffic in farmed fish.

Author Contributions: Conceptualization, I.M. and M.P.A.; writing—Original draft preparation, Y.W., A.R., M.P.A. and I.M.; writing—Review and editing, M.P.A. and I.M.; supervision, I.M.; project administration, I.M.; funding acquisition, I.M. All authors have read and agreed to the published version of the manuscript.

Funding: This research was funded by Ministerio de Economía, Industria y Competitividad (Spain), grant number AGL2016-78124-R, co-funded by the European Regional Development Fund (EC).

Conflicts of Interest: The authors declare no conflict of interest.

References

1. Mahdy Samar, M.; El-Kalyoubi, M.H.; Khalaf, M.M.; Abd El-Razik, M.M. Physicochemical, functional, antioxidant and antibacterial properties of chitosan extracted from shrimp wastes by microwave technique. *Ann. Agric. Sci.* **2013**, *58*, 33–41. [CrossRef]
2. Sun, M.; Wang, T.; Pang, J.; Chen, X.; Liu, Y. Hydroxybutyl chitosan centered biocomposites for potential curative applications: A critical review. *Biomacromolecules* **2020**, *21*, 1351–1367. [CrossRef] [PubMed]
3. Sacco, P.; Cok, M.; Scognamiglio, F.; Pizzolitto, C.; Vecchies, F.; Marfoglia, A.; Marsich, E.; Donati, I. Glycosylated-chitosan derivatives: A systematic review. *Molecules* **2020**, *25*, 1534. [CrossRef] [PubMed]
4. Chuan, D.; Jin, T.; Fan, R.; Zhou, L.; Guo, G. Chitosan for gene delivery: Methods for improvement and applications. *Adv. Colloid Interface Sci.* **2019**, *268*, 25–38. [CrossRef]
5. Ivanova, D.G.; Yaneva, Z.L. Antioxidant properties and redox-modulating activity of chitosan and its derivatives: Biomaterials with application in cancer therapy. *Biores. Open Access* **2020**, *9*, 64–72. [CrossRef]
6. Wang, W.; Meng, Q.; Li, Q.; Liu, J.; Zhou, M.; Jin, Z.; Zhao, K. Chitosan derivatives and their application in biomedicine. *Int. J. Mol. Sci.* **2020**, *21*, 487. [CrossRef]
7. Raftery, R.; O'Brien, F.J.; Cryan, S.-A. Chitosan for gene delivery and orthopedic tissue engineering applications. *Molecules* **2013**, *18*, 5611–5647. [CrossRef]
8. Lostalé-Seijo, I.; Montenegro, J. Synthetic materials at the forefront of gene delivery. *Nat. Rev. Chem.* **2018**, *2*, 258–277. [CrossRef]
9. Cao, Y.; Tan, Y.F.; Wong, Y.S.; Liew, M.W.J.; Venkatraman, S. Recent Advances in Chitosan-Based Carriers for Gene Delivery. *Mar. Drugs* **2019**, *17*, 381. [CrossRef]
10. Santos-Carballal, B.; Fernández Fernández, E.; Goycoolea, F. Chitosan in non-viral gene delivery: Role of structure, characterization methods, and insights in cancer and rare diseases therapies. *Polymers* **2018**, *10*, 444. [CrossRef]
11. Ginn, S.L.; Amaya, A.K.; Alexander, I.E.; Edelstein, M.; Abedi, M.R. Gene therapy clinical trials worldwide to 2017: An update. *J. Gene Med.* **2018**, *20*, e3015. [CrossRef] [PubMed]
12. Picanço-Castro, V.; Pereira, C.G.; Covas, D.T.; Porto, G.S.; Athanassiadou, A.; Figueiredo, M.L. Emerging patent landscape for non-viral vectors used for gene therapy. *Nat. Biotechnol.* **2020**, *38*, 151–157. [CrossRef] [PubMed]
13. Simões, S.; Filipe, A.; Faneca, H.; Mano, M.; Penacho, N.; Düzgüneş, N.; de Lima, M.P. Cationic liposomes for gene delivery. *Expert Opin. Drug Deliv.* **2005**, *2*, 237–254. [CrossRef] [PubMed]
14. Saffari, M.; Moghimi, H.; Dass, C. Barriers to liposomal gene delivery: From application site to the target. *Iran. J. Pharm. Res.* **2016**, *15*, 3–17.
15. Ramamoorth, M.; Narvekar, A. Non viral vectors in gene therapy- an overview. *J. Clin. Diagn. Res.* **2015**, *9*, GE01–GE06. [CrossRef]
16. Patil, S.; Gao, Y.G.; Lin, X.; Li, Y.; Dang, K.; Tian, Y.; Zhang, W.J.; Jiang, S.F.; Qadir, A.; Qian, A.R. The development of functional non-viral vectors for gene delivery. *Int. J. Mol. Sci.* **2019**, *20*, 5491. [CrossRef]
17. Mumper, R.; Wang, J.; Claspell, J.; Rolland, A. Novel polymeric condensing carriers for gene delivery. *Proc. Int. Symp. Control. Release Bioact. Mater.* **1995**, *22*, 178–179.
18. MacLaughlin, F.C.; Mumper, R.J.; Wang, J.; Tagliaferri, J.M.; Gill, I.; Hinchcliffe, M.; Rolland, A.P. Chitosan and depolymerized chitosan oligomers as condensing carriers for in vivo plasmid delivery. *J. Control. Release* **1998**, *56*, 259–272. [CrossRef]
19. Katas, H.; Alpar, H.O. Development and characterisation of chitosan nanoparticles for siRNA delivery. *J. Control. Release* **2006**, *115*, 216–225. [CrossRef]

20. Richardson, S.C.; Kolbe, H.V.; Duncan, R. Potential of low molecular mass chitosan as a DNA delivery system: Biocompatibility, body distribution and ability to complex and protect DNA. *Int. J. Pharm.* **1999**, *178*, 231–243. [CrossRef]
21. Xu, W.; Shen, Y.; Jiang, Z.; Wang, Y.; Chu, Y.; Xiong, S. Intranasal delivery of chitosan-DNA vaccine generates mucosal SIgA and anti-CVB3 protection. *Vaccine* **2004**, *22*, 3603–3612. [CrossRef] [PubMed]
22. Sun, C.-J.; Pan, S.-P.; Xie, Q.-X.; Xiao, L.-J. Preparation of chitosan-plasmid DNA nanoparticles encoding zona pellucida glycoprotein-3alpha and its expression in mouse. *Mol. Reprod. Dev.* **2004**, *68*, 182–188. [CrossRef] [PubMed]
23. Kang, S.H.; Revuri, V.; Lee, S.-J.; Cho, S.; Park, I.-K.; Cho, K.J.; Bae, W.K.; Lee, Y. Oral siRNA delivery to treat colorectal liver metastases. *ACS Nano* **2017**, *11*, 10417–10429. [CrossRef] [PubMed]
24. Lin, F.; Jia, H.-R.; Wu, F.-G. Glycol chitosan: A water-soluble polymer for cell imaging and drug delivery. *Molecules* **2019**, *24*, 4371. [CrossRef] [PubMed]
25. Huang, M.; Fong, C.-W.; Khor, E.; Lim, L.-Y. Transfection efficiency of chitosan vectors: Effect of polymer molecular weight and degree of deacetylation. *J. Control. Release* **2005**, *106*, 391–406. [CrossRef] [PubMed]
26. Huang, M.; Khor, E.; Lim, L.-Y. Uptake and cytotoxicity of chitosan molecules and nanoparticles: Effects of molecular weight and degree of deacetylation. *Pharm. Res.* **2004**, *21*, 344–353. [CrossRef] [PubMed]
27. Liu, X.; Howard, K.A.; Dong, M.; Andersen, M.Ø.; Rahbek, U.L.; Johnsen, M.G.; Hansen, O.C.; Besenbacher, F.; Kjems, J. The influence of polymeric properties on chitosan/siRNA nanoparticle formulation and gene silencing. *Biomaterials* **2007**, *28*, 1280–1288. [CrossRef]
28. Razmi, F.A.; Ngadi, N.; Wong, S.; Inuwa, I.M.; Opotu, L.A. Kinetics, thermodynamics, isotherm and regeneration analysis of chitosan modified pandan adsorbent. *J. Clean. Prod.* **2019**, *231*, 98–109. [CrossRef]
29. Alameh, M.; Lavertu, M.; Tran-Khanh, N.; Chang, C.-Y.; Lesage, F.; Bail, M.; Darras, V.; Chevrier, A.; Buschmann, M.D. siRNA delivery with chitosan: Influence of chitosan molecular weight, degree of deacetylation, and amine to phosphate ratio on in vitro silencing efficiency, hemocompatibility, biodistribution, and in vivo efficacy. *Biomacromolecules* **2018**, *19*, 112–131. [CrossRef]
30. Al-Remawi, M. Application of N-hexoyl chitosan derivatives with high degree of substitution in the preparation of super-disintegrating pharmaceutical matrices. *J. Drug Deliv. Sci. Technol.* **2015**, *29*, 31–41. [CrossRef]
31. Chung, Y.-C.; Kuo, C.-L.; Chen, C.-C. Preparation and important functional properties of water-soluble chitosan produced through Maillard reaction. *Bioresour. Technol.* **2005**, *96*, 1473–1482. [CrossRef] [PubMed]
32. Gullón, B.; Montenegro, M.I.; Ruiz-Matute, A.I.; Cardelle-Cobas, A.; Corzo, N.; Pintado, M.E. Synthesis, optimization and structural characterization of a chitosan-glucose derivative obtained by the Maillard reaction. *Carbohydr. Polym.* **2016**, *137*, 382–389. [CrossRef] [PubMed]
33. Uccello-Barretta, G.; Balzano, F.; Aiello, F.; Senatore, A.; Fabiano, A.; Zambito, Y. Mucoadhesivity and release properties of quaternary ammonium-chitosan conjugates and their nanoparticulate supramolecular aggregates: An NMR investigation. *Int. J. Pharm.* **2014**, *461*, 489–494. [CrossRef] [PubMed]
34. Li, H.; Zhang, Z.; Bao, X.; Xu, G.; Yao, P. Fatty acid and quaternary ammonium modified chitosan nanoparticles for insulin delivery. *Colloids Surf. B. Biointerfaces* **2018**, *170*, 136–143. [CrossRef]
35. Jiang, X.; Dai, H.; Leong, K.W.; Goh, S.-H.; Mao, H.-Q.; Yang, Y.-Y. Chitosan-g-PEG/DNA complexes deliver gene to the rat liver via intrabiliary and intraportal infusions. *J. Gene Med.* **2006**, *8*, 477–487. [CrossRef]
36. Ping, Y.; Liu, C.; Zhang, Z.; Liu, K.L.; Chen, J.; Li, J. Chitosan-graft-(PEI-β-cyclodextrin) copolymers and their supramolecular PEGylation for DNA and siRNA delivery. *Biomaterials* **2011**, *32*, 8328–8341. [CrossRef]
37. Lee, H.; Jeong, J.H.; Park, T.G. PEG grafted polylysine with fusogenic peptide for gene delivery: High transfection efficiency with low cytotoxicity. *J. Control. Release* **2002**, *79*, 283–291. [CrossRef]
38. Strand, S.P.; Issa, M.M.; Christensen, B.E.; Vårum, K.M.; Artursson, P. Tailoring of chitosans for gene delivery: Novel self-branched glycosylated chitosan oligomers with improved functional properties. *Biomacromolecules* **2008**, *9*, 3268–3276. [CrossRef]
39. Thanou, M.; Florea, B.I.; Geldof, M.; Junginger, H.E.; Borchard, G. Quaternized chitosan oligomers as novel gene delivery vectors in epithelial cell lines. *Biomaterials* **2002**, *23*, 153–159. [CrossRef]
40. Kean, T.; Roth, S.; Thanou, M. Trimethylated chitosans as non-viral gene delivery vectors: Cytotoxicity and transfection efficiency. *J. Control. Release* **2005**, *103*, 643–653. [CrossRef]

41. Ren, Y.; Zhao, X.; Liang, X.; Ma, P.X.; Guo, B. Injectable hydrogel based on quaternized chitosan, gelatin and dopamine as localized drug delivery system to treat Parkinson's disease. *Int. J. Biol. Macromol.* **2017**, *105*, 1079–1087. [CrossRef]
42. Raik, S.V.; Andranovitš, S.; Petrova, V.A.; Xu, Y.; Lam, J.K.-W.; Morris, G.A.; Brodskaia, A.V.; Casettari, L.; Kritchenkov, A.S.; Skorik, Y.A. Comparative study of diethylaminoethyl-chitosan and methylglycol-chitosan as potential non-viral vectors for gene therapy. *Polymers* **2018**, *10*, 442. [CrossRef] [PubMed]
43. Calvo, P.; Remuñán-López, C.; Vila-Jato, J.L.; Alonso, M.J. Novel hydrophilic chitosan-polyethylene oxide nanoparticles as protein carriers. *J. Appl. Polym. Sci.* **1997**, *63*, 125–132. [CrossRef]
44. Gan, Q.; Wang, T.; Cochrane, C.; McCarron, P. Modulation of surface charge, particle size and morphological properties of chitosan–TPP nanoparticles intended for gene delivery. *Colloids Surf. B Biointerfaces* **2005**, *44*, 65–73. [CrossRef] [PubMed]
45. Raja, M.A.G.; Katas, H.; Jing Wen, T. Stability, intracellular delivery, and release of siRNA from chitosan nanoparticles using different cross-linkers. *PLoS ONE* **2015**, *10*, e0128963.
46. Vimal, S.; Abdul Majeed, S.; Taju, G.; Nambi, K.S.N.; Sundar Raj, N.; Madan, N.; Farook, M.A.; Rajkumar, T.; Gopinath, D.; Sahul Hameed, A.S. Chitosan tripolyphosphate (CS/TPP) nanoparticles: Preparation, characterization and application for gene delivery in shrimp. *Acta Trop.* **2013**, *128*, 486–493. [CrossRef]
47. Fàbregas, A.; Miñarro, M.; García-Montoya, E.; Pérez-Lozano, P.; Carrillo, C.; Sarrate, R.; Sánchez, N.; Ticó, J.R.; Suñé-Negre, J.M. Impact of physical parameters on particle size and reaction yield when using the ionic gelation method to obtain cationic polymeric chitosan-tripolyphosphate nanoparticles. *Int. J. Pharm.* **2013**, *446*, 199–204. [CrossRef]
48. Serrano-Sevilla, I.; Artiga, Á.; Mitchell, S.G.; De Matteis, L.; de la Fuente, J.M. Natural polysaccharides for siRNA delivery: Nanocarriers based on chitosan, hyaluronic acid, and their derivatives. *Molecules* **2019**, *24*, 2570. [CrossRef]
49. Köping-Höggård, M.; Tubulekas, I.; Guan, H.; Edwards, K.; Nilsson, M.; Vårum, K.; Artursson, P. Chitosan as a nonviral gene delivery system. Structure–property relationships and characteristics compared with polyethylenimine in vitro and after lung administration in vivo. *Gene Ther.* **2001**, *8*, 1108–1121. [CrossRef]
50. Mao, H.Q.; Roy, K.; Troung-Le, V.L.; Janes, K.A.; Lin, K.Y.; Wang, Y.; August, J.T.; Leong, K.W. Chitosan-DNA nanoparticles as gene carriers: Synthesis, characterization and transfection efficiency. *J. Control. Release* **2001**, *70*, 399–421. [CrossRef]
51. Chan, P.; Kurisawa, M.; Chung, J.E.; Yang, Y.-Y. Synthesis and characterization of chitosan-g-poly(ethylene glycol)-folate as a non-viral carrier for tumor-targeted gene delivery. *Biomaterials* **2007**, *28*, 540–549. [CrossRef] [PubMed]
52. Jhaveri, A.; Deshpande, P.; Pattni, B.; Torchilin, V. Transferrin-targeted, resveratrol-loaded liposomes for the treatment of glioblastoma. *J. Control. Release* **2018**, *277*, 89–101. [CrossRef] [PubMed]
53. Gao, S.; Chen, J.; Xu, X.; Ding, Z.; Yang, Y.-H.; Hua, Z.; Zhang, J. Galactosylated low molecular weight chitosan as DNA carrier for hepatocyte-targeting. *Int. J. Pharm.* **2003**, *255*, 57–68. [CrossRef]
54. Park, I.-K.; Yang, J.; Jeong, H.-J.; Bom, H.-S.; Harada, I.; Akaike, T.; Kim, S.-I.; Cho, C.-S. Galactosylated chitosan as a synthetic extracellular matrix for hepatocytes attachment. *Biomaterials* **2003**, *24*, 2331–2337. [CrossRef]
55. Kim, T.H.; Nah, J.W.; Cho, M.-H.; Park, T.G.; Cho, C.S. Receptor-mediated gene delivery into antigen presenting cells using mannosylated chitosan/DNA nanoparticles. *J. Nanosci. Nanotechnol.* **2006**, *6*, 2796–2803. [CrossRef] [PubMed]
56. Negm, N.A.; Hefni, H.H.H.; Abd-Elaal, A.A.A.; Badr, E.A.; Abou Kana, M.T.H. Advancement on modification of chitosan biopolymer and its potential applications. *Int. J. Biol. Macromol.* **2020**, *152*, 681–702. [CrossRef] [PubMed]
57. Shastri, D.H. Thiolated chitosan: A boon to ocular delivery of therapeutics. *MOJ Bioequivalence Bioavailab.* **2017**, *3*, 34–37. [CrossRef]
58. Mahmood, A.; Lanthaler, M.; Laffleur, F.; Huck, C.W.; Bernkop-Schnürch, A. Thiolated chitosan micelles: Highly mucoadhesive drug carriers. *Carbohydr. Polym.* **2017**, *167*, 250–258. [CrossRef]
59. Boateng, J.S.; Ayensu, I. Preparation and characterization of laminated thiolated chitosan-based freeze-dried wafers for potential buccal delivery of macromolecules. *Drug Dev. Ind. Pharm.* **2014**, *40*, 611–618. [CrossRef]
60. Boateng, J.; Mitchell, J.; Pawar, H.; Ayensu, I. Functional characterisation and permeation studies of lyophilised thiolated chitosan xerogels for buccal delivery of insulin. *Protein Pept. Lett.* **2014**, *21*, 1163–1175. [CrossRef]

61. Liu, Q.; Zhang, C.; Zheng, X.; Shao, X.; Zhang, X.; Zhang, Q.; Jiang, X. Preparation and evaluation of antigen/N-trimethylaminoethylmethacrylate chitosan conjugates for nasal immunization. *Vaccine* **2014**, *32*, 2582–2590. [CrossRef] [PubMed]
62. Huang, G.-Q.; Zhang, Z.-K.; Cheng, L.-Y.; Xiao, J.-X. Intestine-targeted delivery potency of O-carboxymethyl chitosan-coated layer-by-layer microcapsules: An in vitro and in vivo evaluation. *Mater. Sci. Eng. C Mater. Biol. Appl.* **2019**, *105*, 110129. [CrossRef] [PubMed]
63. Zhou, P.; Sun, X.; Zhang, Z. Kidney-targeted drug delivery systems. *Acta Pharm. Sin. B* **2014**, *4*, 37–42. [CrossRef] [PubMed]
64. Vasanthakumar, T.; Rubinstein, J.L. Structure and roles of V-type ATPases. *Trends Biochem. Sci.* **2020**, *45*, 295–307. [CrossRef]
65. Li, W.; Nicol, F.; Szoka, F.C. GALA: A designed synthetic pH-responsive amphipathic peptide with applications in drug and gene delivery. *Adv. Drug Deliv. Rev.* **2004**, *56*, 967–985. [CrossRef]
66. El Ouahabi, A.; Thiry, M.; Pector, V.; Fuks, R.; Ruysschaert, J.M.; Vandenbranden, M. The role of endosome destabilizing activity in the gene transfer process mediated by cationic lipids. *FEBS Lett.* **1997**, *414*, 187–192.
67. Ma, Z.; Yang, C.; Song, W.; Wang, Q.; Kjems, J.; Gao, S. Chitosan Hydrogel as siRNA vector for prolonged gene silencing. *J. Nanobiotechnol.* **2014**, *12*, 23. [CrossRef]
68. Shi, B.; Zheng, M.; Tao, W.; Chung, R.; Jin, D.; Ghaffari, D.; Farokhzad, O.C. Challenges in DNA delivery and recent advances in multifunctional polymeric DNA delivery systems. *Biomacromolecules* **2017**, *18*, 2231–2246. [CrossRef]
69. Molinaro, R.; Wolfram, J.; Federico, C.; Cilurzo, F.; Di Marzio, L.; Ventura, C.A.; Carafa, M.; Celia, C.; Fresta, M. Polyethylenimine and chitosan carriers for the delivery of RNA interference effectors. *Expert Opin. Drug Deliv.* **2013**, *10*, 1653–1668. [CrossRef]
70. Boussif, O.; Lezoualc'h, F.; Zanta, M.A.; Mergny, M.D.; Scherman, D.; Demeneix, B.; Behr, J.P. A versatile vector for gene and oligonucleotide transfer into cells in culture and in vivo: Polyethylenimine. *Proc. Natl. Acad. Sci. USA* **1995**, *92*, 7297–7301. [CrossRef]
71. Bae, Y.; Lee, Y.H.; Lee, S.; Han, J.; Ko, K.S.; Choi, J.S. Characterization of glycol chitosan grafted with low molecular weight polyethylenimine as a gene carrier for human adipose-derived mesenchymal stem cells. *Carbohydr. Polym.* **2016**, *153*, 379–390. [CrossRef] [PubMed]
72. Chen, H.; Cui, S.; Zhao, Y.; Zhang, C.; Zhang, S.; Peng, X. Grafting chitosan with polyethylenimine in an ionic liquid for efficient gene delivery. *PLoS ONE* **2015**, *10*, e0121817. [CrossRef] [PubMed]
73. Liu, Q.; Jin, Z.; Huang, W.; Sheng, Y.; Wang, Z.; Guo, S. Tailor-made ternary nanopolyplexes of thiolated trimethylated chitosan with pDNA and folate conjugated cis-aconitic amide-polyethylenimine for efficient gene delivery. *Int. J. Biol. Macromol.* **2020**, *152*, 948–956. [CrossRef]
74. Lee, Y.H.; Park, H.I.; Choi, J.S. Novel glycol chitosan-based polymeric gene carrier synthesized by a Michael addition reaction with low molecular weight polyethylenimine. *Carbohydr. Polym.* **2016**, *137*, 669–677. [CrossRef]
75. Javan, B.; Atyabi, F.; Shahbazi, M. Hypoxia-inducible bidirectional shRNA expression vector delivery using PEI/chitosan-TBA copolymers for colorectal Cancer gene therapy. *Life Sci.* **2018**, *202*, 140–151. [CrossRef]
76. Sun, B.; Zhao, R.; Kong, F.; Ren, Y.; Zuo, A.; Liang, D.; Zhang, J. Phosphorylatable short peptide conjugation for facilitating transfection efficacy of CS/DNA complex. *Int. J. Pharm.* **2010**, *397*, 206–210. [CrossRef]
77. Zhao, R.; Sun, B.; Liu, T.; Liu, Y.; Zhou, S.; Zuo, A.; Liang, D. Optimize nuclear localization and intra-nucleus disassociation of the exogene for facilitating transfection efficacy of the chitosan. *Int. J. Pharm.* **2011**, *413*, 254–259. [CrossRef]
78. Miao, J.; Yang, X.; Gao, Z.; Li, Q.; Meng, T.; Wu, J.; Yuan, H.; Hu, F. Redox-responsive chitosan oligosaccharide-SS-Octadecylamine polymeric carrier for efficient anti-Hepatitis B Virus gene therapy. *Carbohydr. Polym.* **2019**, *212*, 215–221. [CrossRef]
79. Cryan, S.-A.; McKiernan, P.; Cunningham, C.M.; Greene, C. Targeting miRNA-based medicines to cystic fibrosis airway epithelial cells using nanotechnology. *Int. J. Nanomed.* **2013**, *8*, 3907. [CrossRef]
80. González, J.D.; Silva-Marrero, J.I.; Metón, I.; Caballero-Solares, A.; Viegas, I.; Fernández, F.; Miñarro, M.; Fàbregas, A.; Ticó, J.R.; Jones, J.G.; et al. Chitosan-mediated shRNA knockdown of cytosolic alanine aminotransferase improves hepatic carbohydrate metabolism. *Mar. Biotechnol.* **2016**, *18*, 85–97. [CrossRef]

81. Gaspar, C.; Silva-Marrero, J.I.; Fàbregas, A.; Miñarro, M.; Ticó, J.R.; Baanante, I.V.; Metón, I. Administration of chitosan-tripolyphosphate-DNA nanoparticles to knockdown glutamate dehydrogenase expression impairs transdeamination and gluconeogenesis in the liver. *J. Biotechnol.* **2018**, *286*, 5–13. [CrossRef] [PubMed]
82. Acharya, R. The recent progresses in shRNA-nanoparticle conjugate as a therapeutic approach. *Mater. Sci. Eng. C* **2019**, *104*, 109928. [CrossRef] [PubMed]
83. Zheng, H.; Tang, C.; Yin, C. Oral delivery of shRNA based on amino acid modified chitosan for improved antitumor efficacy. *Biomaterials* **2015**, *70*, 126–137. [CrossRef] [PubMed]
84. Wang, S.-L.; Yao, H.-H.; Guo, L.-L.; Dong, L.; Li, S.-G.; Gu, Y.-P.; Qin, Z.-H. Selection of optimal sites for TGFB1 gene silencing by chitosan-TPP nanoparticle-mediated delivery of shRNA. *Cancer Genet. Cytogenet.* **2009**, *190*, 8–14. [CrossRef] [PubMed]
85. Karimi, M.; Avci, P.; Ahi, M.; Gazori, T.; Hamblin, M.R.; Naderi-Manesh, H. Evaluation of chitosan-tripolyphosphate nanoparticles as a p-shRNA delivery vector: Formulation, optimization and cellular uptake study. *J. Nanopharm. Drug Deliv.* **2013**, *1*, 266–278. [CrossRef] [PubMed]
86. Ahmed, F.; Soliman, F.M.; Adly, M.A.; Soliman, H.A.M.; El-Matbouli, M.; Saleh, M. Recent progress in biomedical applications of chitosan and its nanocomposites in aquaculture: A review. *Res. Vet. Sci.* **2019**, *126*, 68–82. [CrossRef]
87. Abdel-Ghany, H.M.; Salem, M.E. Effects of dietary chitosan supplementation on farmed fish; a review. *Rev. Aquac.* **2020**, *12*, 438–452. [CrossRef]
88. Kono, M.; Matsui, T.; Shimizu, C. Effect of chitin, chitosan, and cellulose as deit supplements on the growth of cultured fish. *Bull. Jpn. Soc. Sci. Fish.* **1987**, *53*, 125–129. [CrossRef]
89. Shiau, S.-Y.; Yu, Y.-P. Dietary supplementation of chitin and chitosan depresses growth in tilapia, Oreochromis niloticus×O. aureus. *Aquaculture* **1999**, *179*, 439–446. [CrossRef]
90. Wu, S. The growth performance, body composition and nonspecific immunity of Tilapia (Oreochromis niloticus) affected by chitosan. *Int. J. Biol. Macromol.* **2020**, *145*, 682–685. [CrossRef]
91. Fadl, S.E.; El-Gammal, G.A.; Abdo, W.S.; Barakat, M.; Sakr, O.A.; Nassef, E.; Gad, D.M.; El-Sheshtawy, H.S. Evaluation of dietary chitosan effects on growth performance, immunity, body composition and histopathology of Nile tilapia (*Oreochromis niloticus*) as well as the resistance to *Streptococcus agalactiae* infection. *Aquac. Res.* **2020**, *51*, 1120–1132. [CrossRef]
92. Chen, Y.; Zhu, X.; Yang, Y.; Han, D.; Jin, J.; Xie, S. Effect of dietary chitosan on growth performance, haematology, immune response, intestine morphology, intestine microbiota and disease resistance in gibel carp (*Carassius auratus gibelio*). *Aquac. Nutr.* **2014**, *20*, 532–546. [CrossRef]
93. El-Sayed, H.S.; Barakat, K.M. Effect of dietary chitosan on challenged Dicentrarchus labrax post larvae with Aeromonas hydrophila. *Russ. J. Mar. Biol.* **2016**, *42*, 501–508. [CrossRef]
94. Yan, J.; Guo, C.; Dawood, M.A.O.; Gao, J. Effects of dietary chitosan on growth, lipid metabolism, immune response and antioxidant-related gene expression in *Misgurnus anguillicaudatus*. *Benef. Microbes* **2017**, *8*, 439–449. [CrossRef] [PubMed]
95. Chen, J.; Chen, L. Effects of chitosan-supplemented diets on the growth performance, nonspecific immunity and health of loach fish (Misgurnus anguillicadatus). *Carbohydr. Polym.* **2019**, *225*, 115227. [CrossRef] [PubMed]
96. Kamali Najafabad, M.; Imanpoor, M.R.; Taghizadeh, V.; Alishahi, A. Effect of dietary chitosan on growth performance, hematological parameters, intestinal histology and stress resistance of Caspian kutum (Rutilus frisii kutum Kamenskii, 1901) fingerlings. *Fish Physiol. Biochem.* **2016**, *42*, 1063–1071. [CrossRef] [PubMed]
97. Samarakoon, K.W.; Cha, S.-H.; Lee, J.-H.; Jeon, Y.-J. The growth, innate immunity and protection against H_2O_2-induced oxidative damage of a chitosan-coated diet in the olive flounder Paralichthys olivaceus. *Fish. Aquat. Sci.* **2013**, *16*, 149–158. [CrossRef]
98. Ranjan, R.; Prasad, K.P.; Vani, T.; Kumar, R. Effect of dietary chitosan on haematology, innate immunity and disease resistance of Asian seabass Lates calcarifer (Bloch). *Aquac. Res.* **2014**, *45*, 983–993. [CrossRef]
99. Shanthi Mari, L.S.; Jagruthi, C.; Anbazahan, S.M.; Yogeshwari, G.; Thirumurugan, R.; Arockiaraj, J.; Mariappan, P.; Balasundaram, C.; Harikrishnan, R. Protective effect of chitin and chitosan enriched diets on immunity and disease resistance in Cirrhina mrigala against Aphanomyces invadans. *Fish Shellfish Immunol.* **2014**, *39*, 378–385. [CrossRef]

100. Harikrishnan, R.; Kim, J.-S.; Balasundaram, C.; Heo, M.-S. Immunomodulatory effects of chitin and chitosan enriched diets in Epinephelus bruneus against Vibrio alginolyticus infection. *Aquaculture* **2012**, *326–329*, 46–52. [CrossRef]
101. Harikrishnan, R.; Kim, J.-S.; Balasundaram, C.; Heo, M.-S. Dietary supplementation with chitin and chitosan on haematology and innate immune response in Epinephelus bruneus against Philasterides dicentrarchi. *Exp. Parasitol.* **2012**, *131*, 116–124. [CrossRef] [PubMed]
102. Wang, Y.; Li, J. Effects of chitosan nanoparticles on survival, growth and meat quality of tilapia, Oreochromis nilotica. *Nanotoxicology* **2011**, *5*, 425–431. [CrossRef] [PubMed]
103. Abdel-Tawwab, M.; Razek, N.A.; Abdel-Rahman, A.M. Immunostimulatory effect of dietary chitosan nanoparticles on the performance of Nile tilapia, Oreochromis niloticus (L.). *Fish Shellfish Immunol.* **2019**, *88*, 254–258. [CrossRef] [PubMed]
104. Abd El-Naby, F.S.; Naiel, M.A.E.; Al-Sagheer, A.A.; Negm, S.S. Dietary chitosan nanoparticles enhance the growth, production performance, and immunity in Oreochromis niloticus. *Aquaculture* **2019**, *501*, 82–89. [CrossRef]
105. Naiel, M.A.E.; Ismael, N.E.M.; Abd El-hameed, S.A.A.; Amer, M.S. The antioxidative and immunity roles of chitosan nanoparticle and vitamin C-supplemented diets against imidacloprid toxicity on Oreochromis niloticus. *Aquaculture* **2020**, *523*, 735219. [CrossRef]
106. Abd El-Naby, A.S.; Al-Sagheer, A.A.; Negm, S.S.; Naiel, M.A.E. Dietary combination of chitosan nanoparticle and thymol affects feed utilization, digestive enzymes, antioxidant status, and intestinal morphology of Oreochromis niloticus. *Aquaculture* **2020**, *515*, 734577. [CrossRef]
107. Gao, J.-Q.; Hu, Y.L.; Wang, Q.; Han, F.; Shao, J.Z. Toxicity evaluation of biodegradable chitosan nanoparticles using a zebrafish embryo model. *Int. J. Nanomed.* **2011**, *6*, 3351–3359. [CrossRef]
108. Nikapitiya, C.; Dananjaya, S.H.S.; De Silva, B.C.J.; Heo, G.-J.; Oh, C.; De Zoysa, M.; Lee, J. Chitosan nanoparticles: A positive immune response modulator as display in zebrafish larvae against Aeromonas hydrophila infection. *Fish Shellfish Immunol.* **2018**, *76*, 240–246. [CrossRef]
109. Qin, C.; Zhang, Y.; Liu, W.; Xu, L.; Yang, Y.; Zhou, Z. Effects of chito-oligosaccharides supplementation on growth performance, intestinal cytokine expression, autochthonous gut bacteria and disease resistance in hybrid tilapia Oreochromis niloticus ♀× Oreochromis aureus ♂. *Fish Shellfish Immunol.* **2014**, *40*, 267–274. [CrossRef]
110. Gopalakannan, A.; Arul, V. Immunomodulatory effects of dietary intake of chitin, chitosan and levamisole on the immune system of Cyprinus carpio and control of Aeromonas hydrophila infection in ponds. *Aquaculture* **2006**, *255*, 179–187. [CrossRef]
111. Karlsen, Ø.; Amlund, H.; Berg, A.; Olsen, R.E. The effect of dietary chitin on growth and nutrient digestibility in farmed Atlantic cod, Atlantic salmon and Atlantic halibut. *Aquac. Res.* **2017**, *48*, 123–133. [CrossRef]
112. Lin, S.-M.; Jiang, Y.; Chen, Y.-J.; Luo, L.; Doolgindachbaporn, S.; Yuangsoi, B. Effects of Astragalus polysaccharides (APS) and chitooligosaccharides (COS) on growth, immune response and disease resistance of juvenile largemouth bass, Micropterus salmoides. *Fish Shellfish Immunol.* **2017**, *70*, 40–47. [CrossRef] [PubMed]
113. Nguyen, N.D.; Van Dang, P.; Le, A.Q.; Nguyen, T.K.L.; Pham, D.H.; Van Nguyen, N.; Nguyen, Q.H. Effect of oligochitosan and oligo-β-glucan supplementation on growth, innate immunity, and disease resistance of striped catfish (Pangasianodon hypophthalmus). *Biotechnol. Appl. Biochem.* **2017**, *64*, 564–571. [CrossRef]
114. Meng, X.; Wang, J.; Wan, W.; Xu, M.; Wang, T. Influence of low molecular weight chitooligosaccharides on growth performance and non-specific immune response in Nile tilapia Oreochromis niloticus. *Aquac. Int.* **2017**, *25*, 1265–1277. [CrossRef]
115. Su, P.; Han, Y.; Jiang, C.; Ma, Y.; Pan, J.; Liu, S.; Zhang, T. Effects of chitosan-oligosaccharides on growth performance, digestive enzyme and intestinal bacterial flora of tiger puffer (*Takifugu rubripes* Temminck et Schlegel, 1850). *J. Appl. Ichthyol.* **2017**, *33*, 458–467. [CrossRef]
116. Lin, S.; Mao, S.; Guan, Y.; Luo, L.; Luo, L.; Pan, Y. Effects of dietary chitosan oligosaccharides and Bacillus coagulans on the growth, innate immunity and resistance of koi (Cyprinus carpio koi). *Aquaculture* **2012**, *342–343*, 36–41.
117. Lin, S.; Mao, S.; Guan, Y.; Lin, X.; Luo, L. Dietary administration of chitooligosaccharides to enhance growth, innate immune response and disease resistance of Trachinotus ovatus. *Fish Shellfish Immunol.* **2012**, *32*, 909–913. [CrossRef]

118. Liu, Y.; Xing, R.; Liu, S.; Qin, Y.; Li, K.; Yu, H.; Li, P. Effects of chitooligosaccharides supplementation with different dosages, molecular weights and degrees of deacetylation on growth performance, innate immunity and hepatopancreas morphology in Pacific white shrimp (Litopenaeus vannamei). *Carbohydr. Polym.* **2019**, *226*, 115254. [CrossRef]

119. Luo, L.; Cai, X.; He, C.; Xue, M.; Wu, X.; Cao, H. Immune response, stress resistance and bacterial challenge in juvenile rainbow trouts Oncorhynchus mykiss fed diets containing chitosan-oligosaccharides. *Curr. Zool.* **2009**, *55*, 416–422. [CrossRef]

120. Liu, L.; Zhou, Y.; Zhao, X.; Wang, H.; Wang, L.; Yuan, G.; Asim, M.; Wang, W.; Zeng, L.; Liu, X.; et al. Oligochitosan stimulated phagocytic activity of macrophages from blunt snout bream (Megalobrama amblycephala) associated with respiratory burst coupled with nitric oxide production. *Dev. Comp. Immunol.* **2014**, *47*, 17–24. [CrossRef]

121. Fernández-Díaz, C.; Coste, O.; Malta, E. Polymer chitosan nanoparticles functionalized with Ulva ohnoi extracts boost in vitro ulvan immunostimulant effect in Solea senegalensis macrophages. *Algal Res.* **2017**, *26*, 135–142. [CrossRef]

122. Wisdom, K.S.; Bhat, I.A.; Chanu, T.I.; Kumar, P.; Pathakota, G.-B.; Nayak, S.K.; Walke, P.; Sharma, R. Chitosan grafting onto single-walled carbon nanotubes increased their stability and reduced the toxicity in vivo (catfish) model. *Int. J. Biol. Macromol.* **2020**, *155*, 697–707. [CrossRef] [PubMed]

123. Alishahi, A.; Mirvaghefi, A.; Tehrani, M.R.; Farahmand, H.; Koshio, S.; Dorkoosh, F.A.; Elsabee, M.Z. Chitosan nanoparticle to carry vitamin C through the gastrointestinal tract and induce the non-specific immunity system of rainbow trout (Oncorhynchus mykiss). *Carbohydr. Polym.* **2011**, *86*, 142–146. [CrossRef]

124. Jiménez-Fernández, E.; Ruyra, A.; Roher, N.; Zuasti, E.; Infante, C.; Fernández-Díaz, C. Nanoparticles as a novel delivery system for vitamin C administration in aquaculture. *Aquaculture* **2014**, *432*, 426–433. [CrossRef]

125. Bhat, I.A.; Nazir, M.I.; Ahmad, I.; Pathakota, G.-B.; Chanu, T.I.; Goswami, M.; Sundaray, J.K.; Sharma, R. Fabrication and characterization of chitosan conjugated eurycomanone nanoparticles: In vivo evaluation of the biodistribution and toxicity in fish. *Int. J. Biol. Macromol.* **2018**, *112*, 1093–1103. [CrossRef]

126. Wisdom, K.S.; Bhat, I.A.; Kumar, P.; Pathan, M.K.; Chanu, T.I.; Walke, P.; Sharma, R. Fabrication of chitosan nanoparticles loaded with aromatase inhibitors for the advancement of gonadal development in Clarias magur (Hamilton, 1822). *Aquaculture* **2018**, *497*, 125–133. [CrossRef]

127. Bhat, I.A.; Ahmad, I.; Mir, I.N.; Yousf, D.J.; Ganie, P.A.; Bhat, R.A.H.; Gireesh-Babu, P.; Sharma, R. Evaluation of the in vivo effect of chitosan conjugated eurycomanone nanoparticles on the reproductive response in female fish model. *Aquaculture* **2019**, *510*, 392–399. [CrossRef]

128. Bhat, I.A.; Ahmad, I.; Mir, I.N.; Bhat, R.A.H.; P, G.-B.; Goswami, M.; Sundaray, J.K.; Sharma, R. Chitosan-eurycomanone nanoformulation acts on steroidogenesis pathway genes to increase the reproduction rate in fish. *J. Steroid Biochem. Mol. Biol.* **2019**, *185*, 237–247. [CrossRef]

129. Barakat, K.M.; El-Sayed, H.S.; Gohar, Y.M. Protective effect of squilla chitosan–silver nanoparticles for Dicentrarchus labrax larvae infected with Vibrio anguillarum. *Int. Aquat. Res.* **2016**, *8*, 179–189. [CrossRef]

130. Udayangani, R.M.C.; Dananjaya, S.H.S.; Nikapitiya, C.; Heo, G.-J.; Lee, J.; De Zoysa, M. Metagenomics analysis of gut microbiota and immune modulation in zebrafish (Danio rerio) fed chitosan silver nanocomposites. *Fish Shellfish Immunol.* **2017**, *66*, 173–184. [CrossRef]

131. Xia, I.F.; Cheung, J.S.; Wu, M.; Wong, K.-S.; Kong, H.-K.; Zheng, X.-T.; Wong, K.-H.; Kwok, K.W. Dietary chitosan-selenium nanoparticle (CTS-SeNP) enhance immunity and disease resistance in zebrafish. *Fish Shellfish Immunol.* **2019**, *87*, 449–459. [CrossRef] [PubMed]

132. Victor, H.; Zhao, B.; Mu, Y.; Dai, X.; Wen, Z.; Gao, Y.; Chu, Z. Effects of Se-chitosan on the growth performance and intestinal health of the loach Paramisgurnus dabryanus (Sauvage). *Aquaculture* **2019**, *498*, 263–270. [CrossRef]

133. Zhang, J.; Fu, X.; Zhang, Y.; Zhu, W.; Zhou, Y.; Yuan, G.; Liu, X.; Ai, T.; Zeng, L.; Su, J. Chitosan and anisodamine improve the immune efficacy of inactivated infectious spleen and kidney necrosis virus vaccine in Siniperca chuatsi. *Fish Shellfish Immunol.* **2019**, *89*, 52–60. [CrossRef]

134. Zhu, W.; Zhang, Y.; Zhang, J.; Yuan, G.; Liu, X.; Ai, T.; Su, J. Astragalus polysaccharides, chitosan and poly(I:C) obviously enhance inactivated Edwardsiella ictaluri vaccine potency in yellow catfish Pelteobagrus fulvidraco. *Fish Shellfish Immunol.* **2019**, *87*, 379–385. [CrossRef]

135. Liu, X.; Zhang, H.; Gao, Y.; Zhang, Y.; Wu, H.; Zhang, Y. Efficacy of chitosan oligosaccharide as aquatic adjuvant administrated with a formalin-inactivated Vibrio anguillarum vaccine. *Fish Shellfish Immunol.* **2015**, *47*, 855–860. [CrossRef]
136. Wei, G.; Cai, S.; Wu, Y.; Ma, S.; Huang, Y. Immune effect of Vibrio harveyi formalin-killed cells vaccine combined with chitosan oligosaccharide and astragalus polysaccharides in ♀Epinephelus fuscoguttatus×♂Epinephelus lanceolatus. *Fish Shellfish Immunol.* **2020**, *98*, 186–192. [CrossRef]
137. Halimi, M.; Alishahi, M.; Abbaspour, M.R.; Ghorbanpoor, M.; Tabandeh, M.R. Valuable method for production of oral vaccine by using alginate and chitosan against Lactococcus garvieae/Streptococcus iniae in rainbow trout (Oncorhynchus mykiss). *Fish Shellfish Immunol.* **2019**, *90*, 431–439. [CrossRef]
138. Kole, S.; Qadiri, S.S.N.; Shin, S.-M.; Kim, W.-S.; Lee, J.; Jung, S.-J. Nanoencapsulation of inactivated-viral vaccine using chitosan nanoparticles: Evaluation of its protective efficacy and immune modulatory effects in olive flounder (Paralichthys olivaceus) against viral haemorrhagic septicaemia virus (VHSV) infection. *Fish Shellfish Immunol.* **2019**, *91*, 136–147. [CrossRef]
139. Tandberg, J.; Lagos, L.; Ropstad, E.; Smistad, G.; Hiorth, M.; Winther-Larsen, H.C. The use of chitosan-coated membrane vesicles for immunization against salmonid rickettsial septicemia in an adult zebrafish model. *Zebrafish* **2018**, *15*, 372–381. [CrossRef]
140. Dubey, S.; Avadhani, K.; Mutalik, S.; Sivadasan, S.; Maiti, B.; Girisha, S.; Venugopal, M.; Mutoloki, S.; Evensen, Ø.; Karunasagar, I.; et al. Edwardsiella tarda OmpA encapsulated in chitosan nanoparticles shows superior protection over inactivated whole cell vaccine in orally vaccinated fringed-lipped peninsula carp (Labeo fimbriatus). *Vaccines* **2016**, *4*, 40. [CrossRef]
141. Wang, E.; Wang, X.; Wang, K.; He, J.; Zhu, L.; He, Y.; Chen, D.; Ouyang, P.; Geng, Y.; Huang, X.; et al. Preparation, characterization and evaluation of the immune effect of alginate/chitosan composite microspheres encapsulating recombinant protein of Streptococcus iniae designed for fish oral vaccination. *Fish Shellfish Immunol.* **2018**, *73*, 262–271. [CrossRef] [PubMed]
142. Chen, T.; Hu, Y.; Zhou, J.; Hu, S.; Xiao, X.; Liu, X.; Su, J.; Yuan, G. Chitosan reduces the protective effects of IFN-γ2 on grass carp (Ctenopharyngodon idella) against Flavobacterium columnare infection due to excessive inflammation. *Fish Shellfish Immunol.* **2019**, *95*, 305–313. [CrossRef] [PubMed]
143. Sharma, D.; Maheshwari, D.; Philip, G.; Rana, R.; Bhatia, S.; Singh, M.; Gabrani, R.; Sharma, S.K.; Ali, J.; Sharma, R.K.; et al. Formulation and optimization of polymeric nanoparticles for intranasal delivery of lorazepam using Box-Behnken design: In vitro and in vivo evaluation. *Biomed Res. Int.* **2014**, *2014*, 156010. [CrossRef]
144. Rather, M.A.; Bhat, I.A.; Gireesh-Babu, P.; Chaudhari, A.; Sundaray, J.K.; Sharma, R. Molecular characterization of kisspeptin gene and effect of nano-encapsulted kisspeptin-10 on reproductive maturation in Catla catla. *Domest. Anim. Endocrinol.* **2016**, *56*, 36–47. [CrossRef] [PubMed]
145. Tian, J.; Yu, J.; Sun, X. Chitosan microspheres as candidate plasmid vaccine carrier for oral immunisation of Japanese flounder (Paralichthys olivaceus). *Vet. Immunol. Immunopathol.* **2008**, *126*, 220–229. [CrossRef]
146. Vimal, S.; Taju, G.; Nambi, K.S.N.; Abdul Majeed, S.; Sarath Babu, V.; Ravi, M.; Sahul Hameed, A.S. Synthesis and characterization of CS/TPP nanoparticles for oral delivery of gene in fish. *Aquaculture* **2012**, *358–359*, 14–22. [CrossRef]
147. Li, L.; Lin, S.-L.; Deng, L.; Liu, Z.-G. Potential use of chitosan nanoparticles for oral delivery of DNA vaccine in black seabream Acanthopagrus schlegelii Bleeker to protect from Vibrio parahaemolyticus. *J. Fish Dis.* **2013**, *36*, 987–995. [CrossRef]
148. Vimal, S.; Abdul Majeed, S.; Nambi, K.S.N.; Madan, N.; Farook, M.A.; Venkatesan, C.; Taju, G.; Venu, S.; Subburaj, R.; Thirunavukkarasu, A.R.; et al. Delivery of DNA vaccine using chitosan–tripolyphosphate (CS/TPP) nanoparticles in Asian sea bass, Lates calcarifer (Bloch, 1790) for protection against nodavirus infection. *Aquaculture* **2014**, *420–421*, 240–246. [CrossRef]
149. Zheng, F.; Liu, H.; Sun, X.; Zhang, Y.; Zhang, B.; Teng, Z.; Hou, Y.; Wang, B. Development of oral DNA vaccine based on chitosan nanoparticles for the immunization against reddish body iridovirus in turbots (Scophthalmus maximus). *Aquaculture* **2016**, *452*, 263–271. [CrossRef]
150. Bhat, I.A.; Rather, M.A.; Saha, R.; Pathakota, G.-B.; Pavan-Kumar, A.; Sharma, R. Expression analysis of Sox9 genes during annual reproductive cycles in gonads and after nanodelivery of LHRH in Clarias batrachus. *Res. Vet. Sci.* **2016**, *106*, 100–106. [CrossRef]
151. Valero, Y.; Awad, E.; Buonocore, F.; Arizcun, M.; Esteban, M.Á.; Meseguer, J.; Chaves-Pozo, E.; Cuesta, A. An oral chitosan DNA vaccine against nodavirus improves transcription of cell-mediated cytotoxicity and

interferon genes in the European sea bass juveniles gut and survival upon infection. *Dev. Comp. Immunol.* **2016**, *65*, 64–72. [CrossRef] [PubMed]
152. Sáez, M.I.; Vizcaíno, A.J.; Alarcón, F.J.; Martínez, T.F. Comparison of lacZ reporter gene expression in gilthead sea bream (Sparus aurata) following oral or intramuscular administration of plasmid DNA in chitosan nanoparticles. *Aquaculture* **2017**, *474*, 1–10. [CrossRef]
153. Rathor, P.K.; Bhat, I.A.; Rather, M.A.; Gireesh-Babu, P.; Kumar, K.; Purayil, S.B.P.; Sharma, R. Steroidogenic acute regulatory protein (StAR) gene expression construct: Development, nanodelivery and effect on reproduction in air-breathing catfish, Clarias batrachus. *Int. J. Biol. Macromol.* **2017**, *104*, 1082–1090. [CrossRef] [PubMed]
154. Ahmadivand, S.; Soltani, M.; Behdani, M.; Evensen, Ø.; Alirahimi, E.; Hassanzadeh, R.; Soltani, E. Oral DNA vaccines based on CS-TPP nanoparticles and alginate microparticles confer high protection against infectious pancreatic necrosis virus (IPNV) infection in trout. *Dev. Comp. Immunol.* **2017**, *74*, 178–189. [CrossRef] [PubMed]
155. Sáez, M.I.; Vizcaíno, A.J.; Alarcón, F.J.; Martínez, T.F. Feed pellets containing chitosan nanoparticles as plasmid DNA oral delivery system for fish: In vivo assessment in gilthead sea bream (Sparus aurata) juveniles. *Fish Shellfish Immunol.* **2018**, *80*, 458–466. [CrossRef]
156. Kole, S.; Kumari, R.; Anand, D.; Kumar, S.; Sharma, R.; Tripathi, G.; Makesh, M.; Rajendran, K.V.; Bedekar, M.K. Nanoconjugation of bicistronic DNA vaccine against Edwardsiella tarda using chitosan nanoparticles: Evaluation of its protective efficacy and immune modulatory effects in Labeo rohita vaccinated by different delivery routes. *Vaccine* **2018**, *36*, 2155–2165. [CrossRef]
157. Silva-Marrero, J.I.; Villasante, J.; Rashidpour, A.; Palma, M.; Fàbregas, A.; Almajano, M.P.; Viegas, I.; Jones, J.G.; Miñarro, M.; Ticó, J.R.; et al. The administration of chitosan-tripolyphosphate-DNA nanoparticles to express exogenous SREBP1a enhances conversion of dietary carbohydrates into lipids in the liver of Sparus aurata. *Biomolecules* **2019**, *9*, 297. [CrossRef]
158. Rao, B.M.; Kole, S.; Gireesh-Babu, P.; Sharma, R.; Tripathi, G.; Bedekar, M.K. Evaluation of persistence, bio-distribution and environmental transmission of chitosan/PLGA/pDNA vaccine complex against Edwardsiella tarda in Labeo rohita. *Aquaculture* **2019**, *500*, 385–392. [CrossRef]
159. Ramos, E.A.; Relucio, J.L.V.; Torres-Villanueva, C.A.T. Gene expression in tilapia following oral delivery of chitosan-encapsulated plasmid DNA incorporated into fish feeds. *Mar. Biotechnol.* **2005**, *7*, 89–94. [CrossRef]
160. Rajesh Kumar, S.; Ishaq Ahmed, V.P.; Parameswaran, V.; Sudhakaran, R.; Sarath Babu, V.; Sahul Hameed, A.S. Potential use of chitosan nanoparticles for oral delivery of DNA vaccine in Asian sea bass (Lates calcarifer) to protect from Vibrio (Listonella) anguillarum. *Fish Shellfish Immunol.* **2008**, *25*, 47–56. [CrossRef]
161. Kumari, R.; Gupta, S.; Singh, A.R.; Ferosekhan, S.; Kothari, D.C.; Pal, A.K.; Jadhao, S.B. Chitosan nanoencapsulated exogenous trypsin biomimics zymogen-like enzyme in fish gastrointestinal tract. *PLoS ONE* **2013**, *8*, e74743. [CrossRef] [PubMed]
162. Naylor, R.L.; Hardy, R.W.; Bureau, D.P.; Chiu, A.; Elliott, M.; Farrell, A.P.; Forster, I.; Gatlin, D.M.; Goldburg, R.J.; Hua, K.; et al. Feeding aquaculture in an era of finite resources. *Proc. Natl. Acad. Sci. USA* **2009**, *106*, 15103–15110. [CrossRef] [PubMed]
163. Polakof, S.; Panserat, S.; Soengas, J.L.; Moon, T.W. Glucose metabolism in fish: A review. *J. Comp. Physiol. B* **2012**, *182*, 1015–1045. [CrossRef]
164. Rashidpour, A.; Silva-Marrero, J.I.; Seguí, L.; Baanante, I.V.; Metón, I. Metformin counteracts glucose-dependent lipogenesis and impairs transdeamination in the liver of gilthead sea bream (*Sparus aurata*). *Am. J. Physiol. Integr. Comp. Physiol.* **2019**, *316*, R265–R273. [CrossRef]
165. Metón, I.; Egea, M.; Anemaet, I.G.; Fernández, F.; Baanante, I.V. Sterol regulatory element binding protein-1a transactivates 6-phosphofructo-2-kinase/fructose-2,6-bisphosphatase gene promoter. *Endocrinology* **2006**, *147*, 3446–3456. [CrossRef]
166. Egea, M.; Metón, I.; Córdoba, M.; Fernández, F.; Baanante, I.V. Role of Sp1 and SREBP-1a in the insulin-mediated regulation of glucokinase transcription in the liver of gilthead sea bream (Sparus aurata). *Gen. Comp. Endocrinol.* **2008**, *155*, 359–367. [CrossRef]

© 2020 by the authors. Licensee MDPI, Basel, Switzerland. This article is an open access article distributed under the terms and conditions of the Creative Commons Attribution (CC BY) license (http://creativecommons.org/licenses/by/4.0/).

Article

Design and Preparation of New Multifunctional Hydrogels Based on Chitosan/Acrylic Polymers for Drug Delivery and Wound Dressing Applications

Ioana A. Duceac [1,2,*], Liliana Verestiuc [2,*], Cristina D. Dimitriu [3], Vasilica Maier [4] and Sergiu Coseri [1]

1. "Petru Poni" Institute of Macromolecular Chemistry of Romanian Academy, 41 A Gr. Ghica Voda Alley, 700487 Iasi, Romania; coseris@icmpp.ro
2. Department of Biomedical Sciences, Faculty of Medical Bioengineering, "Grigore T. Popa" University of Medicine and Pharmacy, 9-13 M. Kogalniceanu Street, 700454 Iasi, Romania
3. Department of Morpho-Functional Sciences, Faculty of Medicine, "Grigore T. Popa" University of Medicine and Pharmacy, 16 Universitatii Street, 700115 Iasi, Romania; daniela.dimitriu@umfiasi.ro
4. Department of Textiles and Leather Chemical Engineering, "Gheorghe Asachi" Technical University of Iasi, 700050 Iasi, Romania; vmaier@tuiasi.ro
* Correspondence: duceac.ioana@icmpp.ro (I.A.D.); liliana.verestiuc@bioinginerie.ro (L.V.); Tel.: +40-232-217454 (I.A.D.); Fax: +40-232-211299 (I.A.D.)

Received: 18 May 2020; Accepted: 26 June 2020; Published: 30 June 2020

Abstract: The dynamic evolution of materials with medical applications, particularly for drug delivery and wound dressing applications, gives impetus to design new proposed materials, among which, hydrogels represent a promising, powerful tool. In this context, multifunctional hydrogels have been obtained from chemically modified chitosan and acrylic polymers as cross-linkers, followed by subsequent conjugation with arginine. The hydrogels were finely tuned considering the variation of the synthetic monomer and the preparation conditions. The advantage of using both natural and synthetic polymers allowed porous networks with superabsorbent behavior, associated with a non-Fickian swelling mechanism. The in vitro release profiles for ibuprofen and the corresponding kinetics were studied, and the results revealed a swelling-controlled release. The biodegradability studies in the presence of lysozyme, along with the hemostatic evaluation and the induced fibroblast and stem cell proliferation, have shown that the prepared hydrogels exhibit characteristics that make them suitable for local drug delivery and wound dressing.

Keywords: superabsorbent hydrogel; *N*-citraconyl-chitosan; poly(acrylic acid)/poly(methacrylic acid)

1. Introduction

Hydrogels are a momentous collection of materials with highly diverse applications in engineering, biomedical and pharmaceutical sciences [1]. Among these, drug delivery and wound dressing are specifically of interest for scientists due to the increasing number of patients having various types of acute or chronic wounds (surgical, ulcers or burns that need emergency or constant, long-term medical assistance) combined with a considerable economic burden. The significance of this problem is easily illustrated by the growing wound dressing market share and size, from USD 7.1 bln. of the global market in 2019 to an estimated USD 12.5 bln. by 2022 [2].

An effective treatment is a constant challenge which leads to a careful selection of materials in medical practice and requests an imperative development of new advanced wound dressings with combined properties. These smart materials are able to absorb blood and wound fluids, protect the injury, accelerate the healing process by one or more mechanisms: promoting fibroblast proliferation and keratinocyte migration, which are both necessary for complete epithelialization; prevention of the wound contamination with opportunistic pathogen species; efficient transport of biologically active molecules (e.g., antimicrobial agents and other pharmaceuticals) [2].

Aiming at products that could be marketed, one of the research strategies focuses on multifunctional advanced dressings. These smart, high performance materials in the form of superabsorbent hydrogels, are expected to show various characteristics and fulfil several demands for application as wound dressing, such as to provide a porous structure required to absorb exudates, while maintaining a moist environment; a high swelling ratio and fluid retention, which entail the ability to rehydrate dry wounds (e.g., eschars); transparency to allow wound monitoring; offer the possibility to endow bioactive molecules in their matrix; good biodegradability, biocompatibility, and hemocompatibility [3,4].

Furthermore, the wound healing process is improved by using the same hydrogel system for the additional delivery of a therapeutic payload, such as antibiotics, anti-inflammatory drugs or growth factors [5]. Given the premises that hydrogels enable large amounts of bioactive molecules to be loaded into the polymeric network, they permit controlled release at the desired site as the hydrogels are placed directly at the targeted location (like a dressing or injectable/in situ gelling material), and the drug release occurs through a specific mechanism or strategy (swelling, network relaxation, temperature-induced transition, etc.), it can be hypothesized that hydrogels can be used as dressings able to perform controlled drug delivery [6].

The challenge of multifunctional hydrogel design and preparation consists of finding the optimal formulation that yields in the material with the best performance as both a drug delivery system and wound dressing. Such a macromolecular structure can be obtained by using a wise selection from the large variety of natural and synthetic polymers able to provide good biological interactions and modulated mechanical resistance and biochemical interactions [7].

Chitosan, a copolymer which consists of β-(1→4)-linked 2-acetamido-2-deoxy-D-glucopyranose and 2-amino-2-deoxy-D-glucopyranose units, is one of the most abundant polysaccharides in nature and possesses specific properties highly required for materials suitable in drug delivery, wound healing, and, ultimately, in regenerative medicine. This natural macromolecule is widely used for hydrogel fabrication due to its intrinsic properties such as excellent biocompatibility, low toxicity and immune-stimulatory activity [8], leading to positive effects on wound healing [9]. However, chitosan is poorly soluble in water and in the common organic solvents, except in aqueous acidic medium. To overcome this drawback, chitosan is often subjected to chemical modifications [10]. Chitosan is already the major component of various commercial wound dressings with hemostatic activity, such as ChitoSAM™ (SAM Medical, Wilsonville, OR, USA, 2018), ChitoGauze XR pro (North American Rescue, Greer, SC, USA, 2018), ChitoFlex (H.M.T. Inc., The Woodlands, TX, USA, 2007), and Axiostat®(AXIO, New York, NY, USA, 2018), which recommends this polymer to be further investigated for this type of application [11].

On the other hand, poly(acrylic acid) (PAA) and poly(methacrylic acid) (PAM) are synthetic polymers with biocompatible and antibacterial properties, and, therefore, are widely used as (bio)adhesives or superabsorbent materials [12–14]. Polymers grafted with acrylic acid yield in highly hydrophilic materials and tunable scaffolds for drug delivery systems due to the presence of pendant carboxylic groups [15–18]. In other words, acrylic acid-based hydrogels make excellent wound dressings due to their ability to retain large volumes of water and water-soluble drugs loaded into their matrix [19].

This study focused on the design, synthesis and advanced characterization of novel multifunctional hydrogels, having a specific composition optimized according to the targeted applications. Thus, a series of hydrogels based on N-citraconyl-chitosan and acrylates with various structures has been obtained, and their level of performance was comparatively assessed. The variation of two parameters was considered for the optimization: (i) the nature of the cross-linking agent, either PAA or PAM, and (ii) the initiator ratio used for free radical polymerization. In addition, L-arginine, an amino acid which is a key factor in accelerated wound healing and in cellular recognition and adhesion, was conjugated with the hydrogels network in order to modulate their interaction with various fluids and contact with cells. The influence of different components on the ability of hydrogels to respond to relevant biological media was also investigated. Moreover, the hydrogels capacity to act as a drug carrier and delivery system was evaluated by studying the release profiles and kinetics of ibuprofen, a nonsteroidal anti-inflammatory drug.

2. Materials and Methods

2.1. Materials and Hydrogel Preparation

Chitosan (MW 80 kDa, Sigma-Aldrich, Darmstadt, Germany) was purified by the solubilization–precipitation method. Briefly, chitosan was solubilized in 5% aqueous acetic acid solution, and then, precipitated with 20% NaOH solution, centrifuged and dialyzed against distilled water. Citraconic anhydride, acrylic acid (AA), methacrylic acid (AM), ammonium persulfate (APS), N,N,N′,N′-tetramethylethylenediamine (TEMED), L-arginine, 1-ethyl-3-(3-dimethyl-aminopropyl)carbodiimide (EDC), N-hydroxysuccinimide (NHS) and ibuprofen (Ib) were purchased from Sigma-Aldrich, Germany. AA and AM were purified by passing through an inhibitor removal column (HQ, Sigma-Aldrich, Germany) and APS was recrystallized from methanol. For the cytotoxicity assays, the following were used: Hank's Balanced Salt Solution (HBSS), Dulbecco's Modified Eagle's Medium (DMEM), thiazolyl blue tetrazolium bromide (MTT) and calcein AM (Sigma-Aldrich, Germany). All other solvents and chemicals were obtained from Sigma-Aldrich, Germany, and were used as received.

2.1.1. Chitosan Chemical Modification

Chitosan was chemically modified with citraconic anhydride by the N-acylation reaction using a previously described method. [20] Briefly, 27.5 mL citraconic anhydride were solubilized in 40 mL acetone and the solution was added dropwise under stirring to a chitosan solution (1% w/wt. in 5% acetic acid aqueous solution) in the presence of 80 mL methanol, to prevent chitosan precipitation in the presence of acetone, and has been allowed to react at room temperature (25 °C), for 18 h. The chitosan/anhydride molar ratio was 1:15, chosen based on our previous studies [20]. The obtained product was further purified through dialysis against distilled water for 7 days and then freeze-dried.

2.1.2. Hydrogels Preparation

The chemical method of hydrogel preparation is the copolymerization of citraconyl-chitosan with AA or AM by free radical polymerization with the formation of cross-linked networks. Thereby, a stock solution of 1% citraconyl-chitosan in distilled water was prepared and the necessary volume for each hydrogel was transferred. AA or AM, followed by APS and TEMED, were added to the chitosan solution under vigorous stirring at 800 rpm and allowed to homogenize for 10 min. In Table 1, the hydrogels composition is presented in terms of synthetic monomer/initiator ratio, which is the variable for the hydrogels. The AA and AM volumes are not given in the table, as it is a constant ratio between the natural and synthetic polymers of 1:7 *w/w* citraconyl-chitosan/synthetic monomer. The APS/TEMED thermolabile initiator system was chosen due to its low toxicity and it was added to the polymerization mixture in different ratios, as presented in Table 1. The mixtures were transferred into molds (10 mL glass cylinders) and the radical polymerization reaction was carried out for 2 h at 70 °C. The obtained hydrogels have been thoroughly washed in distilled water until constant pH (pH ≈ 6.5) to ensure the removal of all unreacted monomers, APS, TEMED, oligomers and other residues. The purified materials were freeze-dried for further characterization, thus, ensuring porous network conservation.

Table 1. Hydrogels composition, the cross-linking reaction yield, and final chitosan and synthetic polymer content present in 100% hydrogel, as determined by elemental analysis.

Code *	Cross-Linker	Molar Ratio (%) AA/AM: APS/TEMED	Cross-Linking Reaction Yield (%)	N-Citraconyl-Chitosan (%)	Synthetic Polymer (%)
CA6	PAA	0.6	79.13	25.89	74.11
CA8	PAA	0.8	76.26	23.09	76.91
CA10	PAA	1	59.12	29.74	70.26
CA12	PAA	1.2	63.96	25.95	74.05
CA14	PAA	1.4	60.23	20.30	79.70
CA16	PAA	1.6	58.5	34.42	65.58
CM6	PAM	0.6	59.21	17.99	82.01
CM8	PAM	0.8	51.44	20.88	79.12
CM10	PAM	1	48.03	22.16	77.84
CM12	PAM	1.2	44.17	26.88	73.12
CM14	PAM	1.4	41.36	33.56	66.44

* an additional A was added to hydrogels code after the reaction with arginine.

2.1.3. L-Arginine Coupling with the Hydrogels Network

The amino acid coupling reaction evolved in two stages. Firstly, the weighted samples were immersed in specific volumes of EDC/NHS solution buffered at pH = 5.4 for 6 h, for the activation of carboxylic groups. The pH value was chosen to ensure the availability of COOH groups from the synthetic polymer chains for the coupling reaction. The EDC/NHS solution volumes were determined for each sample based on the molar ratios COOH:EDC = 1:2, EDC:NHS = 1:1, and on the concentration of COOH groups for each hydrogel calculated from the elemental analysis data. Then, all hydrogels were washed three times with PBS, and then, put in contact with the arginine solution, at pH = 10 for 24 h, to enable the formation of the amide bond between the activated carboxylic groups from hydrogels and the α-amino moieties in the structure of arginine. Finally, the functionalized scaffolds were washed thoroughly with PBS and distilled water, then freeze-dried.

2.2. Hydrogels Characterization

2.2.1. Structural and Morphological Analysis

The chemical composition was evaluated by elemental analysis, quantifying the nitrogen content in hydrogels by the Kjeldahl method. All FTIR spectra were recorded on dried samples in KBr pellets using a Bruker Vertex 70 spectrophotometer (Berlin, Germany) and scanned within the range 400–4000 cm^{-1} in transmittance mode. Gold coated cross-sections of hydrogels were examined with a SEM Tescan-Vega microscope (Brno, Czech Republic), at ambient temperature, with an operating voltage of 30 kV, and the morphology images were analyzed with ImageJ software.

2.2.2. Fluid Retention Study

The swelling behavior of all hydrogels was tested in phosphate-buffered solution (PBS) and simulated wound fluid (SWF), using the gravimetric method (50 mg aliquots immersed in 10 mL solution, incubated at 37 °C). The PBS solution had precise parameters: pH = 7.4, density 1.02 g/mL and concentration 0.01 M. The SWF solution was prepared using albumin (2%), NaCl (0.4 M), CaCl$_2$ (0.02 M) and was finally buffered at pH = 7.4 with Trisbase (0.08 M); all components were solubilized in deionized water. Measurements were made for all aliquots at regular intervals and the swelling degree was calculated using the following equation:

$$SD = \frac{W_t - W_0}{W_0} \times 100 \qquad (1)$$

where SD—swelling degree (%), W_t—hydrogel weight at different times and W_0—initial hydrogel weight.

Before each gravimetric measurement, the swollen aliquots were gently tapped on absorbing paper to eliminate the surface liquid excess. The tests were performed in triplicate.

2.2.3. Drug Release Study

Ibuprofen was used as a model for drug release study. Initially, hydrogels were loaded by the swelling-diffusion method [21,22]. Weighted dry aliquots of 80 mg were immersed in 10 mL of 8 mg/mL ibuprofen solution in water/ethanol (1:1 v/v). After 24 h, the excess was removed and the samples were freeze-dried (using a freeze-dryer with pump for solvents, Labconco, USA). The in vitro release study was performed at 37 °C and neutral pH, chosen to mimic the extracellular environment. The aliquots were introduced in dialysis membrane and immersed in 500 mL PBS (pH = 7.2, 0.035 M), under dynamic conditions, and the samples were analyzed. The release was studied for 600 min and readings were performed automatically at preset time intervals. The absorbance was recorded at 220 nm, using an Erweka Dissolution tester DT 700 (Germany) coupled with a PharmaSpec UV-1700 spectrophotometer (Shimadzu, Japan). Blank experiments using drug-free hydrogels confirmed that there was no interfering background absorption. The released amount of ibuprofen was calculated using a standard calibration curve measured for the drug at 220 nm on a dilution series. Finally, the cumulative amount of released drug was plotted against time. Data analysis was done by fitting various kinetic models to the curves.

2.2.4. In Vitro Degradation Study

Aliquots of the materials (30 mg) were immersed in 10 mL PBS, pH = 7.4, 0.01 M, containing 1200 µg/mL lysozyme, in a dialysis membrane (MWCO = 14,000 Da) that was further immersed in 30 mL PBS and incubated at 37 °C. For measurements, 1 mL of solution was sampled and evaluated. The concentration in saccharide reducing units was measured by the potassium ferricyanide method [23]: the extracted solution was mixed with 4 mL of 0.05% potassium ferricyanide solution in 0.5M Na_2CO_3 solution, and boiled for 15 min, allowed to cool and then, diluted with 5 mL PBS. The sample absorbance was measured at 420 nm, employing a PharmaSpec UV-1700 spectrophotometer (Shimadzu, Japan) and the amount of chitosan reducing ends was calculated using a calibration curve for N-acetyl-D-glucosamine.

2.2.5. In Vitro Cytocompatibility Study

Selected hydrogels (synthesized with 0.8%, 1.2%, and 1.4% APS) were sterilized by immersion in 5 mL of 70% ethanol for 15 min and then, washed several times with HBSS and DMEM. The circular samples with 5 mm diameter cut with a sterile biopsy circular scalpel were incubated with normal human dermal fibroblasts (NHDF, Lonza, Basel, Switzerland), in 48-well plates and, as a reference, cells were seeded under the same conditions. Cell viability was assessed at 24, 48 and 72 h using a direct contact MTT assay. The absorbance of the obtained solutions was measured at 570 nm using a Tescan Sunrise Plate Reader. Cell viability was calculated by Equation (2), where RMA is the relative metabolic activity, As is absorbance of the sample and Ac is absorbance of the negative control. Each result represents the mean viability ± standard deviation of four independent experiments.

$$RMA = \frac{A_S}{A_C} \times 100 \qquad (2)$$

In addition to the MTT test, a live/dead staining assay was performed. Sterile hydrogel samples were put in direct contact with stem cells. At pre-set time intervals, the calcein solution was used to color the live cells by incubation for 20 min at 37 °C, followed by a microscopy analysis. Images were taken in phase contrast and fluorescence using a Leica DM IL LED Inverted Microscope with a Phase Contrast System and Fluorescence (Leica Microsystems GmbH, Wetzlar, Germany).

2.2.6. Hemostatic Properties

The prothrombin time (PT) of blood and fibrinogen concentration in blood after contact with the prepared hydrogels were measured. Integral blood from healthy, non-smoker volunteers (venous puncture) was incubated with anticoagulant (aqueous sodium citrate 3.8% *w/v*; ratio 1/9 *v/v*). The hydrogels (5 × 10 × 1 mm) were added to 5 mL blood, and the control sample was considered the free integral blood, and both the sample and control were incubated at room temperature for 30 min. After that, the hydrogels were separated from blood by centrifugation (2500 rpm, 10 min). PT in blood plasma was determined (as a mean of three values) using a semi-automatic coagulometer Helena with 2 channels and photo-optical technique coagulation systems and PT-Fibrinogen kit (International Sensitivity Index (ISI) = 1.07). The International Normalized Ratio (INR) was calculated as the ratio of prothrombin times recorded for the hydrogels (PTH) and control (PTC) samples, raised to the power of the ISI value:

$$INR = \left(\frac{PTH}{PTC}\right)^{ISI} \quad (3)$$

3. Results and Discussions

3.1. Hydrogels Structure

The design of chitosan-based hydrogels considered the parameters that influence the formation of the cross-linked polymeric networks. A schematic illustration of the hydrogel preparation is illustrated in Figure 1. In the first stage, chitosan was chemically modified with citraconic anhydride. The anhydride reacted with the amino groups available on the chitosan backbone and yielded in amide bonds, which induced novel useful properties. The C=C bonds from the anhydride moieties are available for further copolymerization.

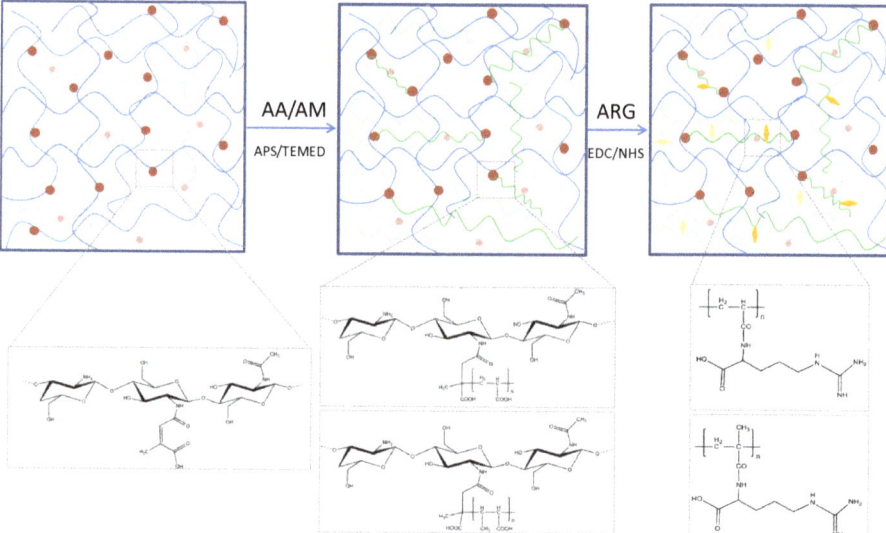

Figure 1. Schematic representation of the hydrogel network.

In order to observe and compare the hydrogels structure and properties depending on the acrylate nature, a constant 1:7 ratio (*w/w*) between citraconyl-chitosan and the synthetic monomer was chosen for both AA- and AM-containing materials.

The free radical polymerization in the second stage was initiated by the thermal decomposition of APS molecules, in the presence of TEMED, at 70 °C, when the free radicals reacted with either monomer molecules, AA or AM, or with citraconyl sequences in the chitosan structure. Consequently, it can be assumed that the network was finally formed of (1) PAA or PAM cross-links between N-citraconyl-chitosan chains, as designed, (2) N-citraconyl-chitosan/PAA or PAM graft copolymer, and (3) a semi-IPN of citraconyl-chitosan and PAA or PAM. All polymeric chains have been stabilized through hydrogen bonds, ionic interactions between –COOH and –NH_2 groups, or other physical interactions.

From the material science point of view, the aim of this study was to assess the influence of APS concentration on the hydrogel cross-linking process. The APS percentage is reflected in the cross-linking density of the polymeric network, due to the occurrence of more or less reaction centers. The acrylic chain length further affects the hydrogel morphology, pore dimension, fluid absorption and retention etc. The cross-linking reaction yield was between 58 and 79% for hydrogel with PAA and in the range 41–59% for those with PAM (data in Table 1). A correlation can be observed between the polymerization yield and the initiator ratio: the yield decreased along with the increase in APS concentration in the reaction mixture. Moreover, the values were lower for PAM polymerization, but the value range was similar for both synthetic polymers.

Finally, in order to achieve improved biological interactions, the hydrogels were optimized by the bioconjugation of L-arginine, an amino acid that could be coupled with the carboxylic groups existing on the synthetic polymer chains in both PAA and PAM. L-arginine is an amino acid precursor of proline which is converted to hydroxyproline and then, to collagen; it has a positive influence on the insulin-like growth factor (IGF-1), a hormone which promotes wound healing. Given its properties, L-arginine is expected to impart valuable characteristics to the conjugated materials. Consequently, the properties were assessed prior and after this reaction in order to compare the hydrogels performance and arginine influence.

The elemental analysis was performed on chitosan, citraconyl-chitosan and all hydrogel samples, before the arginine conjugation reaction. The composition was calculated and the data are shown in Figure 2.

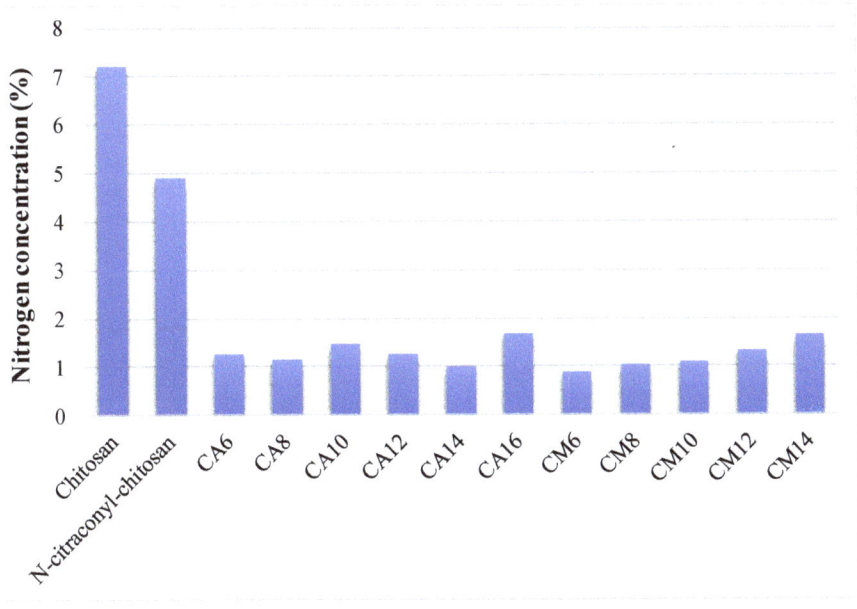

Figure 2. Nitrogen concentration of chitosan, N-citraconyl-chitosan and prepared hydrogels.

Compared to raw chitosan, the modified polysaccharide contained less nitrogen, thus, confirming the presence of citraconic anhydride in its structure. The decrease in nitrogen percentage is more significant in hydrogel samples due to the acrylic cross-links. The amount of nitrogen increased along with APS ratio in the hydrogels prepared with AM, indicating a higher reaction yield when less initiator was used. These data allowed the determination of the final composition in each hydrogel and the content of chitosan and synthetic polymer in 100% hydrogel is presented in Table 1. Based on these results, the carboxylic groups available for the final arginine coupling reaction could be determined.

The comparative FTIR spectra recorded for chitosan and citraconyl-chitosan, and for the selected scaffolds CA10, CA10A and CM10, respectively, are presented in Figure 3.

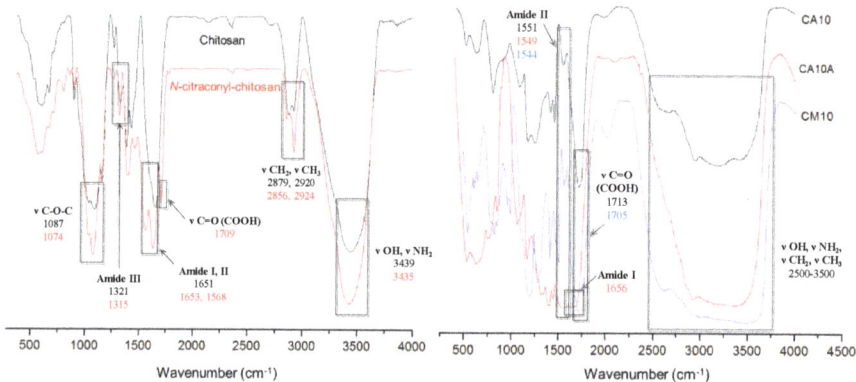

Figure 3. FTIR spectra for chitosan, N-citraconyl-chitosan (left), and selected hydrogels (right).

Chitosan exhibits specific absorption bands in the FTIR spectrum, as follows: one intense absorption peak at 3439 cm^{-1} assigned to the OH and NH$_2$ functional groups present in the polysaccharides structure; the amide I band was present at 1651 cm^{-1}, shifted at higher wavenumber and combined with OH. The specific peak for C-O-C glycoside bond appeared at 1087 cm^{-1}.

By comparison, the spectra of citraconyl-chitosan displayed new peaks correlated to the new amide bonds and to the structure of citraconyl residue. Hence, a large intense band was recorded between 1709 and 1568 cm^{-1} as a result of the fusion of the peaks specific to amide I and the carboxylic group, due to C=O stretching. Furthermore, C-O-H in plane bending from the COOH group led to the peak at 1461 cm^{-1}, while the peaks at 2924 and 2856 cm^{-1} were correlated to the methyl groups symmetric and asymmetric stretching vibrations.

Scaffolds with 1% APS, cross-linked with AA or AM, before and after arginine conjugation, are presented in Figure 3 (right). The chemical cross-linking was confirmed by the presence of intense peaks at 1713 cm^{-1} (CA10) and 1705 cm^{-1} (CM10), correlated with the C=O bond from the carboxylic group and the intense frequency bands in the interval 2500–3500 cm^{-1} specific to PAA–O–H bond stretching vibrations (from the carboxylic group); in the range 2600–3600 cm^{-1}, for the hydrogels with PAM, the bands were more intense due to, and shifted by, the symmetric and asymmetric stretch specific to the methyl group. In addition, the presence of the amide bond previously formed between chitosan and the anhydride was confirmed by the signals at 1551 cm^{-1} (CA10) and 1544 cm^{-1} (CM10), bands assigned to N–H bending vibrations and C–N stretching specific to amide II, hence, confirming polymer modification.

Arginine coupling with hydrogels yielded in new peaks that appeared in the FTIR spectrum of sample CA10A, corresponding to the amide bond vibrations, as follows: at 1656 cm^{-1} a new peak related to the stretching vibrations of C=O bond (amide I); at 1549 and 1323 cm^{-1}, respectively, novel bands assigned to C–N stretching and N–H bending vibrations (amide II, III).

3.2. Hydrogels Morphology

SEM images from Figure 4 were recorded for samples with AA and AM, synthesized in the presence of 0.6% or 1% APS, before and after arginine coupling, in order to compare and determine the influence of the considered preparation parameters. All samples displayed a porous morphology with interconnected pores of variable dimensions. At a microscopic level, there are areas that differ in aspect, which may be explained by the fact that the radical polymerization did not occur uniformly in the whole reaction volume, thus, leading to regions where more synthetic polymer was formed.

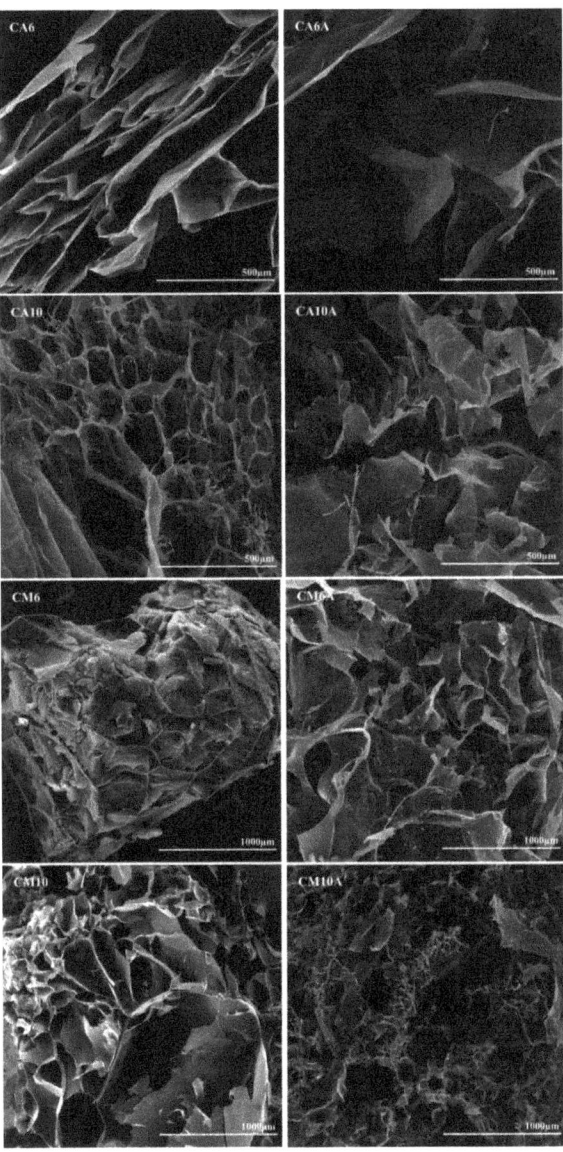

Figure 4. SEM images of hydrogels with acrylic and methacrylic acid, obtained with 0.6% or 1% APS, before and after arginine immobilization.

Particularly, for the AA-based hydrogels, well defined pores were obtained and the arginine coupling caused the formation of larger pores and high pore dispersity (see Table 2). However, in the case of hydrogels prepared with AM, the pore size increased with the increasing amount of APS and the presence of the amino acid favored the formation of pores with an even higher size (the average pore dimension varied between 40 and 550 µm). In the case of the arginine-containing hydrogels (sample CM10A in Figure 4), a significant decrease in pore size was noticed. The explanation resides in the direct bonding of arginine to the synthetic polymer moieties and, thus, arginine molecules are pendant toward the interior of the pores, which finally entailed the reduction in pore diameter.

Table 2. Pore size variations with hydrogel composition.

Hydrogel	Monomer Type	APS (%)	Presence of Arginine	Pore Dimension (µm)		
				Min	Average	Max
CA6	AA	0.6%	✗	55.2	148.0	448.4
CA6A		0.6%	✓	145.3	318.6	526.0
CA10		1%	✗	91.4	185.9	387.5
CA10A		1%	✓	150.4	240.1	372.6
CM6	AM	0.6%	✗	149.7	383.9	907.8
CM6A		0.6%	✓	287.6	539.8	899.4
CM10		1%	✗	251.0	548.5	1885.0
CM10A		1%	✓	13.6	41.8	73.1

3.3. Swelling Behavior in Simulated Physiological Conditions

The hydrogels' ability to absorb fluids is essential in skin treatment applications, as it concerns, for example, the wound exudates or a hemorrhage. The swelling behavior of the hydrogels can be affected by multiple factors: the hydrophilic or hydrophobic functional groups (–COOH, –OH, –CO–NH–) existing on the polymer chains, the internal morphology, the network parameters and its elasticity, temperature, pH and swelling medium [7]. The experimental data obtained during swelling assays are shown in Figure 5. In order to test the hydrogels, two types of fluids were chosen according to various methods described in the literature: phosphate-buffered solution (PBS) at pH 7.4 and simulated extracellular or wound fluid (SWF) [24].

Hydrogels with AA showed SDs of up to 30,000% due to the hydrophilicity of the synthetic polymer and the maximum swelling degree, associated to the plateau, was reached 20–60 min after contact, depending on the polymer chain length and network morphology, but mainly due to the carboxylic groups that provide an important hydrophilic behavior. By comparison, the hydrogels with AM hydrated slower and absorbed up to 16,000%. APS influence can also be noted: a higher initiator concentration is associated with a higher swelling capacity and a slower absorption rate. Such a result is considered to be determined by a higher cross-linking density and shorter polymeric chains, which led to pores with smaller size. Hence, the network's elasticity was altered, and the hydrogels became less flexible. [7] The presence of arginine in the polymeric networks determined a fast fluid absorption evident in all hydrogel samples, but with a smaller retained volume, up to 19,000% PBS. The influence of the swelling media was also investigated. The obtained experimental data suggested that SWF is absorbed in lower quantities as compared to PBS, presumably due to the high viscosity of the wound fluid and to existing albumin large molecules which do not access the domains with small pores, thus, leading to swelling degrees of a maximum of 12,000%.

Upon contact with any of the fluids, the polymer network started to swell progressively and the solution accessed the pores. The swelling mechanism may be either Fickian and the swelling is diffusion-controlled, or non-Fickian, when the swelling kinetics are based mostly on network relaxation, according to Ritger–Peppas model (Equation (4), data in Table 3) [25–28].

$$\frac{M_t}{M_\infty} = kt^n \quad (4)$$

where M_t is the material weight at time t, M_∞ is the initial material weight, k is the process rate constant, t is time and n is the diffusional exponent. The model was fit to experimental data for all hydrogels, before arginine conjugation, and tested in PBS.

Table 3. Kinetic parameters for hydrogels without arginine, immersed in PBS.

Hydrogel	k	n	Hydrogel	k	n
CA6	0.028	0.790096	CM6	0.0338	0.720515
CA8	0.0397	0.759804	CM8	0.0409	0.665587
CA10	0.0093	0.938351	CM10	0.0478	0.614805
CA12	0.0132	0.893475	CM12	0.0375	0.715687
CA14	0.0153	0.879649	CM14	0.0339	0.732957
CA16	0.0187	0.868381			

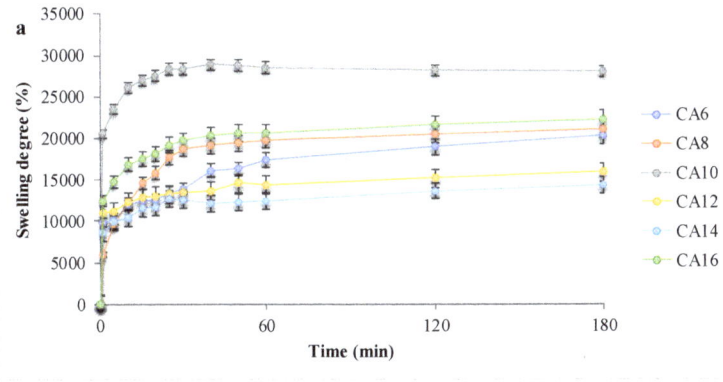

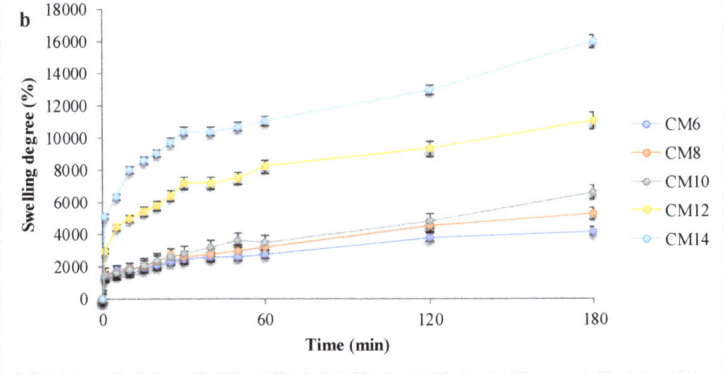

Figure 5. Cont.

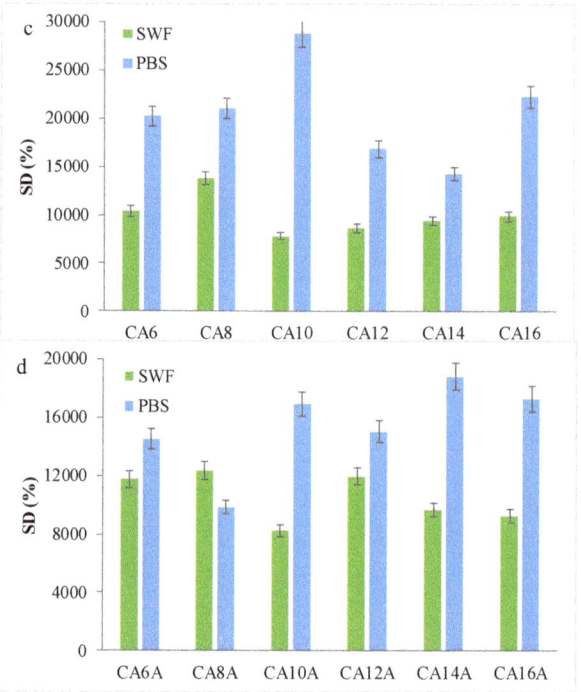

Figure 5. Swelling behavior of the hydrogels: (**a**,**b**) Kinetic swelling degree data, in PBS for hydrogels with acrylic and methacrylic acid; (**c**,**d**) Maximum swelling degrees in PBS or SWF, before and after arginine coupling.

In the present case, the diffusional exponent n had values between 0.75 and 0.93 for hydrogels with AA and between 0.61 and 0.73 for those with AM; hence, for all hydrogel samples the value of the diffusional exponent n fell within the range 0.45–1, a fact that indicates a non-Fickian swelling mechanism and substantiated that the water retention was driven by the relaxation of the network.

The fast swelling, favored by the presence of arginine, and the large volumes retained in the hydrogel networks, due to the chitosan/acrylate cross-linking, indicated a superabsorbent behavior [12,29–34] for the prepared hydrogels, which is a highly advantageous characteristic for applications such as drug delivery and wound dressing, as intended.

3.4. Ibuprofen Release Profiles and Kinetics

Ibuprofen (Ib), a nonsteroidal anti-inflammatory drug, is used for the treatment of pain, fever, and inflammation. It is insoluble in water, but has a high solubility in most organic solvents, including ethanol. The mechanism that leads to the analgesic, antipyretic, and anti-inflammatory effects evolves through a non-selective inhibition of cyclooxygenase (COX) enzymes. Normally, COX enzymes convert arachidonic acid to a prostaglandin, which further mediates pain, inflammation, and fever [35]. It is well known that Ib systemic administration causes severe issues, such as stomach and intestine injuries, circulatory problems (heart failure, AVC), and kidney lesions. To limit these drawbacks, a local delivery of Ib is preferred, such as a dressing. Moreover, in this case, the therapeutic effects are enhanced.

The in vitro release of Ib from the hydrogels was analyzed in PBS, pH = 7.2, at 37 °C, mimicking the physiological environment. It is important to emphasize that the pH was maintained at the level recorded for the extracellular environment in physiological conditions, in order to be fit for the medical applications taken into consideration in the present article. The readings expressed as cumulative release were plotted versus time in order to obtain the drug release profiles and the results are shown in Figure 6.

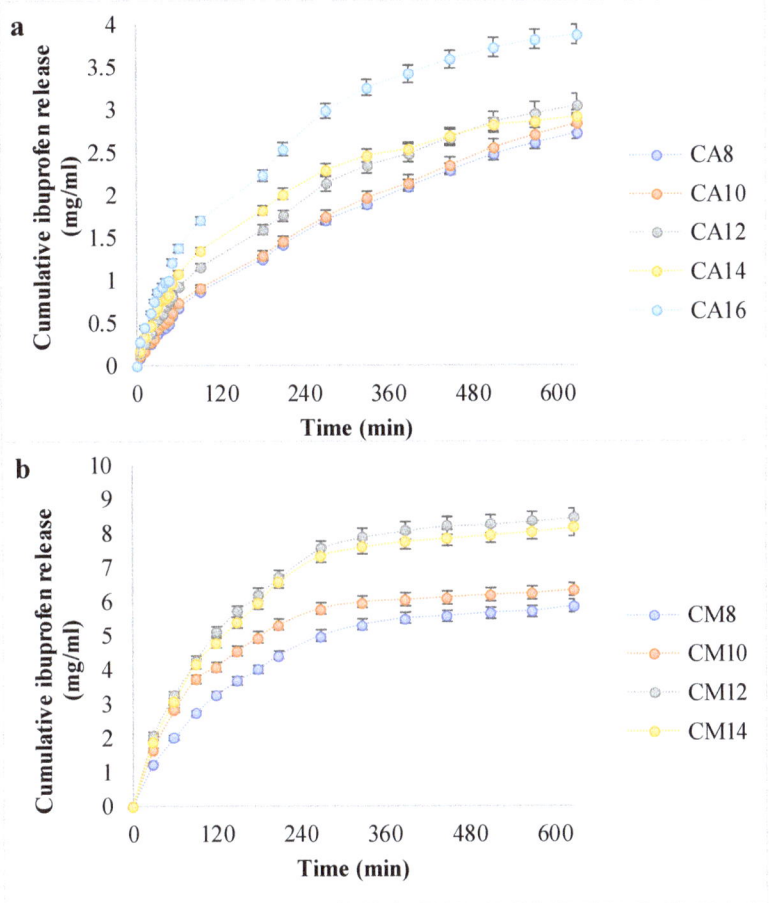

Figure 6. The in vitro ibuprofen release kinetic data (**a**,**b**).

The release profiles for the hydrogels with AA indicated a steady, controlled release of a maximum of 4 mg ibuprofen over 10 h. However, in the case of hydrogels with AM, Ib was released in double the amount and in a shorter time, similar to a burst effect. In addition, AA-based hydrogels did not reach a plateau after 600 min, while hydrogels with AM reached the plateau after 300–360 min, meaning that the equilibrium in drug concentration was reached between the hydrogel and the environment. Moreover, as the APS ratio grows from 0.8% to 1.4 or 1.6%, respectively, the amount of released Ib is greater. This can be explained by the fact that more APS led to larger pores, according to SEM micrographs, which favored drug diffusion in both directions—at loading and at release.

The drug loading and release capacity depend on the polymeric matrices' structure and morphology, their ability to absorb and retain the drug solution, and the drug–hydrogel interaction [36]. The Ib molecule is amphiphilic and able to participate in various molecular/supramolecular associations. Ionic interactions between the COOH group from Ib and the NH_2 moiety from chitosan were generally present in all hydrogels, since the same amount of chitosan was present in all matrices. It should be emphasized that there was no arginine in the hydrogels tested for drug delivery. The significant difference in hydrogel–drug interactions was correlated with the type of synthetic polymer.

Interestingly, the hydrogels with PAM released higher amounts of drug and in shorter time intervals, results that were in contradiction with the swelling degrees, since the PAA favored a faster absorption and the retention of larger fluid volumes. It is worth mentioning that the drug loading took place in a water/ethanol solution, which consequently may have been favorable for the swelling in chitosan/PAM networks, rather than in matrices with PAA. Moreover, it can be assumed that hydrophobic interactions appeared between Ib and the methyl groups in PAM structure.

It is known that the drug release from hydrogels is governed by several phenomena: diffusion, erosion, network relaxation, all in various proportions, depending on the polymer nature, network stability and parameters, morphology, hydrophilicity, the nature of the drug and its interaction with the matrix, the release medium etc., [37–40]. In order to study the drug release kinetics and mechanism, the drug release data were fitted into four models, using the following equations [21,41–43]:

$$\text{Zero order}: \frac{Q_t}{Q_0} = k_0 t \tag{5}$$

$$\text{First order}: \ln \frac{Q_t}{Q_0} = k_1 t \tag{6}$$

$$\text{Higuchi model}: \frac{Q}{Q_0} = k_H t^{1/2} \tag{7}$$

$$\text{Korsmeyer–Peppas model}: \frac{Q_t}{Q_0} = k t^n \tag{8}$$

where Q_t is the amount of drug released at time t, Q_0 is the original drug concentration in the material (40 mg), n is the release exponent and K is the release rate constant.

The models were fitted to the curves with the highest cumulative drug release in each series, namely CA16 (which had the highest APS ratio in the PAA-based hydrogels series) and CM12. The latter had a cumulative amount close to but higher than CA14, which had the highest APS ratio of the materials cross-linked with PAM. The results obtained for the correlation coefficient are shown in Table 4 and suggest that for the hydrogels with PAA, the release mechanism followed the Korsmeyer–Peppas kinetics [41], while the hydrogels with PAM fitted the Korsmeyer–Peppas and the Higuchi models well [42,43]. In addition to these data, it was important to determine the diffusional exponent n, which is a parameter that indicates the mechanism of drug release and varies depending on the geometry of the release device. Fickian diffusion is confirmed for $n < 0.45$; when $0.45 < n < 0.89$, the drug transport is anomalous (non-Fickian)—in other words, the diffusion and relaxation rates are similar and the physical phenomena are diffusion and erosion of the matrix; if $n > 0.89$, then the key mechanism of drug release is Case II transport, thus, the drug release is determined by polymer relaxation [26,44]. The values obtained for the exponent n suggested anomalous (non-Fickian) transport, which indicated that the drug is released from the hydrogels by diffusion associated with the materials erosion. The drug release can be controlled by selecting the acrylate nature for the cross-linking and the initiator ratio. The drug is incorporated in a hydrophilic, initially glassy material, and the release is basically swelling-controlled [26].

Table 4. Drug release correlation coefficient values from different kinetic models.

	Correlation Coefficient (r^2)				Release Rate Constant, k	Release Exponent, n
	Zero Order	First Order	Higuchi	Korsmeyer–Peppas		
CA16	0.9545	0.8342	0.9838	0.9936	0.0158	0.6234
CM12	0.9236	0.858	0.9954	0.9947	0.0025	0.5907

3.5. The In Vitro Hydrogels Degradation

The biodegradation assays have proven that the chitosan-based scaffolds were susceptible to lysozyme attack under simulated physiological conditions of pH and temperature, as shown in Figure 7.

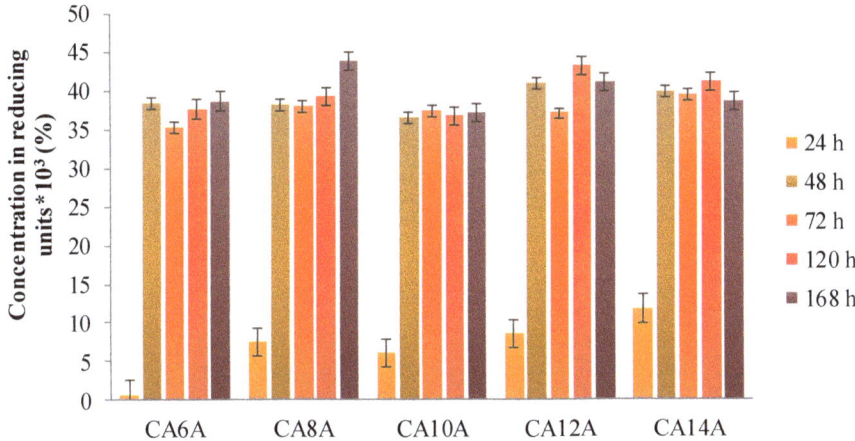

Figure 7. Degradation behavior of hydrogels in medium with enzyme.

The presence of the arginine molecules in the polymeric network caused the rapid absorption of the enzyme solution into the hydrogel networks, followed by their expansion, which entailed a fast degradation rate in all investigated samples, similar to a burst effect after 24–48 h. This phenomenon can be explained by the enhanced accessibility of lysozyme to the glycoside bonds within chitosan chains.

Lysozyme is present in any type of wound due to neutrophils secretion, in both infected and non-infected wounds. This phenomenon is associated with the inflammation phase of wound evolution and has the crucial role of cleaning the wound bed before any regeneration processes are triggered. The enzyme levels determined in the wound fluids were of 0.4 mg/mL for a wound with inflammation and at least 0.5 mg/mL lysozyme in an ulcer wound fluid. [45] Compared to the literature data, the 1.2 mg/mL lysozyme concentration used in this experiment was indicating that the hydrogels had a good stability in a more aggressive simulated environment, being able to maintain the cross-linked network for up to 48 h.

3.6. Hydrogels Cytocompatibility

All materials intended to be used for medical applications must be tested by means of biocompatibility. One of the standard assays is the MTT study for in vitro evaluation of cytocompatibility by the direct contact method. In addition to the obtained MTT data, a live/dead staining was performed and the results are shown in Figures 8 and 9.

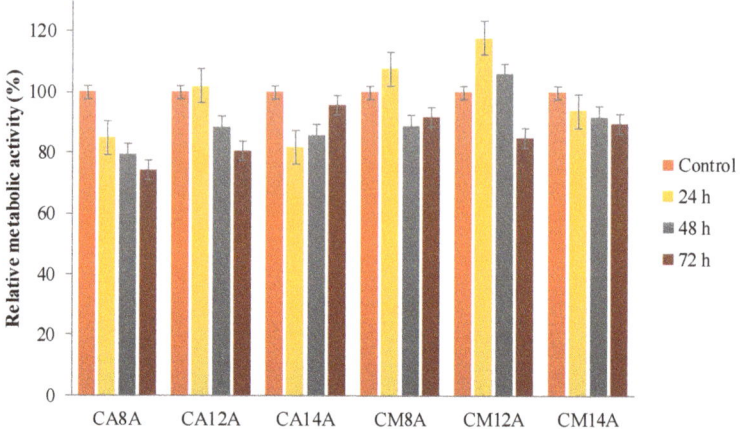

Figure 8. Relative metabolic activity data from the MTT assays for hydrogels with AA or AM, obtained with 0.8%, 1.2% and 1.4% APS, respectively, tested for different time intervals.

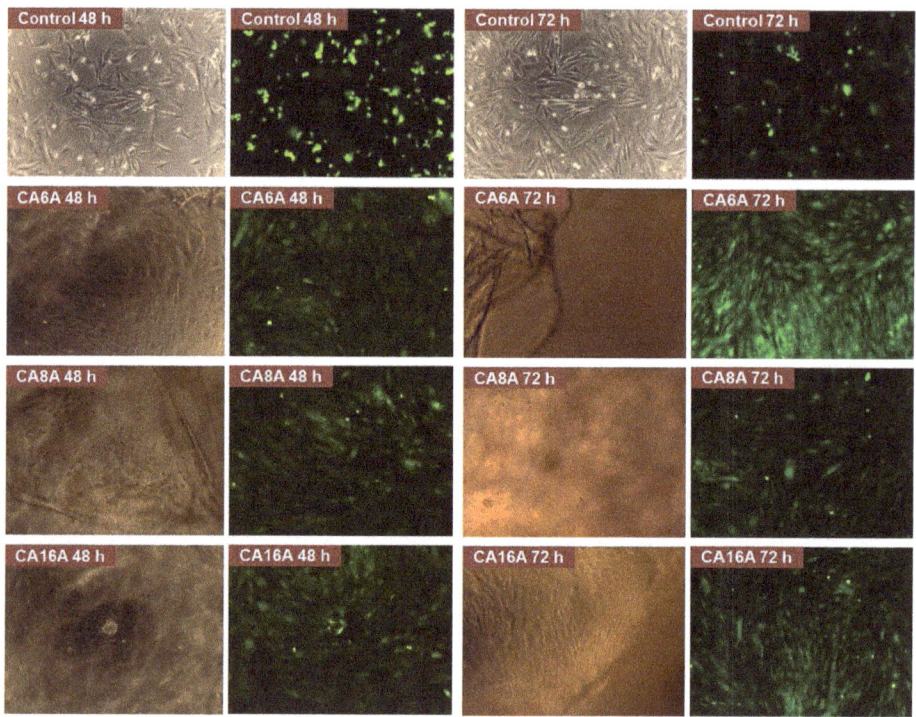

Figure 9. Live/dead staining assay images of negative control and cells after incubation with different hydrogels (magnification 10×).

The relative metabolic activity induced by the hydrogels was between 76 and 100%, depending on the materials' composition. All tested materials were conjugated with arginine, an amino acid well-known for its key role in accelerated wound healing and in cellular recognition and adhesion. Therefore, arginine is a valuable element in multicomponent systems used for drug delivery and wound dressing. [46] After the first 24 h, the best results were for materials with 1.2% APS, followed by a decrease in time, while the hydrogels with 1.4% APS showed a better cell response after a lag period of 72 h.

The fluorescence microscopy images were analyzed and several observations can be made: all hydrogels are transparent in the hydrated state, which can be an advantage for wound monitoring; the hydrogels induced an intense proliferation compared to the control with the formation of a confluent monolayer after 48 h; based on cell morphology, the stem cells appeared to have undergone differentiation. The cell response is the result of various cell–hydrogel interactions determined by the materials' structure and properties on the one hand, and, on the other hand, by cell biology and biochemistry. Thus, it is possible that the inclusion of L-arginine in the matrices induced this cell behavior—the intense proliferation and the differentiation. Due to the fact that the differentiation is associated with a lower metabolic effect and that it may have begun after a fast proliferation—a process that lasted less than 24 h—it may explain the low RMA values.

3.7. Hemostatic Properties

The thrombogenic character is an important property for wound dressing materials. The protein adhesion initiates the coagulation cascade, and thus, the material can accelerate the thrombus formation, stop the hemorrhage, and stimulate the healing process. The hydrogels porosity and their rough surface are known to be beneficial to blood coagulation [47]. Moreover, the colloid formed on the pore's surface upon the blood fluid sorption favors the adhesion of blood cells, and the dressing material presses the wound area and efficiently limits the bleeding as a physical barrier. The prothrombin time (PT) test is an efficient method to evidence the thrombotic or antithrombotic activity of biomaterials. Generally, the lower the value of PT, the faster the clotting rate is and the better the antithrombotic activity of these materials. [48] The experimental data recorded for the selected samples in terms of hemostatic activity are given in Table 5.

Table 5. PT, INR and fibrinogen values.

Parameter	Blood	CA6A	CA10A	CM6A	CM10A
PT (s)	13.1 ± 1.2	10. 2 ± 0.5	10. 2 ± 0.5	11.3 ± 0.4	11.0 ± 0.5
INR	1.07 ± 0.12	0.70 ± 0.13	0.70 ± 0.16	0.79 ± 0.18	0.76 ± 0.21
Fibrinogen (mg/dl)	390 ± 15	460 ± 21	421 ± 12	447 ± 21	410 ± 33

The excellent hemostatic properties of the prepared hydrogels were due to their strong swelling capacity and porous structure. For all tested hydrogels, the PT and INR values decreased in comparison with the control sample, while higher values for fibrinogen concentration were recorded. The PAA-based hydrogels exhibited the most intense hemostatic activity, compared with those with PAM, and this cumulative effect can be attributed to the synergic action of their specific characteristics: porosity, high hydrophilicity, and the positively charged moieties from chitosan and arginine immobilized onto networks, as was separately reported for chitosan-based materials and arginine [49–54]. Furthermore, at a lower APS ratio, the hemostatic effect was more intense for both hydrogel series. As superabsorbent hydrogels, these materials absorbed the plasma in the blood and, hence, promoted the local accumulation of the coagulation factors, erythrocytes and platelets, and accelerate the adhesion of the blood components to their surface, and thus, the blood coagulation was sped up. Therefore, taking into consideration these results and the literature data in the field, it can be stated that these new hydrogels successfully respond to the performance requirements for wound dressing materials.

4. Conclusions

In the present study, two series of hydrogels based on chemically modified chitosan grafted with acrylic polymers as cross-linkers were compared. The advantages of both natural and synthetic polymers led to hydrogels with combined desirable characteristics, and the networks were further conjugated with arginine in order to achieve enhanced properties. The results obtained from FTIR spectra, elemental analysis and SEM images confirmed the formation of networks with interconnected pores and evidenced a great influence on the hydrogel properties of the monomer and of the initiator ratio used during the synthesis. It has been confirmed that hydrogels have a high fluid absorption behavior after their interaction with buffer solution and simulated wound fluid, under physiological conditions. The ibuprofen release profiles were studied in vitro and the kinetics fitted the Korsmeyer–Peppas model best. The hydrogels have proved to be biodegradable in the presence of lysozyme with hemostatic properties and the cytocompatibility tests indicated the hydrogels' ability to induce cell proliferation and differentiation. In conclusion, these superabsorbent hydrogels with tunable properties may be considered suitable materials for both drug delivery and wound dressing applications.

Author Contributions: Conceptualization, I.A.D., L.V. and S.C.; Data curation, I.A.D.; Formal analysis, I.A.D.; Investigation, I.A.D., C.D.D. and V.M.; Methodology, I.A.D., L.V. and S.C.; Visualization, I.A.D.; Writing—original draft, I.A.D.; Writing—review and editing, I.A.D., L.V. and S.C. All authors have read and agreed to the published version of the manuscript.

Funding: This research received no external funding.

Conflicts of Interest: The authors declare no conflict of interest.

References

1. Ullah, F.; Othman, M.B.; Javed, F.; Ahmad, Z.; Akil, H. Classification, processing and application of hydrogels: A review. *Mater. Sci. Eng. C* **2015**, *57*, 414–433. [CrossRef]
2. Gupta, A.; Kowalczuk, M.; Heaselgrave, W.; Britland, S.T.; Martin, C.; Radecka, I. The Production and Application of Hydrogels for Wound Management: A Review. *Eur. Polym. J.* **2019**, *111*, 134–151. [CrossRef]
3. Kamoun, E.A.; Kenawy, E.R.S.; Chen, X. A Review on Polymeric Hydrogel Membranes for Wound Dressing Applications: PVA-Based Hydrogel Dressings. *J. Adv. Res.* **2017**, *8*, 217–233. [CrossRef]
4. Baron, R.I.; Culica, M.E.; Biliuta, G.; Bercea, M.; Gherman, S.; Zavastin, D.; Ochiuz, L.; Avadanei, M.; Coseri, S. Physical Hydrogels of Oxidized Polysaccharides and Poly(Vinyl Alcohol) For Wound Dressing Applications. *Materials* **2019**, *12*, 1569. [CrossRef]
5. Bhattarai, N.; Gunn, J.; Zhang, M. Chitosan-based hydrogels for controlled, localized drug delivery. *Adv. Drug Deliv. Rev.* **2010**, *62*, 83–99. [CrossRef]
6. Matthews, K.H. Drug delivery dressings. In *Advanced Wound Repair Therapies*; Farrar, D., Ed.; Woodhead Publishing: Cambridge, UK, 2011; pp. 361–394.
7. Slaughter, B.V.; Khurshid, S.S.; Fisher, O.Z.; Khademhosseini, A.; Peppas, N.A. Hydrogels in Regenerative Medicine. *Adv. Mater.* **2009**, *21*, 3307–3329. [CrossRef]
8. Liu, H.; Wang, C.; Li, C.; Qin, Y.; Wang, Z.; Yang, F.; Li, Z.; Wang, J. A functional chitosan-based hydrogel as a wound dressing and drug delivery system in the treatment of wound healing. *RSC Adv.* **2018**, *8*, 7533–7549. [CrossRef]
9. Ahmed, S.; Ikram, S. Chitosan Based Scaffolds and Their Applications in Wound Healing. *Achiev. Life Sci.* **2016**, *10*, 27–37. [CrossRef]
10. Croisier, F.; Jérôme, C. Chitosan-Based Biomaterials for Tissue Engineering. *Eur. Polym. J.* **2013**, *49*, 780–792. [CrossRef]
11. Hamedi, H.; Moradi, S.; Hudson, S.M.; Tonelli, A.E. Chitosan Based Hydrogels and Their Applications for Drug Delivery in Wound Dressings: A Review. *Carbohydr. Polym.* **2018**, *199*, 445–460. [CrossRef]
12. Ferfera-Harrar, H.; Aiouaz, N.; Dairi, N. Synthesis and Properties of Chitosan Graft-Polyacrylamide Gelatin Superabsorbent Composites for Wastewater. *Int. J. Chem. Mol. Eng.* **2015**, *9*, 849–856.
13. Sakthivel, M.; Franklin, D.S.; Guhanathan, S. PH-Sensitive Itaconic Acid Based Polymeric Hydrogels for Dye Removal Applications. *Ecotoxicol. Environ. Saf.* **2016**, *134*, 427–432. [CrossRef] [PubMed]
14. Wang, Y.; Zeng, L.; Ren, X.; Song, H.; Wang, A. Removal of Methyl Violet from Aqueous Solutions Using Poly (Acrylic Acid-Co-Acrylamide)/Attapulgite Composite. *J. Environ. Sci.* **2010**, *22*, 7–14. [CrossRef]
15. Lee, J.W.; Kim, S.Y.; Kim, S.S.; Lee, Y.M.; Lee, K.H.; Kim, S.J. Synthesis and characteristics of interpenetrating polymer network hydrogel composed of chitosan and poly(acrylic acid). *J. Appl. Polym. Sci.* **1999**, *73*, 113–120. [CrossRef]

16. Lee, J.S.; Kumar, R.N.; Rozman, H.D.; Azemi, B.M.N. Pasting, Swelling and Solubility Properties of UV Initiated Starch-Graft-Poly(AA). *Food Chem.* **2005**, *91*, 203–211. [CrossRef]
17. Athawale, V.; Lele, V. Recent Trends in Hydrogels Based on Starch-Graft-Acrylic Acid: A Review. *Starch/Stärke* **2001**, *53*, 7–13. [CrossRef]
18. Prabaharam, M. Graft Copolymerization of Acrylic Monomers on Chitosan and Its Derivatives: Recent Developments and Applications. In *Chitin and Chitosan Derivatives: Advances in Drug Discovery and Developments*; Se-Kwon, K., Ed.; CRC Press: Boca Raton, FL, USA, 2014; p. 167.
19. Dos Santos, K.S.C.R.; Coelho, J.F.J.; Ferreira, P.; Pinto, I.; Lorenzetti, S.G.; Ferreira, E.I.; Higa, O.Z.; Gil, M.H. Synthesis and Characterization of Membranes Obtained by Graft Copolymerization of 2-Hydroxyethyl Methacrylate and Acrylic Acid onto Chitosan. *Int. J. Pharm.* **2006**, *310*, 37–45. [CrossRef]
20. Duceac, I.A.; Lobiuc, A.; Coseri, S.; Verestiuc, L. Tunable Hydrogels Based on Chitosan, Collagen and Poly(Acrylic Acid) for Regenerative Medicine. In Proceedings of the EHB 2019: IEEE International Conference on e-Health and Bioengineering, Iasi, Romania, 21–23 November 2019; Volume 2, pp. 3–6.
21. Dwivedi, R.; Singh, A.K.; Dhillon, A. pH-Responsive Drug Release from Dependal-M Loaded Polyacrylamide Hydrogels. *J. Sci. Adv. Mater.* **2017**, *2*, 45–50. [CrossRef]
22. Chuah, C.; Wang, J.; Tavakoli, J.; Tang, Y. Novel Bacterial Cellulose-Poly(Acrylic Acid) Hybrid Hydrogels with Controllable Antimicrobial Ability as Dressings for Chronic Wounds. *Polymers* **2018**, *10*, 1323. [CrossRef]
23. Hoffman, W.S. A rapid photoelectric method for the determination of glucose in blood and urine. *J. Biol. Chem.* **1937**, *120*, 51.
24. Singh, B.; Sharma, A.; Sharma, A.; Dhiman, A. Design of Antibiotic Drug Loaded Carbopol-Hydrogel for Wound Dressing Applications. *Am. J. Drug Deliv. Ther.* **2017**, *4*, 1–9.
25. Peppas, N.A.; Bures, P.; Leobandung, W.; Ichikawa, H. Hydrogels in Pharmaceutical Formulations. *Eur. J. Pharm. Biopharm.* **2000**, *50*, 27–46. [CrossRef]
26. Ritger, P.L.; Peppas, N.A. A Simple Equation for Description of Solute Release II. Fickian and Anomalous Release from Swellable Devices. *J. Control. Release* **1987**, *5*, 37–42. [CrossRef]
27. Blanco, A.; González, G.; Casanova, E.; Pirela, M.E.; Briceño, A. Mathematical Modeling of Hydrogels Swelling Based on the Finite Element Method. *Appl. Math.* **2013**, *4*, 161–170. [CrossRef]
28. Ganji, F.; Vasheghani-Farahani, S.; Vasheghani-Farahani, E. Theoretical Description of Hydrogel Swelling: A Review. *Iran. Polym. J. (Eng. Ed.)* **2010**, *19*, 375–398.
29. Mignon, A.; De Belie, N.; Dubruel, P.; Van Vlierberghe, S. Superabsorbent Polymers: A Review on the Characteristics and Applications of Synthetic, Polysaccharide-Based, Semi-Synthetic and 'Smart' Derivatives. *Eur. Polym. J.* **2019**, *117*, 165–178. [CrossRef]
30. Cheng, W.M.; Hu, X.M.; Wang, D.M.; Liu, G.H. Preparation and Characteristics of Corn Straw-*co*-AMPS-*co*-AA Superabsorbent Hydrogel. *Polymers* **2015**, *7*, 2431–2445. [CrossRef]
31. Khan, H.; Chaudhary, J.P.; Meena, R. Anionic Carboxymethylagarose-Based PH-Responsive Smart Superabsorbent Hydrogels for Controlled Release of Anticancer Drug. *Int. J. Biol. Macromol.* **2019**, *124*, 1220–1229. [CrossRef]
32. Spagnol, C.; Rodrigues, F.H.A.; Pereira, A.G.B.; Fajardo, A.R.; Rubira, A.F.; Muniz, E.C. Superabsorbent hydrogel composite made of cellulose nanofibrils and chitosan-*graft*-poly(acrylic acid). *Carbohydr. Polym.* **2012**, *87*, 2038–2045. [CrossRef]
33. Guilherme, M.R.; Aouada, F.A.; Fajardo, A.R.; Martins, A.F.; Paulino, A.T.; Davi, M.F.T.; Rubira, A.F.; Muniz, E.C. Superabsorbent Hydrogels Based on Polysaccharides for Application in Agriculture as Soil Conditioner and Nutrient Carrier: A Review. *Eur. Polym. J.* **2015**, *72*, 365–385. [CrossRef]
34. Kollár, J.; Mrlik, M.; Moravcikova, D.; Ivan, B.; Mosnacek, J. Effect of monomer content and external stimuli on properties of renewable Tulipalin A-based superabsorbent hydrogels. *Eur. Polym. J.* **2019**, *115*, 99–106. [CrossRef]
35. Rao, P.N.P.; Knaus, E.E. Evolution of Nonsteroidal Anti-Inflammatory Drugs (NSAIDs): Cyclooxygenase (COX) Inhibition and Beyond. *J. Pharm. Pharm. Sci.* **1997**, *11*, 81–110. [CrossRef]
36. Rodríguez, R.; Alvarez-Lorenzo, C.; Concheiro, A. Interactions of Ibuprofen with Cationic Polysaccharides in Aqueous Dispersions and Hydrogels: Rheological and Diffusional Implications. *Eur. J. Pharm. Sci.* **2003**, *20*, 429–438. [CrossRef]

37. Si, H.; Xing, T.; Ding, Y.; Zhang, H.; Yin, R.; Zhang, W. 3D Bioprinting of the Sustained Drug Release Wound Dressing with Double-Crosslinked Hyaluronic-Acid-Based Hydrogels. *Polymers* **2019**, *11*, 1584. [CrossRef] [PubMed]
38. Tzereme, A.; Christodoulou, E.; Kyzas, G.Z.; Kostoglou, M.; Bikiaris, D.N.; Lambropoulou, D.A. Chitosan Grafted Adsorbents for Diclofenac Pharmaceutical Compound Removal from Single-Component Aqueous Solutions and Mixtures. *Polymers* **2019**, *11*, 497. [CrossRef] [PubMed]
39. Sattari, S.; Tehrani, A.D.; Adeli, M. pH-Responsive Hybrid Hydrogels as Antibacterial and Drug Delivery Systems. *Polymers* **2018**, *10*, 660. [CrossRef] [PubMed]
40. Pistone, A.; Iannazzo, D.; Celesti, C.; Scolaro, C.; Giofré, S.V.; Romeo, R.; Visco, A. Chitosan/PAMAM/Hydroxyapatite Engineered Drug Release Hydrogels with Tunable Rheological Properties. *Polymers* **2020**, *12*, 754. [CrossRef]
41. Korsmeyer, R.W.; Gurny, R.; Doelker, E.; Buri, P.; Peppas, N.A. Mechanisms of Solute Release from Porous Hydrophilic Polymers. *Int. J. Pharm.* **1983**, *15*, 25–35. [CrossRef]
42. Higuchi, T. Rate of Release of Medicaments from Ointment Bases Containing Drugs in Suspension. *J. Pharm. Sci.* **1961**, *50*, 874–875. [CrossRef]
43. Higuchi, T. Mechanism of Sustained- Action Medication. *J. Pharm. Sci.* **1963**, *52*, 1145–1149. [CrossRef]
44. Mankotia, P.; Choudhary, S.; Sharma, K.; Kumar, V.; Kaur Bhatia, J.; Parmar, A.; Sharma, S.; Sharma, V. Neem Gum Based PH Responsive Hydrogel Matrix: A New Pharmaceutical Excipient for the Sustained Release of Anticancer Drug. *Int. J. Biol. Macromol.* **2019**, *142*, 742–755. [CrossRef] [PubMed]
45. Frohm, M.; Gunne, H.; Bergman, A.C.; Agerberth, B.; Bergman, T.; Boman, A.; Lidén, S.; Jörnvall, H.; Boman, H.G. Biochemical and Antibacterial Analysis of Human Wound and Blister Fluid. *Eur. J. Biochem.* **1996**, *237*, 86–92. [CrossRef] [PubMed]
46. Iacob, A.T.; Drăgan, M.; Ghețu, N.; Pieptu, D.; Vasile, C.; Buron, F.; Routier, S.; Giusca, S.E.; Caruntu, I.D.; Profire, L. Preparation, Characterization and Wound Healing Effects of New Membranes Based on Chitosan, Hyaluronic Acid and Arginine Derivatives. *Polymers* **2018**, *10*, 607. [CrossRef]
47. Hong, Y.; Zhou, F.; Hua, Y.; Zhang, X.; Ni, C.; Pan, D.; Zhang, Y.; Jiang, D.; Yang, L.; Lin, Q.; et al. A strongly adhesive hemostatic hydrogel for the repair of arterial and heart bleeds. *Nat. Commun.* **2019**, *10*, 1–11. [CrossRef] [PubMed]
48. Lee, W.; Lee, J.I.; Kulkarni, R.; Kim, M.A.; Hwang, J.S.; Na, M.K.; Bae, J.S. Antithrombotic and antiplatelet activities of small-molecule alkaloids from Scolopendra subspinipes mutilans. *Sci. Rep.* **2016**, *6*, 1–12. [CrossRef]
49. Behrens, A.M.; Sikorski, M.J.; Kofinas, P. Hemostatic strategies for traumatic and surgical bleeding. *J. Biomed. Mater. Res. Part A* **2014**, *102*, 4182–4194. [CrossRef]
50. Hu, Z.; Zhang, D.Y.; Lu, S.T.; Li, P.W.; Li, S.D. Chitosan-based composite materials for prospective hemostatic applications. *Mar. Drugs* **2018**, *16*, 273. [CrossRef]
51. Yang, J.; Tian, F.; Wang, Z.; Wang, Q.; Zeng, Y.-J.; Chen, S.-Q. Effect of chitosan molecular weight and deacetylation degree on hemostasis. *J. Biomed. Mater. Res. B. Appl. Biomater.* **2008**, *84*, 131–137. [CrossRef]
52. Kang, P.L.; Chang, S.J.; Manousakas, I.; Lee, C.W.; Yao, C.H.; Lin, F.H.; Kuo, S.M. Development and assessment of hemostasis chitosan dressings. *Carbohydr. Polym.* **2011**, *85*, 565–570. [CrossRef]
53. Muzzarelli, R.A.A. Chitins and chitosans for the repair of wounded skin, nerve, cartilage and bone. *Carbohydr. Polym.* **2009**, *76*, 167–182. [CrossRef]
54. Witte, M.B.; Barbul, A. Arginine physiology and its implication for wound healing. *Wound Repair Regen.* **2003**, *11*, 419–423. [CrossRef] [PubMed]

© 2020 by the authors. Licensee MDPI, Basel, Switzerland. This article is an open access article distributed under the terms and conditions of the Creative Commons Attribution (CC BY) license (http://creativecommons.org/licenses/by/4.0/).

Article

Developed Chitosan/Oregano Essential Oil Biocomposite Packaging Film Enhanced by Cellulose Nanofibril

Shunli Chen [1], Min Wu [1,2,*], Caixia Wang [1], Shun Yan [1], Peng Lu [1,2] and Shuangfei Wang [1,2,*]

1. College of Light Industry and Food Engineering, Guangxi University, Nanning 530004, China; chenshunli@st.gxu.edu.cn (S.C.); wangcx@st.gxu.edu.cn (C.W.); yanshun2020gxu@163.com (S.Y.); lupeng@gxu.edu.cn (P.L.)
2. Guangxi Key Laboratory of Clean Pulp & Papermaking and Pollution Control, Nanning 530004, China
* Correspondence: wumin@gxu.edu.cn (M.W.); wangsf@gxu.edu.cn (S.W.); Tel.: +86-0771-323-7305 (M.W.); +86-0771-323-7301 (S.W.)

Received: 8 July 2020; Accepted: 8 August 2020; Published: 9 August 2020

Abstract: The use of advanced and eco-friendly materials has become a trend in the field of food packaging. Cellulose nanofibrils (CNFs) were prepared from bleached bagasse pulp board by a mechanical grinding method and were used to enhance the properties of a chitosan/oregano essential oil (OEO) biocomposite packaging film. The growth inhibition rate of the developed films with 2% (w/w) OEO against *E. coli* and *L. monocytogenes* reached 99%. With the increased levels of added CNFs, the fibrous network structure of the films became more obvious, as was determined by SEM and the formation of strong hydrogen bonds between CNFs and chitosan was observed in FTIR spectra, while the XRD pattern suggested that the strength of diffraction peaks and crystallinity of the films slightly increased. The addition of 20% CNFs contributed to an oxygen-transmission rate reduction of 5.96 cc/m^2·day and water vapor transmission rate reduction of 741.49 g/m^2·day. However, the increase in CNFs contents did not significantly improve the barrier properties of the film. The addition of 60% CNFs significantly improved the barrier properties of the film to light and exhibited the lowest light transmittance (28.53%) at 600 nm. Addition of CNFs to the chitosan/OEO film significantly improved tensile strength and the addition of 60% CNFs contributed to an increase of 16.80 MPa in tensile strength. The developed chitosan/oregano essential oil/CNFs biocomposite film with favorable properties and antibacterial activity can be used as a green, functional material in the food-packaging field. It has the potential to improve food quality and extend food shelf life.

Keywords: cellulose nanofibrils; chitosan; oregano essential oil; antimicrobial; oxygen barrier properties

1. Introduction

Petroleum-based plastic films have been widely used in recent decades in the packaging field due to their low cost, good chemical stability and excellent barrier performance [1]. However, the use of non-biodegradable materials in packaging applications has raised concerns about environmental pollution. The demand for advanced and eco-friendly packaging materials owing to their excellent physical, mechanical, and barrier properties and antimicrobial activity is significantly increasing, especially with increased consumer awareness of environmental protection and with increased attention being paid to food quality and safety. Recently, biodegradable materials, such as chitosan [2–4], starch [5], pectin [6], gelatine [7] and cellulose [8], as matrices to develop biologic composite packaging materials with good oxygen, water vapor barrier properties, antibacterial properties and mechanical properties, have become a focus of scholars. However, biodegradable packaging films prepared by a single biomass material often cannot simultaneously possess a variety of favorable properties which limits the application of biomass packaging materials in the field of food packaging. Therefore, two or more kinds of material are blended together, and the functional components are added to prepare biocomposite packaging materials with good mechanical properties, oxygen and water vapor barrier properties and antibacterial or antioxidant properties.

Chitosan, which consists of (1,4)-linked-2-amino-deoxy-b-D-glucan, is a kind of cationic polysaccharide and is the deacetylated form of chitin [9]. As a renewable natural biopolymer, chitosan is derived from various sources and exhibits non-toxicity, biocompatibility, biodegradability and excellent film-forming properties. Meanwhile, chitosan has broad-spectrum antimicrobial activity against both Gram-positive and Gram-negative bacteria as well as fungi. Films prepared from chitosan were extensively used in food packaging as a degradable packaging material and inhibited bacterial reproduction, prolonged the shelf life of food and improved food quality and safety [10,11]. However, the poor mechanical and barrier properties of chitosan-based packaging materials compared to those of non-biodegradable materials have limited its widespread usage [12]. In order to develop the application of chitosan-based films in the field of food packaging, physical or chemical modification strategies have been tried. Poverenov et al. [13] prepared the alginate–chitosan coating by layer-by-layer electrostatic deposition, the coating showed excellent gas-exchange and water vapor permeability properties that protected the appearance of the fresh-cut melon. Caseinate and chitosan by ion interaction was carried out to obtain the composite films, which showed improved water vapor barrier properties [14]. The chitosan-grafted salicylic acid films have been found to maintain better quality in cucumber, making it a promising material for food packaging applications [15]. Cellulose nanofibrils (CNFs) from natural resources are recognized as the most abundant, renewable and biodegradable polymeric materials [16] and have been widely used due to their good biocompatibility, chemical stability, mechanical properties and oxygen barrier properties [17–20]. The addition of 3% (w/w) of CNFs increased the tensile strength, elongation at breaking and Young's modulus of corn starch film [21]. The water vapor permeability of bio-nanocomposite films was reduced by incorporating 4% (w/w) CNFs [22].

The antibacterial properties of chitosan are mainly related to its degree of deacetylation, molecular weight, types of microorganisms and other factors. Sanchez-Gonzalez et al. [23] showed that pure chitosan films had obvious antibacterial effects against *E. coli* and *L. monocytogenes*, but could not inhibit *S. aureus* growth. Song et al. [24] reported that chitosan film exhibited only slight inhibitory activity against *E. coli* and *S. aureus*. Some research showed that the antibacterial properties of chitosan were still controversial and unstable when chitosan was used as a single antibacterial agent when preparing film. Essential oil as a natural antibacterial agent was often added to the film to improve the antibacterial properties of the films. Oregano essential oil (OEO) mainly consists of phenolic compounds [25] that can effectively prevent spoilage and prolong the shelf life of food and is widely used in food packaging due to its favorable antioxidant and antibacterial properties. OEO can be directly incorporated into the matrix of packaging film and is released during transportation and/or storage of food and thereby contributes to reducing food spoilage [26]. However, to our knowledge, no information has been reported regarding the enhancement of cellulose nanofibers on the mechanical, thermal and barrier properties of chitosan/OEO packaging film.

The main objective of this study was to develop chitosan/OEO biocomposite packaging film enhanced by CNFs to obtain favorable physical characteristics and antimicrobial properties. The morphology, chemical structure, physical characteristics and antimicrobial properties of the developed biocomposite packaging films were measured and analyzed. The film will have potential applicability for food packaging and high-value goods.

2. Materials and Methods

2.1. Chemicals and Materials

Chitosan powder (80%–95%, deacetylated) was purchased from Sinopharm Chemical Reagent Co., Ltd., Shanghai, China. Pure OEO and Tween-80 were obtained from Shanghai Aladdin Bio-Chem

Technology Co., Ltd., Shanghai, China. *E. coli* (ATCC 25,922) and *L. monocytogenes* (ATCC 19,115) strains were obtained from the China Center of Industrial Culture Collection, Beijing, China. Bleached bagasse pulp board was purchased from Guangxi GuiTang (Group) Co., Ltd, Guigang, Guangxi, China. All other chemicals were of analytical grade.

2.2. Preparation of CNFs

CNFs were prepared using a mechanical grinding method according to the process described by Nie et al. [27]. Initially, the bleached bagasse pulp board was soaked in deionized water overnight at room temperature and was then disintegrated by a fiber disintegrator (AG 04, Estanit GmbH, Muhlheim, Germany) for 30 min to obtain disintegrated pulp with 1% (w/w) concentration. The water in the disintegrated pulp was dehydrated and the disintegrated pulp was placed in a refrigerator at 4 °C overnight to balance the moisture levels. Then, the disintegrated pulp with a solid content of 2% (w/w) was ground using an ultrafine grinder (MKZA10-15J, Masuko Sangyo, kawaguchi, Japan) with -100-μm disc spacing at a speed of 1500 rpm to obtain MFC suspensions. After 10 grinding cycles, a CNFs suspension with a solid content of 2.56% (w/w) was obtained and stored in a 4 °C refrigerator for further use.

2.3. Characterization of CNFs

The morphology of CNFs was observed using a transmission electron microscope (TEM). The CNFs suspensions were diluted to concentrations of 0.008% (w/w) with deionized water and were ultrasonically dispersed for 30 min. A drop of the dispersed CNFs suspension was deposited on a carbon-coated grid and was then stained with 1.5% (w/w) phosphotungstic acid water for 15 min in a dark place. The dried grid was observed by a TEM (HT7700, Hitachi, Tokyo, Japan) with an acceleration voltage of 100 kV.

2.4. Preparation Chitosan/OEO Films Enhanced by CNFs

Chitosan (2%, w/v) was dispersed in a glacial acetic acid solution (1%, v/v) and magnetically stirred at 250 rpm for 8 h at room temperature to completely dissolve the chitosan. The pure OEO with addition amounts of 0%, 1%, 2%, 3% (w/w, chitosan-based) and Tween-80 (40% w/w, OEO-based) were added to the chitosan solution and stirred by a high-shear homogenizer (Unidrive-Model×1000D, CAT M.Zipperer GmbH, Ballrechten-Dottingen, Germany) at 8000 rpm for 10 min to obtain film-forming solutions. Tween-80 was used as surfactant to facilitate emulsion formation and stability. After ultrasonic deaeration for 1 h, 30-g film-forming solutions were cast onto Teflon plates (150 mm × 150 mm) and dried in an oven at 35 °C for 2 days. All dried film samples were removed from the molds and were stored at 25 °C and 50% RH until the antibacterial activity was evaluated.

To evaluate the enhancement of the CNFs on the properties of the chitosan/OEO biocomposite film with the best antimicrobial activity, the optimal added OEO amount was first determined. Two percent (w/w, chitosan-based) OEO was added in the follow-up experiments, based on the results (Section 3.2). The CNFs with a concentration of 2.56 wt % were dispersed in distilled water by a high-shear homogenizer at 13,000 rpm for 6 min to obtain a CNFs suspension with a concentration of 1.0 wt %. The 0%, 20%, 40%, 60% CNFs (w/w, chitosan-based) were added to the chitosan solution and stirred by a high-shear homogenizer at high speed (13,000 rpm) for 6 min to obtain mixtures of chitosan and CNFs. In addition, the 2% OEO (w/w, chitosan- and CNFs-based) and Tween-80 (40% w/w, OEO-based) were added to the mixture and homogenized using a high-shear homogenizer at 8000 rpm for 10-min to obtain film-forming solutions which contained CNFs. The film-forming solutions were deaerated, cast and dried as described above to obtain the chitosan/OEO/0% CNFs (COC_0), chitosan/OEO/20% CNFs (COC_{20}), chitosan/OEO/40% CNFs (COC_{40}) and chitosan/OEO/60% CNFs (COC_{60}) films.

2.5. Antimicrobial Properties

Antimicrobial activities of the sample films were evaluated using growth inhibition rates and disk inhibition zone assays. The *Escherichia coli* and *Listeria monocytogenes* cultures were regenerated through the exponential growth phase (24 h) in nutrient broth and brain heart infusion (BHI) broth in incubators at 37 °C and 75% RH to obtain a bacterial suspension with a concentration of 10^8 CFU/mL.

Before the inhibition rate test, the bacterial suspension was diluted to 10^5 CFU/mL by phosphate buffer saline (PBS) and the sample film was chopped into fragments and placed under ultraviolet sterilization for 1 h. Then, 100 mg of sample film fragments were placed into 5 mL of an *E. coli* and *L. monocytogenes* suspension of 10^5 CFU/mL and then shaken at 250 rpm in a water bath shaker at 37 °C for 2 h. After 2 h of contact time, 0.1 mL of *E. coli* and *L. monocytogenes* suspension diluted to 10^2 CFU/mL (with the sample film) was uniformly coil-coated on the nutrient surfaces and BHI agar plates, respectively. The plates were incubated for 24 h at 37 °C and 75% RH. The inhibition rates of *E. coli* and *L. monocytogenes* growth were calculated by the following equation:

$$\text{Growth inhibition rate (\%)} = (A - B)/A \times 100\% \tag{1}$$

where A and B are the bacterial counts from the control and the sample films, respectively. All values were averaged from three parallel experiments.

The bacterial suspension was diluted by 100 times with PBS to obtain an inoculum which contained approximately 10^6 CFU/mL for disk inhibition zone assays. All sample films were cut into circular discs of 10-mm diameter. All culture media were double-layered in which the concentration of the upper agar was 0.5 times that of the underlying agar (BHI agar was used as the medium for *L. monocytogenes* and nutrient agar as the medium for *E. coli*). One hundred microliters bacterial cultures with 10^5 CFU/mL were uniformly coil-coated on the surface of the BHI and nutrient agar plates and the film discs were placed on plates. The plates were incubated for 24 h at 37 °C and 75% RH. The diameters of the inhibition zones were measured with a vernier caliper.

2.6. SEM Analysis

The cross-sectional morphologies of the sample films were observed with a scanning electron microscope (F16502, Phenom, Eindhoven, Netherlands) at 5 kV. The cross-sections of the sample films were exposed by fracturing the films in liquid nitrogen and sprayed with a thin layer of gold under vacuum.

2.7. FTIR Spectrum

The chemical structures of sample films were characterized using a Fourier-transform infrared spectrometer (TENSEOR 27, Bruker, Ettlingen, Germany) over a range of 400–4000 cm^{-1} that was operating in attenuated total reflection (ATR) mode and with a resolution of 4 cm^{-1}.

2.8. X-ray Diffraction (XRD)

X-ray diffraction (XRD) spectra of the sample films were measured by an X-ray diffractometer (MiniFlex600, Rigaku Corporation, Tokyo, Japan) with Cu Kα radiation ($\lambda = 0.15418$ nm) that was generated at 40 kV and 30 mA. The diffraction patterns of the films were recorded over an angular range of $2\theta = 5°–50°$ at a constant rate of 5°/min.

2.9. Thermal Stability

The thermal stability of sample films was measured using a thermal gravimetric (TG) analyzer (STA449F5, NETZSCH, Bayern, Germany) in a nitrogen atmosphere and the films were heated from 30 to 600 °C at a heating rate of 10 °C/min and nitrogen flow rate of 20 mL/min.

2.10. Mechanical Property

Sample film thicknesses were measured using a digital micrometer (model 11,248-001, TMI, New Castle, DE, USA). Elongations at the breaking and tensile strengths of the sample films were determined by an electronic universal material testing machine (MODEL 3367, Instron, MA, USA).

The films were cut into strips (100 mm × 15 mm). Stretching rates and initial grip separations were set to 10 mm/min and 50 mm, respectively.

2.11. Light Transmittance

The light transmittance of the developed biocomposite films was measured using a UV-visible spectrophotometer (Specord 50 Plus, Analytik Jena, Jena, Germany) in a wavelength range from 380 to 800 nm. An empty quartz cuvette was used as the blank. Each film sample was cut to 9 mm × 40 mm and was attached to the wall of the cuvette before measurement.

2.12. Barrier Properties

The oxygen transmission rate (OTR) values of the developed biocomposite films were measured by an automated oxygen permeability testing instrument (OX-TRAN 2/21, MOCON, Inc., Minneapolis, MN, USA) with a coulometric oxygen sensor method and followed the ASTM D3985 standard [28]. OTR is the volume of permeant oxygen passing through a film per unit surface area and time under equilibrium with testing conditions, and the unit of OTR was expressed as $cc/m^2 \cdot day$ [29]. The test area of the samples was 5 cm^2 and the tests were performed at 23 °C and 50% RH. The test gas was oxygen with a flow rate of 20 mL/min while a mixture of nitrogen (98%) and hydrogen (2%) was used as the carrier gas with a flow rate of 10 mL/min. The test mode was convergence by cycles.

The water vapor transmission rate (WVTR) of the developed biocomposite films was measured using a water vapor permeability testing instrument (TSY-T1, Labthink, Jinan, China). WVTR is the weight of permeant moisture passing through a film per unit surface area and time under equilibrium with testing conditions, and the unit of WVTR was expressed as $g/m^2 \cdot day$ [29]. The film was cut into Ø100 mm circular pieces and was placed onto a permeability cup with a 63.58 cm^2 testing area. The cup was previously filled with 10 mL of distilled water (RH 100%). The cup was sealed and placed into the dry chamber of the instrument at 38 ± 0.6 °C and 10% RH. The sealed cup was weighed periodically (0.001 g) until testing was complete. Water vapor amounts transported into the dry chamber were determined by the weight loss of the cup.

2.13. Statistical Analysis

The data were reported as mean ± standard deviation and analyzed by one-way analysis of variance (ANOVA) and Duncan's multiple range tests using the SPSS 22.0 statistical package for Windows (IBM SPSS Statistical software, Inc., Chicago, IL, USA). The significance level was always set to $p < 0.05$.

3. Results and Discussion

3.1. Characteristic of CNFs

The morphology of the CNFs is shown in Figure 1. The TEM image shows that the lengths of the prepared CNFs ranged from 200 nm to several microns, diameters ranged from 20 to 50 nm, length–diameter ratios were greater than 50, and there were intertwinements between the long fibrils [30].

3.2. Antimicrobial Properties

To determine the optimal addition amount of OEO in the following experiments, 1%, 2% and 3% OEO were added to the chitosan/OEO films. The antimicrobial activities of the chitosan/OEO films with different amounts of OEO against *E. coli* (Gram-negative) and *L. monocytogenes* (Gram-negative) were evaluated and the results are shown in Figures 2 and 3. As is shown in Figure 2, there was no clear inhibition zone either on *E. coli* or on *L. monocytogenes* around the pure chitosan film (0% OEO). With increased OEO addition amounts, the areas of the inhibition zones against both *E. coli* and *L. monocytogenes* gradually increased. The antibacterial properties of the chitosan/OEO films significantly improved by adding OEO and demonstrated that OEO had good inhibition properties against *E. coli* and *L. monocytogenes*. As shown in Figure 3, it is worth noting that the inhibition rates against *E. coli* and *L. monocytogenes* of pure chitosan film (0% OEO) reached 40% and 43%, respectively. This suggested that pure chitosan film exhibited certain antibacterial properties toward *E. coli* and *L. monocytogenes*, but that the antibacterial properties were not obvious. However, the poor antibacterial properties of pure chitosan film cannot meet the requirements for packaging materials for some perishable foods. When the addition amounts of OEO were 2% and 3%, the growth inhibition rates of the developed films against *E. coli* and *L. monocytogenes* reached 99%. However, essential oils usually have a strong pungent, are volatile and excessive levels of essential oils in food packaging as antibacterial agents may affect the original food flavor [31]. Therefore, the optimal addition amount of OEO was chosen to be 2% to ensure a high antibacterial rate. The excellent antibacterial properties indicated that chitosan/OEO biocomposite films have the potential to be used as antimicrobial packaging materials to extend food shelf life.

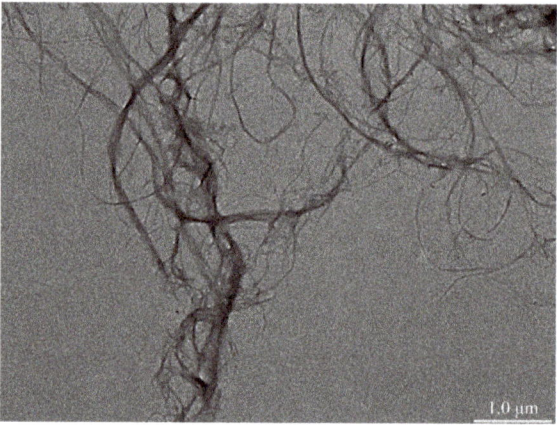

Figure 1. Transmission electron microscopy (TEM) image of cellulose nanofibrils (CNFs).

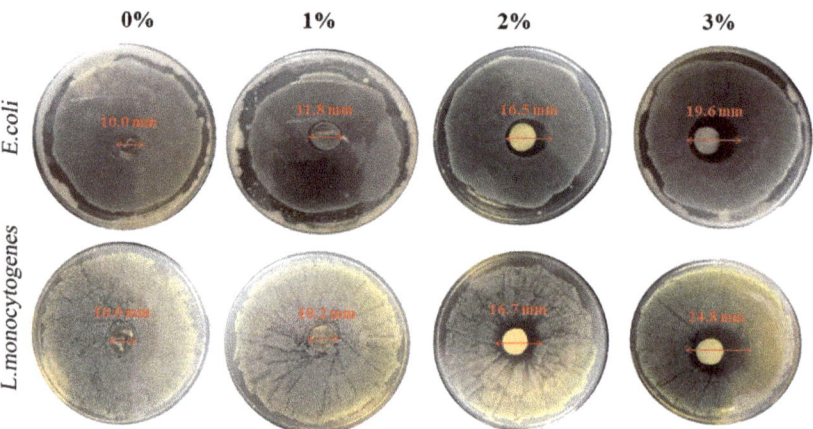

Figure 2. Area of inhibition zone of the films with the different addition amount of oregano essential oil (OEO) against *E. coli* and *L. monocytogenes*.

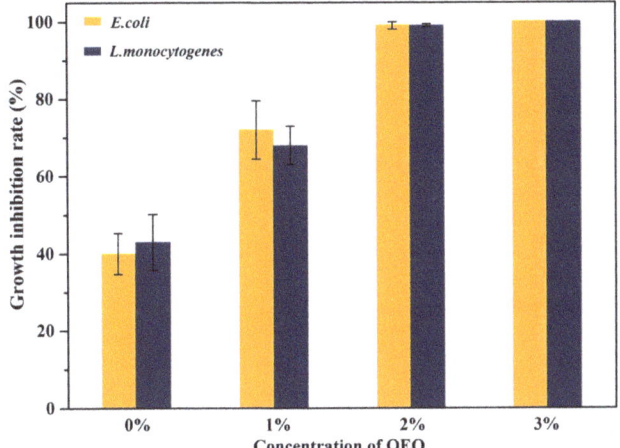

Figure 3. Growth inhibition rate of the films with the different addition amount of OEO against *E. coli* and *L. monocytogenes*.

3.3. SEM Analysis

SEM was used to observe the microstructures of the developed chitosan/OEO/CNFs films to analyze the influence of adding CNFs on the film morphologies. Figure 4 shows the cross-sectional morphologies of the COC_0, COC_{20}, COC_{40} and COC_{60} films. The COC_0 film showed a tight and homogenous structure and few pores may be caused by the volatilization of the OEO [32]. The fibrous-network structure of the films became more obvious with increased addition amounts of CNFs. This behavior was due to the hydroxyl group on the CNFs chains through hydrogen bonding interactions with chitosan. Moreover, the fibers overlapped each other and formed a dense three-dimensional network structure.

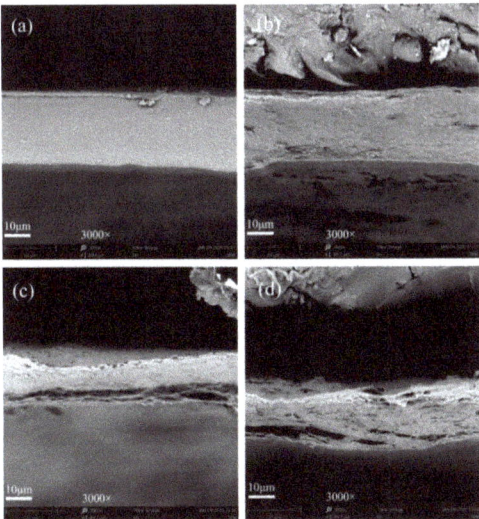

Figure 4. SEM images of the cross-section of the (**a**) chitosan/OEO/0% CNFs (COC_0) film; (**b**) chitosan/OEO/20% CNFs (COC_{20}) film; (**c**) chitosan/OEO/40% CNFs (COC_{40}) film and (**d**) chitosan/OEO/60% CNFs (COC_{60}) film.

3.4. FTIR Spectrum

FTIR spectra are widely used to analyze changes in chemical structure and components of co-composites. The FTIR spectra of all sample films with different addition amounts of CNFs are shown in Figure 5. The characteristic peaks at 1629, 1543 and 1411 cm^{-1} were assigned to C=O stretching (amide I), N–H bending (amide II) and C–N stretching (amide III), respectively [33]. These are the characteristic peaks of chitosan which appeared in all spectra of all films and confirmed that chitosan was the matrix material for all films. In the spectra of the COC$_0$ films, the broad peak at 3276 cm^{-1} was attributed to O–H and N–H stretching of chitosan. After the CNFs were incorporated with the chitosan, the position of the broad peak of the COC$_{20}$, COC$_{40}$ and COC$_{60}$ films was shifted to around 3341 cm^{-1} which was due to overlapping of the O–H bonds in both CNFs and chitosan. The peak at 1070 cm^{-1} was related to the C–O–C stretching vibration of chitosan in the COC$_0$ film spectra [34]. However, the C–O–C stretching vibration of the films containing CNFs was shifted to around 1059 cm^{-1} which was due to the overlap of the C–O–C bonds in both CNFs and chitosan [35]. These results demonstrated that there are strong hydrogen bonds between CNFs and chitosan in the molecular chain. The peaks in the region of 2921 and 2863 cm^{-1} were attributed to symmetric and asymmetric methylene stretching vibrations, respectively. Wu et al. [36] also found that the peaks at 2928 and 2864 cm^{-1} became stronger in a gelatine–chitosan film with 4% OEO. These results indicated that OEO was successfully introduced into the films.

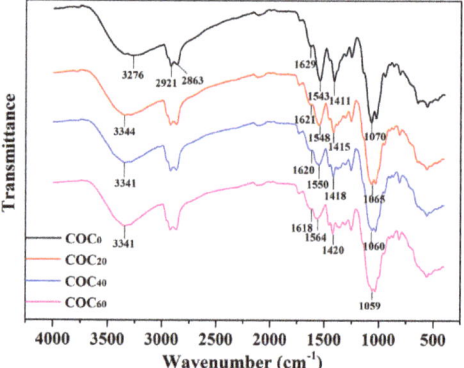

Figure 5. FTIR spectrum of the COC$_0$ film, COC$_{20}$ film, COC$_{40}$ film and COC$_{60}$ film; (COC$_0$: chitosan/OEO/0% CNFs; COC$_{20}$: chitosan/OEO/20% CNFs; COC$_{40}$: chitosan/OEO/40% CNFs; COC$_{60}$: chitosan/OEO/60% CNFs).

3.5. X-ray Diffraction (XRD)

The XRD patterns of the chitosan/OEO/CNFs films with different added amounts of CNFs are shown in Figure 6. Soni et al. [37] reported that the characteristic peaks for pure chitosan films were near 2θ = 9.77° and 19.88°. However, the positions of these characteristic peaks of chitosan-based films with OEO were slightly shifted (e.g., 2θ = 8.48° and 18.32°) which indicated that the original crystalline structure of the chitosan was destroyed [38]. After the addition of CNFs, the characteristic peaks of the CNFs appeared in the region of 2θ = 16.5° and 22°. The strengths of these peaks and the crystallinity of the films slightly increased with increased CNFs content and could be due to the ordered accumulation of chitosan chains on the surface of the crystalline domains of the CNFs [39]. Fernandes et al. [39] also reported that increased bacterial cellulose contents promoted crystallization of chitosan chains as observed in the diffractograms of water soluble chitosan/bacterial cellulose nanocomposite films. The strength of the crystalline peak increased with increased CNFs content which was due to the high biocompatibility between chitosan and cellulose [40].

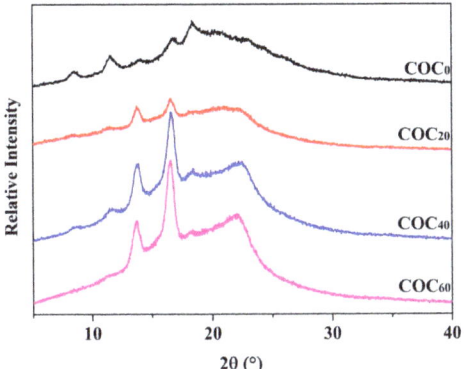

Figure 6. X-ray diffraction patterns of the COC_0 film, COC_{20} film, COC_{40} film and COC_{60} film; (COC_0: chitosan/OEO/0% CNFs; COC_{20}: chitosan/OEO/20% CNFs; COC_{40}: chitosan/OEO/40% CNFs; COC_{60}: chitosan/OEO/60% CNFs).

3.6. Thermal Stability

The thermal stabilities of the sample films were examined by TG to evaluate the effect of CNFs addition on the thermal degradation behavior of the films. Figure 7 shows the TG curves of the COC_0, COC_{20}, COC_{40} and COC_{60} films. The TG curves of all sample films showed the first stage of weight loss occurred between 90 and 250 °C which was associated with water evaporation. Similarly, the weight loss of cassara starch/chitosan/gallic acid films reinforced by CNFs occurred in the range of 90–225 °C and was also due to the weight loss of the absorbed moisture in the films [41]. The second degradation stage consisted of the disaggregation of chitosan molecules or/and disaggregation of cellulose chains which occurred between 250–340 °C. The third stage, between 340 and 500 °C, was due to oxidation of the char or/and breakdown of glucose units in CNFs. All sample films showed similar thermal behavior between 250 and 500 °C and therefore, the effect of CNFs on the thermal stability of chitosan/OEO/CNFs biocomposite films was insignificant [3]. However, the total residues of the COC_{60} films at 600 °C were the highest which indicated that 60% addition of CNFs reduced the rate of char oxidation and shifted the char oxidation to higher temperatures [42]. Therefore, the COC_{60} films are more suitable for food packaging applications even when used at a relatively high temperature.

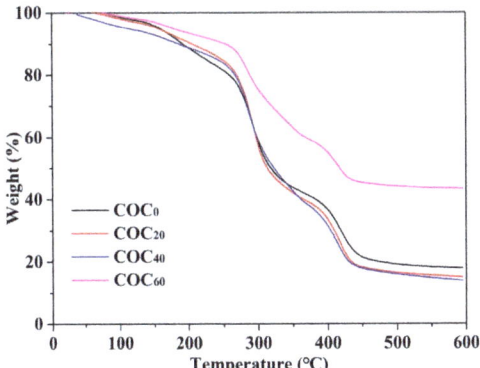

Figure 7. Thermal gravimetric (TG) curve of the COC_0 film, COC_{20} film, COC_{40} film and COC_{60} film; (COC_0: chitosan/OEO/0% CNFs; COC_{20}: chitosan/OEO/20% CNFs; COC_{40}: chitosan/OEO/40% CNFs; COC_{60}: chitosan/OEO/60% CNFs).

3.7. Mechanical Property

Tensile strength (TS) and elongation at break (EB) are fundamental properties for food-packaging films to resist the stresses and strains that the material may endure during food storage and transportation. Xu et al. [43] shown that the carboxylated CNF significantly enhanced the tensile strength of plasticized hemicelluloses/chitosan-based edible films. The mechanical properties of the COC_0, COC_{20}, COC_{40} and COC_{60} films are shown in Table 1. The TS of the COC_0 film was determined to be 7.71 MPa and the TS of the COC_{20}, COC_{40} and COC_{60} films increased to 10.24, 13.79 and 16.80 MPa, respectively ($p < 0.05$). The high TS of the films containing CNFs may result from the large aspect ratio of CNFs [43] and the stronger interfacial interaction between the chitosan and chains of CNFs [22]. Moreover, the result was also related to the mentioned in Section 3.5, the strength of crystalline peak increased in the chitosan amorphous matrix after addition of CNFs [39]. This may be because the intermolecular hydrogen bonding of chitosan was replaced by the new, strong hydrogen bonding between the hydroxyl groups in the CNFs and the hydroxyl groups in chitosan. Therefore, CNFs can be used as a good filler to enhance the mechanical strength of chitosan films. Khan et al. [3] observed a decrease in the EB values of chitosan films from 8.58% to 6.28% due to the addition of nanocrystal cellulose (NCC). In this work, the EB value was determined to be 31.31% for the COC_0 film. Compared to the COC_0 film, the EB was significantly reduced by 5.14%, 5.63% and 4.48% for the COC_{20}, COC_{40} and COC_{60} films, respectively ($p < 0.05$). These results are attributed to the strong hydrogen bonding and electrostatic interactions between CNFs and the chitosan matrix [37].

Table 1. Mechanical property of the COC_0 film, COC_{20} film, COC_{40} film and COC_{60} film; (COC_0: chitosan/OEO/0% CNFs; COC_{20}: chitosan/OEO/20% CNFs; COC_{40}: chitosan/OEO/40% CNFs; COC_{60}: chitosan/OEO/60% CNFs).

Film	Thickness (μm)	TS (MPa)	EB (%)
COC_0	58.80 ± 10.13	7.71 ± 0.62	31.31 ± 1.52
COC_{20}	58.60 ± 8.73	10.24 ± 0.44	5.14 ± 0.86
COC_{40}	57.60 ± 3.71	13.79 ± 0.29	5.63 ± 0.73
COC_{60}	57.40 ± 4.88	16.80 ± 0.66	4.48 ± 0.80

3.8. Optical Properties

The transparency of packaging films is important because light can lead to oxidation of nutrients including vitamins, fats and oils and can affect food quality. At the same time, packaging materials also need to have a certain amount of light transmittance to enable consumers to view the packaged products. The light transmittance spectra of the sample films are shown in Figure 8. The light transmittance of the COC_0 film was the highest of all films which suggested that the light barrier effect of the COC_0 film was poor. In addition, the light transmittance of the COC_0 film was 39.73% at 600 nm (center of visible light spectrum) which was lower than most pure chitosan films mentioned in other research [34]; this indicated that the presence of OEO reduced the transmittance of the films [43]. With increasing CNFs contents, the light transmittance of the films decreased, and the opacity increased. This indicated that the addition of CNFs decreased the transparency of the films. The addition of 60% CNFs improved the light barrier effect the COC_{60} film which had the lowest light transmittance (28.53%) at 600 nm. These results suggested that CNFs were densely packed in the chitosan matrix and with compact lap between the fibers, light scattering was prevented by the small interstices between the fibers [44]. All of the results implied that the COC_{60} film has good prospects for food packaging because it has excellent shading properties.

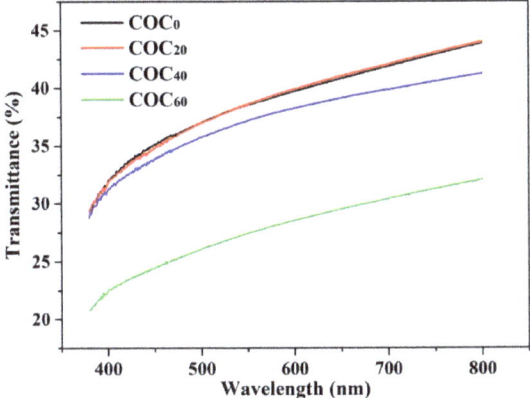

Figure 8. Light transmittance of the COC_0 film, COC_{20} film, COC_{40} film and COC_{60} film; (COC_0: chitosan/OEO/0% CNFs; COC_{20}: chitosan/OEO/20% CNFs; COC_{40}: chitosan/OEO/40% CNFs; COC_{60}: chitosan/OEO/60% CNFs).

3.9. Barrier Properties

Oxygen and water vapor are the important environmental factors that cause spoilage and deterioration of food during storage. Hence, there is concern about the barrier properties of oxygen and water vapor in food packaging materials. Figure 9 shows the barrier properties of the COC_0, COC_{20}, COC_{40} and COC_{60} films. As shown in Figure 9, the WVTR of the COC_{20}, COC_{40} and COC_{60} films significant ($p < 0.05$) decreased when compared to the COC_0 films (861.26 g/m²·day). The reduction in WVTR was due to the physicochemical interactions between CNFs and chitosan which led to reduced numbers of hydrophilic groups (–OH) [8]. As mentioned in Section 3.3, there was good biocompatibility between CNFs and the chitosan matrix and a three-dimensional network structure formed between the fibers by producing winding paths for the water vapor molecules and thus led to reduction of WVTR [22].

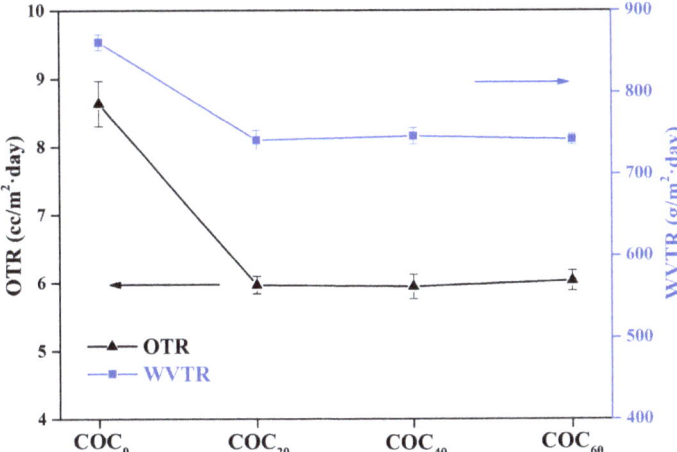

Figure 9. Oxygen transmission rate and water vapor transmission rate of the COC_0 film, COC_{20} film, COC_{40} film and COC_{60} film; (COC_0: chitosan/OEO/0% CNFs; COC_{20}: chitosan/OEO/20% CNFs; COC_{40}: chitosan/OEO/40% CNFs; COC_{60}: chitosan/OEO/60% CNFs).

If the OTR value is in the region of 1–10 cc/m^2·day, the packaging material is considered to have good oxygen barrier performance [45]. All sample films in this study had OTR < 10 cc/m^2·day. However, the OTR of the COC$_{20}$ films was 5.97 cc/m^2·day which were lower than that of the COC$_0$ film (8.64 cc/m^2·day). Their decreased oxygen permeability could be due to the presence of a more tortuous path between the fibers for penetration by oxygen molecules [46]. Compared with the OTRs of the COC$_{40}$ films (5.94 cc/m^2·day) and COC$_{60}$ films (6.03 cc/m^2·day) there were comparable. Oxygen permeability was related to the addition of CNFs and was not affected by the CNFs content. Overall, the addition of CNFs to chitosan films shows good oxygen barrier performance and thus indicates that these composite films can be used as barrier packaging for food.

4. Conclusions

In this study, a novel biocomposite packaging film with good antibacterial activities in addition to good mechanical and barrier properties was successfully developed based on chitosan as the film matrix, CNFs as a reinforcing filler and OEO as an antibacterial agent. The chitosan film, which contained 2% OEO, exhibited significant antimicrobial activity and its growth inhibition rates against *L. monocytogenes* and *E. coli* reached 99%. The fibers overlapped with each other in the chitosan matrix to form a dense three-dimensional network structure that was observed by SEM. The FTIR spectrum showed that there were strong hydrogen bonds between CNFs and chitosan in the molecular chain. The TS of the chitosan/OEO film increased with the addition of CNFs. CNFs improved the barrier performance of the chitosan/OEO film to light, oxygen and water vapor by reducing light transmittance, oxygen permeability and water vapor permeability. However, the effect of CNFs on the thermal stability of the chitosan/OEO film was insignificant. The developed chitosan/oregano essential oil/CNFs biocomposite film can be used as an antibacterial and barrier materials in the field of food packaging. It has the potential to improve food quality and extend food shelf life.

Author Contributions: Conceptualization, M.W., P.L. and S.W.; methodology, S.C. and C.W.; formal analysis, S.C.; investigation, S.C., C.W. and S.Y.; data curation, M.W. and P.L.; writing—original draft preparation, S.C.; writing—review and editing, M.W. All authors have read and agreed to the published version of the manuscript.

Funding: This research was funded by the Natural Science Foundation of Guangxi (No. 2019GXNSFAA185002) and by the Dean Project of Guangxi Key Laboratory of Clean Pulp & Papermaking and Pollution Control (ZR201708).

Conflicts of Interest: The authors declare no conflicts of interest.

References

1. Ye, Q.; Han, Y.; Zhang, J.; Zhang, W.; Xia, C.; Li, J. Bio-based films with improved water resistance derived from soy protein isolate and stearic acid via bioconjugation. *J. Clean. Prod.* **2019**, *214*, 125–131. [CrossRef]
2. Lago, M.A.; Sendón, R.; de Quirós, A.R.-B.; Sanches-Silva, A.; Costa, H.S.; Sánchez-Machado, D.I.; Valdez, H.S.; Angulo, I.; Aurrekoetxea, G.P.; Torrieri, E.; et al. Preparation and characterization of antimicrobial films based on chitosan for active food packaging applications. *Food Bioprocess Technol.* **2014**, *7*, 2932–2941. [CrossRef]
3. Khan, A.; Khan, R.A.; Salmieri, S.; Le Tien, C.; Riedl, B.; Bouchard, J.; Chauve, G.; Tan, V.; Kamal, M.R.; Lacroix, M. Mechanical and barrier properties of nanocrystalline cellulose reinforced chitosan based nanocomposite films. *Carbohydr. Polym.* **2012**, *90*, 1601–1608. [CrossRef] [PubMed]
4. Zhang, H.; Jung, J.; Zhao, Y. Preparation and characterization of cellulose nanocrystals films incorporated with essential oil loaded β-chitosan beads. *Food Hydrocoll.* **2017**, *69*, 164–172. [CrossRef]
5. Fang, Y.; Fu, J.; Tao, C.; Liu, P.; Cui, B. Mechanical properties and antibacterial activities of novel starch-based composite films incorporated with salicylic acid. *Int. J. Biol. Macromol.* **2019**. [CrossRef] [PubMed]
6. Chaichi, M.; Badii, F.; Mohammadi, A.; Hashemi, M. Water resistance and mechanical properties of low methoxy-pectin nanocomposite film responses to interactions of Ca(2+) ions and glycerol concentrations as crosslinking agents. *Food Chem.* **2019**, *293*, 429–437. [CrossRef] [PubMed]
7. Zhao, J.; Wei, F.; Xu, W.; Han, X. Enhanced antibacterial performance of gelatin/chitosan film containing capsaicin loaded MOFs for food packaging. *Appl. Surf. Sci.* **2020**, *510*. [CrossRef]

8. Deng, Z.; Jung, J.; Zhao, Y. Development, characterization, and validation of chitosan adsorbed cellulose nanofiber (CNF) films as water resistant and antibacterial food contact packaging. *LWT Food Sci. Technol.* **2017**, *83*, 132–140. [CrossRef]
9. Yuan, G.; Chen, X.; Li, D. Chitosan films and coatings containing essential oils: The antioxidant and antimicrobial activity, and application in food systems. *Food Res. Int.* **2016**, *89*, 117–128. [CrossRef]
10. Kim, K.W.; Min, B.J.; Kim, Y.-T.; Kimmel, R.M.; Cooksey, K.; Park, S.I. Antimicrobial activity against foodborne pathogens of chitosan biopolymer films of different molecular weights. *LWT Food Sci. Technol.* **2011**, *44*, 565–569. [CrossRef]
11. Kaya, M.; Khadem, S.; Cakmak, Y.S.; Mujtaba, M.; Ilk, S.; Akyuz, L.; Salaberria, A.M.; Labidi, J.; Abdulqadir, A.H.; Deligöz, E. Antioxidative and antimicrobial edible chitosan films blended with stem, leaf and seed extracts of Pistacia terebinthus for active food packaging. *RSC Adv.* **2018**, *8*, 3941–3950. [CrossRef]
12. Liu, J.; Liu, S.; Wu, Q.; Gu, Y.; Kan, J.; Jin, C. Effect of protocatechuic acid incorporation on the physical, mechanical, structural and antioxidant properties of chitosan film. *Food Hydrocoll.* **2017**, *73*, 90–100. [CrossRef]
13. Poverenov, E.; Danino, S.; Horev, B.; Granit, R.; Vinokur, Y.; Rodov, V. Layer-by-layer electrostatic deposition of edible coating on fresh cut melon model: Anticipated and unexpected effects of alginate–chitosan combination. *Food Bioprocess Technol.* **2013**, *7*, 1424–1432. [CrossRef]
14. Khwaldia, K.; Basta, A.H.; Aloui, H.; El-Saied, H. Chitosan-caseinate bilayer coatings for paper packaging materials. *Carbohydr. Polym.* **2014**, *99*, 508–516. [CrossRef] [PubMed]
15. Zhang, Y.; Zhang, M.; Yang, H. Postharvest chitosan-g-salicylic acid application alleviates chilling injury and preserves cucumber fruit quality during cold storage. *Food Chem.* **2015**, *174*, 558–563. [CrossRef]
16. Willberg-Keyrilainen, P.; Vartiainen, J.; Pelto, J.; Ropponen, J. Hydrophobization and smoothing of cellulose nanofibril films by cellulose ester coatings. *Carbohydr. Polym.* **2017**, *170*, 160–165. [CrossRef]
17. Azeredo, H.M.C.; Rosa, M.F.; Mattoso, L.H.C. Nanocellulose in bio-based food packaging applications. *Ind. Crops Prod.* **2017**, *97*, 664–671. [CrossRef]
18. Abdul Khalil, H.P.S.; Davoudpour, Y.; Saurabh, C.K.; Hossain, M.S.; Adnan, A.S.; Dungani, R.; Paridah, M.T.; Islam Sarker, M.Z.; Fazita, M.R.N.; Syakir, M.I.; et al. A review on nanocellulosic fibres as new material for sustainable packaging: Process and applications. *Renew. Sustain. Energy Rev.* **2016**, *64*, 823–836. [CrossRef]
19. Sasikala, M.; Umapathy, M.J. Preparation and characterization of pineapple leaf cellulose nanocrystal reinforced gelatin bio-nanocomposite with antibacterial banana leaf extract for application in food packaging. *New J. Chem.* **2018**, *42*, 19979–19986. [CrossRef]
20. Nie, S.; Zhang, K.; Lin, X.; Zhang, C.; Yan, D.; Liang, H.; Wang, S. Enzymatic pretreatment for the improvement of dispersion and film properties of cellulose nanofibrils. *Carbohydr. Polym.* **2018**, *181*, 1136–1142. [CrossRef]
21. Li, J.; Zhou, M.; Cheng, G.; Cheng, F.; Lin, Y.; Zhu, P.X. Fabrication and characterization of starch-based nanocomposites reinforced with montmorillonite and cellulose nanofibers. *Carbohydr. Polym.* **2019**, *210*, 429–436. [CrossRef] [PubMed]
22. Jahed, E.; Khaledabad, M.A.; Bari, M.R.; Almasi, H. Effect of cellulose and lignocellulose nanofibers on the properties of Origanum vulgare ssp. gracile essential oil-loaded chitosan films. *React. Funct. Polym.* **2017**, *117*, 70–80. [CrossRef]
23. Sánchez-González, L.; Cháfer, M.; Hernández, M.; Chiralt, A.; González-Martínez, C. Antimicrobial activity of polysaccharide films containing essential oils. *Food Control* **2011**, *22*, 1302–1310. [CrossRef]
24. Song, J.; Feng, H.; Wu, M.; Chen, L.; Xia, W.; Zhang, W. Preparation and characterization of arginine-modified chitosan/hydroxypropyl methylcellose antibacterial film. *Int. J. Biol. Macromol.* **2020**, *145*, 750–758. [CrossRef]
25. Ramos, M.; Jiménez, A.; Peltzer, M.; Garrigós, M.C. Characterization and antimicrobial activity studies of polypropylene films with carvacrol and thymol for active packaging. *J. Food Eng.* **2012**, *109*, 513–519. [CrossRef]
26. Ribeiro-Santos, R.; Andrade, M.; Melo, N.R.d.; Sanches-Silva, A. Use of essential oils in active food packaging: Recent advances and future trends. *Trends Food Sci. Technol.* **2017**, *61*, 132–140. [CrossRef]
27. Nie, S.; Zhang, C.; Zhang, Q.; Zhang, K.; Zhang, Y.; Tao, P.; Wang, S. Enzymatic and cold alkaline pretreatments of sugarcane bagasse pulp to produce cellulose nanofibrils using a mechanical method. *Ind. Crops Prod.* **2018**, *124*, 435–441. [CrossRef]
28. ASTM D3985-17. *Standard Test Method for Oxygen Gas Transmission Rate Through Plastic Film and Sheeting Using a Coulometric Sensor*; ASTM International: West Conshohocken, PA, USA, 2017. [CrossRef]

29. Wang, J.; Gardner, D.J.; Stark, N.M.; Bousfield, D.W.; Tajvidi, M.; Cai, Z. Moisture and oxygen barrier properties of cellulose nanomaterial-based films. *ACS Sustain. Chem. Eng.* **2017**, *6*, 49–70. [CrossRef]
30. Abdul Khalil, H.P.S.; Bhat, A.H.; Ireana Yusra, A.F. Green composites from sustainable cellulose nanofibrils: A review. *Carbohydr. Polym.* **2012**, *87*, 963–979. [CrossRef]
31. Han, J.-W.; Ruiz-Garcia, L.; Qian, J.-P.; Yang, X.-T. Food packaging: A comprehensive review and future trends. *Compr. Rev. Food Sci. Food Saf.* **2018**, *17*, 860–877. [CrossRef]
32. do Evangelho, J.A.; da Silva Dannenberg, G.; Biduski, B.; El Halal, S.L.M.; Kringel, D.H.; Gularte, M.A.; Fiorentini, A.M.; da Rosa Zavareze, E. Antibacterial activity, optical, mechanical, and barrier properties of corn starch films containing orange essential oil. *Carbohydr. Polym.* **2019**, *222*, 114981. [CrossRef] [PubMed]
33. Arafa, M.G.; Mousa, H.A.; Afifi, N.N. Preparation of PLGA-chitosan based nanocarriers for enhancing antibacterial effect of ciprofloxacin in root canal infection. *Drug Deliv.* **2020**, *27*, 26–39. [CrossRef] [PubMed]
34. Xu, J.; Xia, R.; Yuan, T.; Sun, R. Use of xylooligosaccharides (XOS) in hemicelluloses/chitosan-based films reinforced by cellulose nanofiber: Effect on physicochemical properties. *Food Chem.* **2019**, *298*, 125041. [CrossRef] [PubMed]
35. Xu, J.; Xia, R.; Zheng, L.; Yuan, T.; Sun, R. Plasticized hemicelluloses/chitosan-based edible films reinforced by cellulose nanofiber with enhanced mechanical properties. *Carbohydr. Polym.* **2019**, *224*, 115164. [CrossRef] [PubMed]
36. Wu, J.; Ge, S.; Liu, H.; Wang, S.; Chen, S.; Wang, J.; Li, J.; Zhang, Q. Properties and antimicrobial activity of silver carp (Hypophthalmichthys molitrix) skin gelatin-chitosan films incorporated with oregano essential oil for fish preservation. *Food Packag. Shelf Life* **2014**, *2*, 7–16. [CrossRef]
37. Soni, B.; Hassan, E.B.; Schilling, M.W.; Mahmoud, B. Transparent bionanocomposite films based on chitosan and TEMPO-oxidized cellulose nanofibers with enhanced mechanical and barrier properties. *Carbohydr. Polym.* **2016**, *151*, 779–789. [CrossRef]
38. Jahed, E.; Khaledabad, M.A.; Almasi, H.; Hasanzadeh, R. Physicochemical properties of Carum copticum essential oil loaded chitosan films containing organic nanoreinforcements. *Carbohydr. Polym.* **2017**, *164*, 325–338. [CrossRef]
39. Fernandes, S.C.M.; Oliveira, L.; Freire, C.S.R.; Silvestre, A.J.D.; Neto, C.P.; Gandini, A.; Desbriéres, J. Novel transparent nanocomposite films based on chitosan and bacterial cellulose. *Green Chem.* **2009**, *11*. [CrossRef]
40. Cabanas-Romero, L.V.; Valls, C.; Valenzuela, S.V.; Roncero, M.B.; Pastor, F.I.J.; Diaz, P.; Martinez, J. Bacterial cellulose-chitosan paper with antimicrobial and antioxidant activities. *Biomacromolecules* **2020**, *21*, 1568–1577. [CrossRef]
41. Zhao, Y.; Huerta, R.R.; Saldaña, M.D.A. Use of subcritical water technology to develop cassava starch/chitosan/gallic acid bioactive films reinforced with cellulose nanofibers from canola straw. *J. Supercrit. Fluids* **2019**, *148*, 55–65. [CrossRef]
42. Uddin, K.M.A.; Ago, M.; Rojas, O.J. Hybrid films of chitosan, cellulose nanofibrils and boric acid: Flame retardancy, optical and thermo-mechanical properties. *Carbohydr. Polym.* **2017**, *177*, 13–21. [CrossRef] [PubMed]
43. Chen, S.; Wu, M.; Lu, P.; Gao, L.; Yan, S.; Wang, S. Development of pH indicator and antimicrobial cellulose nanofibre packaging film based on purple sweet potato anthocyanin and oregano essential oil. *Int. J. Biol. Macromol.* **2020**, *149*, 271–280. [CrossRef] [PubMed]
44. Nogi, M.; Iwamoto, S.; Nakagaito, A.N.; Yano, H. Optically transparent nanofiber paper. *Adv. Mater.* **2009**, *21*, 1595–1598. [CrossRef]
45. Abdellatief, A.; Welt, B.A. Comparison of new dynamic accumulation method for measuring oxygen transmission rate of packaging against the steady-state method described by ASTM D3985. *Packag. Technol. Sci.* **2013**, *26*, 281–288. [CrossRef]
46. Fazeli, M.; Keley, M.; Biazar, E. Preparation and characterization of starch-based composite films reinforced by cellulose nanofibers. *Int. J. Biol. Macromol.* **2018**, *116*, 272–280. [CrossRef]

© 2020 by the authors. Licensee MDPI, Basel, Switzerland. This article is an open access article distributed under the terms and conditions of the Creative Commons Attribution (CC BY) license (http://creativecommons.org/licenses/by/4.0/).

Article

Increased Cytotoxic Efficacy of Protocatechuic Acid in A549 Human Lung Cancer Delivered via Hydrophobically Modified-Chitosan Nanoparticles As an Anticancer Modality

Cha Yee Kuen [1], Tieo Galen [1], Sharida Fakurazi [2], Siti Sarah Othman [1] and Mas Jaffri Masarudin [1,3,*]

[1] Department of Cell and Molecular Biology, Faculty of Biotechnology and Biomolecular Sciences, Universiti Putra Malaysia, Selangor 43400, Malaysia; yeekuen910416cha@gmail.com (C.Y.K.); galenofficial94@gmail.com (T.G.); sarahothman@upm.edu.my (S.S.O.)
[2] Department of Human Anatomy, Faculty of Medicine and Health Sciences, Universiti Putra Malaysia, Selangor 43400, Malaysia; sharida@upm.edu.my
[3] UPM-MAKNA Cancer Research Laboratory, Institute of Biosciences, Universiti Putra Malaysia, Selangor 43400, Malaysia
* Correspondence: masjaffri@upm.edu.my

Received: 10 July 2020; Accepted: 31 July 2020; Published: 28 August 2020

Abstract: The growing incidence of global lung cancer cases against successful treatment modalities has increased the demand for the development of innovative strategies to complement conventional chemotherapy, radiation, and surgery. The substitution of chemotherapeutics by naturally occurring phenolic compounds has been touted as a promising research endeavor, as they sideline the side effects of current chemotherapy drugs. However, the therapeutic efficacy of these compounds is conventionally lower than that of chemotherapeutic agents due to their lower solubility and consequently poor intracellular uptake. Therefore, we report herein a hydrophobically modified chitosan nanoparticle (pCNP) system for the encapsulation of protocatechuic acid (PCA), a naturally occurring but poorly soluble phenolic compound, for increased efficacy and improved intracellular uptake in A549 lung cancer cells. The pCNP system was modified by the inclusion of a palmitoyl group and physico-chemically characterized to assess its particle size, Polydispersity Index (PDI) value, amine group quantification, functional group profiling, and morphological properties. The inclusion of hydrophobic palmitoyl in pCNP-PCA was found to increase the encapsulation of PCA by 54.5% compared to unmodified CNP-PCA samples whilst it only conferred a 23.4% larger particle size. The single-spherical like particles with uniformed dispersity pCNP-PCA exhibited IR bands, suggesting the successful incorporation of PCA within its core, and a hydrophobic layer was elucidated via electron micrographs. The cytotoxic efficacy was then assessed by using an MTT cytotoxicity assay towards A549 human lung cancer cell line and was compared with traditional chitosan nanoparticle system. Fascinatingly, a controlled release delivery and enhanced therapeutic efficacy were observed in pCNP-PCA compared to CNP, which is ascribed to lower IC_{50} values in the 72-h treatment in the pCNP system. Using the hydrophobic system, efficacy of PCA was significantly increased in 24-, 48-, and 72-h treatments compared to a single administration of the compound, and via the unmodified CNP system. Findings arising from this study exhibit the potential of using such modified nanoparticulate systems in increasing the efficacy of natural phenolic compounds by augmenting their delivery potential for better anti-cancer responses.

Keywords: hydrophobically modified-chitosan nanoparticle; protocatechuic acid; nanobiotechnology

1. Introduction

Despite improvements in the medical field nowadays, cancer remains one of the most studied diseases due to its complexity and continually increasing incidence rate throughout the decades. Global cancer statistics show lung cancer is the top cause for cancer-related deaths worldwide, with non-small cell lung cancer (NSCLC) accounting for approximately 84% of this statistic [1]. Current treatment for NSCLC includes surgery, chemotherapy, targeted therapies, immunotherapy, radiation therapy, and radiofrequency ablation therapy depending on the stage of cancer and other factors. However, even in curable NSCLC cases, death in patients can occur due to the onset of extensive distant metastases after initial treatment exercises [2]. The most common treatment regime for NSCLC includes chemotherapy utilizing therapeutic agents such as Carboplatin, Cisplatin, Paclitaxel (Taxol), and Gemcitabine (Gemzar). According to the National Cancer Institute, the use of such drugs in anticancer therapy will frequently confer unwanted side effects, such as hair loss, fatigue, anemia, appetite changes and nausea, and vomiting, which further hinder the recovery of patients. These Food and Drug administration (FDA)-approved drugs can give rise to these side effects frequently due to the high dose administration and non-specific destruction of these chemotherapy agents, where the non-specific cytotoxicity often resulted in low tumor specificity and high toxicity, leading to various concomitant side effects [3]. Therefore, alternatives for these chemotherapy agents are needed to avoid unpleasant side effects for the patients.

In lieu of this, extensive research has been conducted in search of derivative- and natural-based compounds that can be used as alternatives for traditional chemotherapy drugs [4,5]. It is more beneficial to use natural anti-cancer compounds as compared to chemotherapy agents as they can aid in reducing these concomitant side effects and reduce discomfort in patients [6]. This was suggested by Demain and Vaishnav as natural compounds, including curcumin, isoflavone genistine, and resveratrol, induced apoptosis death of cancer cells without any adverse effect on normal cells [7]. Previous studies have shown that phenolic compounds, such as green tea extract [8,9], curcumin [10,11], and caffeic acid [12,13], are among natural compounds that possess potent anticancer effects against lung carcinomas. Among these alternates, protocatechuic acid (PCA) is a natural phenolic compound broadly distributed in most edible plants utilized for folk medicine [14,15]. PCA has been reported to be anti-bacterial [16], anti-oxidative [17], anti-cancer [18], anti-diabetic [19], anti-ageing [20], and anti-inflammatory [21]. Yin et al. have suggested that PCA has revealed an anticancer effect towards human cancer cells, including lung, breast, liver, and prostate cancer cells through apoptosis or the suppression of invasion and metastasis of the cancer cells [22]. Moreover, evidence from Hu et al. shown that PCA at 25 µM concentration has significantly inhibited vascular endothelial growth factor (VEGF)-induced cell proliferation of human umbilical vein endothelial cell (HUVECs) by 22.68 ± 5.6% assessed by an MTT assay which further suggested PCA as a candidate treatment for cancer tumors [23]. Apart from that, a previous study of Tsao et al. has reported that PCA treatments at 2–8 µM were able to inhibit the cell growth of lung cancer A549, H3255, and Calu-6 cells in a dose-dependent manner through modulation of FAK, MAPK, and NF-kB pathways, and downregulation of the protein production of growth factors proposed PCA as a good candidate for lung cancer therapeutics [24]. However, PCA, which is also commonly known as 3,4-dihydroxybenzoic acid, possesses sparingly a solubility of 1:50 ratio in water. This hampers its use in the medical field, including in cancer treatment, since the solubility may directly affect its absorption and bioavailability [25]. Consequently, the efficiency of PCA as a therapeutic agent can be potentiated by improving its cellular delivery and uptake. By increasing its accumulation through higher cellular uptake, the efficacy of PCA can be potentially increased while minimizing adverse responses associated with the many side-effects of more potent chemotherapy drugs.

One possible strategy to increase its efficiency is by optimizing its uptake and delivery into cancer cells. Oral delivery is the most popular and economical administration route for therapeutics but requires overcoming biological barriers such as absorption, solubility, and dissolution, pre-systemic metabolism, and excretion [26]. Parenteral delivery is the most simple and convenient drug delivery system but involves the application of specialized tools and techniques to arrange and administer parenteral formulations [26]. Subsequently, poor cellular uptake of the phenolic compounds has also led to a high dose of therapeutic administration, thus conforming to a restricted therapeutic value due to issues of dose-dependent morbidities [27]. To overcome these problems, research has focused on attempts in assisting or enhancing current drug delivery systems. This has included the adaptation of nanoparticulate delivery systems to complement established oral and parenteral delivery systems [28]. The utilization of nanoparticles for the encapsulation of cargos such as various therapeutic drugs or compounds and genetic materials have been reported by innumerable researchers over the years. Various nanoparticle systems have been formulated by the researchers, including metallic, liposome, carbon nanotube, solid lipid, and polymeric nanoparticle systems. These nanoparticle systems vary in their physical and surface properties due to the features of their respective building materials [29]. Nonetheless, they have shared some crucial common characteristics where they have sizes of less than 100 nm at least in one of the three dimensions and are capable of encapsulating cargos [30]. In recent decades, nanotechnology has emerged as one of the promising tools in various sectors, including cosmetics, electronics, food, and agriculture, as well as biomedical and pharmaceutical fields [31]. The application of nanobiotechnology in cancer therapy has been incorporated into several treatments such as hyperthermia, gene therapy, and targeted cancer therapy. A previous study of Giustini et al. showed that utilization of magnetic nanoparticles in hyperthermia cancer treatment was advantageous in achieving an enhanced permeability and retention (EPR) effect and achieved targeted delivery [32]. Additionally, Wu et al. described a SP94 peptide-conjugated PEGylated liposomal doxorubicin towards human hepatocellular carcinoma cell lines and revealed a significant drug accumulation increment in tumors in comparison to non-targeted PEGylated liposomal doxorubicin by about 8.8-fold greater cellular uptake in SK-HEP-1 cells, and revealed greater therapeutic effects in both in vitro and in vivo studies [33].

Several nanoparticle formulation systems have been shown to aid in delivery purposes, including metal nanoparticles, carbon-based nanoparticles, polymeric nanoparticles, as well as lipid-based nanoparticles [34,35]. However, their eventual adaptation for anticancer modalities are often constricted to issues of inherent toxicity and robust synthesis regimes. For example, titanium dioxide (TiO_2) nanoparticles have been demonstrated to profusely accumulate in the mouse hippocampus post-administration to affect hippocampal apoptosis and damage in spatial recognition memory [36]. Carbon-based nanoparticles potentially induce oxidative stress, as shown by Wang et al. in that single-walled carbon nanotubes exerted significant cytotoxicity towards rat PC12 cells in a wide dose range of 5–600 µg/mL for 24 and 48 h [37]. Previous studies have described that the presence of transition metals in carbon nanotubes induces the formation of molecular oxygen-dependent superoxide anion radicals, hydroxyl radicals, and hydrogen peroxide, which have high redox potentials and reactivities [38,39]. On the other hand, biodegradable polymeric nanoparticles serve as a good candidate vector to develop anticancer modalities with additional sustained release properties while being biocompatible with cells and tissues [40]. Chitosan nanoparticles (CNP) constitute a polymeric nanoparticle system that has been commonly reported for drug delivery applications. Both the amine ($-NH_2$) and hydroxyl ($-OH$) groups of chitosan are active spots for modification to initiate different modification requirements [41]. CNP has been modified using emulsification solvent diffusion methods to increase the entrapment of hydrophobic drugs [41,42]. Glycol-chitosan has been hydrophobically modified through chemical conjugation using hydrophobic 5β-cholanic acid moieties and the hydrophilic glycol chitosan backbone to encapsulate water-insoluble camptothecin (CPT) into the hydrophobically modified glycol chitosan nanoparticles with high loading efficiency with a sustained release property [43]. These previous modifications have therefore led to the modification of chitosan in this current study using palmitic acid to synthesize hydrophobically modified-chitosan nanoparticles (pCNP) to increase the encapsulation of PCA in pCNP through hydrophobic–hydrophobic interactions, and in turn to enhance the therapeutic efficacy of PCA in A549 lung cancer cell treatment.

This current study describes the enhanced therapeutic response and controlled release property of the phenolic acid PCA in the A549 human lung cancer cell line mediated through its encapsulation in hydrophobically modified chitosan nanoparticles (pCNP), as compared with conventional CNP systems. The overview of this study is shown in Figure 1, where a hydrophobic anchor based on palmitoyl was conjugated to chitosan polymer via NHS-ester bridges, and the resulting nanoparticles were characterized via various physicochemical analyses. This novel pCNP system is suggested as a safe and effective alternative nanocarrier system for the enhanced therapeutic delivery of PCA, which could be a potential nanocarrier system for other poorly soluble therapeutics. The findings from this research are expected to aid in the enhancement of PCA for anti-cancer applications.

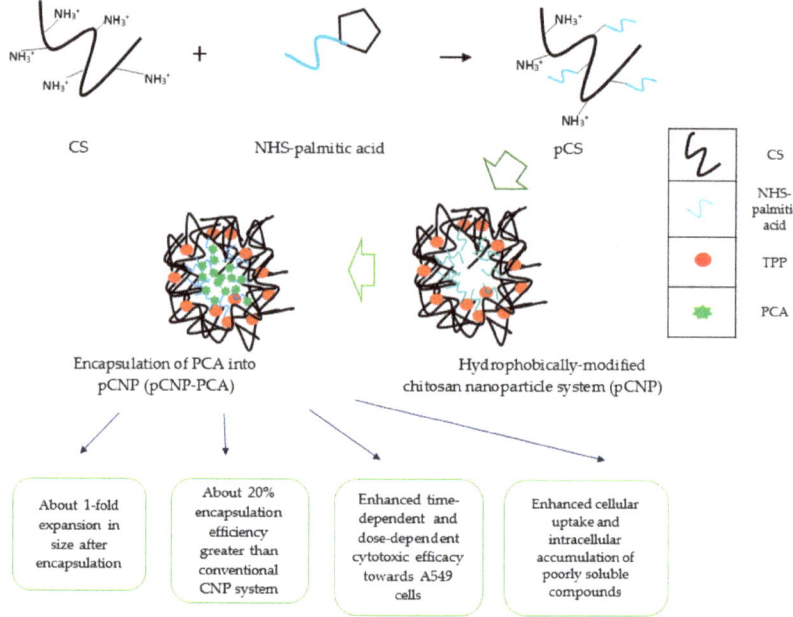

Figure 1. Overview of the study.

2. Materials and Methods

Chitosan (CS, low molecular weight), sodium tripolyphosphate (TPP), palmitic acid N-hydroxy-succinimide ester (NHS-palmitate), protocatechuic acid (PCA), and dimethyl sulfoxide (DMSO) were acquired in powder form from Sigma-Aldrich (St. Louis, MO, USA). Roswell Park Memorial Institute-1640 medium (RPMI-1640), fetal bovine serum (FBS), 0.25% trypsin-EDTA (1×), and Antibiotic-Antimycotic (100×) were purchased from Gibco Life Technologies (Grand island, NY, USA). Glacial acetic acid, sodium hydroxide, and hydrochloric acid (analytical grade) were obtained from Friendemann Schmidt Chemicals (Parkwood, Western Australia). All reagents, unless otherwise stated, were used without further purification.

2.1. Formation of Chitosan Nanoparticles (CNP)

CNPs were prepared by ionic gelation route as previously described by Masarudin et al. [44]. Chitosan (CS) and Tripolyphosphate (TPP) were prepared to a concentration of 1.0 mg/mL in 50 mL centrifuge tubes and further diluted to 0.5 and 0.7 mg/mL respectively and adjusted to pH 5 and pH 2 using 1 M NaOH and 1 M HCl. Subsequently, nanoparticles were formed by adding increasing volumes of TPP solution (0 to 300 µL) to 600 µL of CS solution. The CNPs were purified by centrifugation at

13,000 rpm for 20 min. After that, 40% of the total CNPs supernatant volume were mixed with 60% of deionized water (dH$_2$O) corresponding to the 40% supernatant volume and used for further analyses.

2.2. Hydrophobic Modification of Chitosan Nanoparticles (pCNP)

Hydrophobic modification was performed by the spontaneous conjugation of palmitoyl groups to the CS backbone prior to nanoparticle formation. Initially, 1.0 mg/mL CS solution was adjusted to pH 6. Separately, NHS-palmitate was prepared in absolute ethanol to a concentration of 0.9 mg/mL. The NHS-palmitate solution was subsequently added to the CS solution by dropwise additions at a 2:1 volume ratio and the conjugation reaction was left to occur a further 20 h at 50 °C. Following incubation, hydrophobically modified chitosan (pCS) was precipitated from the mixture by adjusting the pH to 9. It was then centrifuged at 4500 rpm for 45 min to separate the precipitate from the solution. The precipitate was washed once with an acetone: ethanol (50:50) solution, and successively thrice with dH$_2$O before being dried in oven at 50 °C. The pCNP was prepared therewith using similar methods as described for CNP.

2.3. Synthesis of Protocatechuic Acid-Encapsulated Nanoparticles (CNP-PCA and pCNP-PCA)

Approximately 1.5 mg of PCA was dissolved in 2 mL dH$_2$O and allowed to stir at 60 °C for approximately 10 min using a magnetic stirrer to prepare a 5 mM master stock. To form PCA-encapsulated nanoparticles, 200 µL of PCA was mixed with 600 µL of CS/pCS followed by 200 µL of TPP. The resulting CNP-PCA and pCNP-PCA were then directly used for consequent physicochemical analyses.

2.4. Physicochemical Characterization of Nanoparticles

Particle size by intensity and polydispersity index (PDI) of nanoparticle samples (CNP, pCNP, CNP-PCA and pCNP-PCA) were determined using dynamic light scattering on a Malvern Zetasizer Nano S Instrument (Malvern Instruments, Malvern, UK). Approximately 1000 µL of sample was aliquoted into a disposable cuvette and analyzed in triplicate to ensure the stability of the samples. All the data were recorded as mean ± standard error of mean (SEM). Surface morphology of nanoparticles were assessed using field emission-scanning electron microscopy (FESEM). The samples were first diluted prior to analysis by mixing 100 µL of the samples with 500 µL of dH$_2$O. Then, a single drop of each diluted samples was coated onto an aluminum stub and left to dry in an oven for at least 3 days. Next, vacuum gold-coating was performed for the sample-loaded stubs before observation under a FEI NOVA nanoSEM 230 electron microscope. Internal surface of the nanoparticle samples was examined using transmission electron microscopy (TEM). For TEM analysis, the diluted samples were drop-coated directly onto copper grids and dried under a hot light bulb before observation under a TECNAI G2 F20, FEI TEM. Determination of characteristic functional groups in samples were performed using a Spectrum 100 Perkin-Elmer FTIR instrument. Prior to analysis, all samples were freeze dried in a Coolsafe 95-15 PRO freeze drier (SCANVAC, Lynge, Denmark) for 48 h. The samples were analysed using attenuated total reflectance (ATR) at an infrared frequency range of 200–4000 cm^{-1}.

2.5. Determination of Free Amine Groups Using Trinitrobenzene Sulfonic Acid Assay (TNBS)

Free amine groups in chitosan was determined to ascertain conjugation reactions with NHS-palmitate, and successful formation of nanoparticle samples. Precedingly, solutions of 0.05% (v/v) TNBS reagent, 1.0 M HCl, 10% (w/v) SDS and 0.1 M (w/v) NaHCO$_3$ were separately prepared in 15 mL centrifuge tubes. Then, a chitosan standard solution was prepared by serially diluting 50 µL CS solution (0.5 mg/mL) using 0.1 M NaHCO$_3$. About 50µL of 0.05% (v/v) TNBS solution was then added to each CS/pCS sample in 0.5 mL centrifuge tubes. For sample solutions, 100µL of nanoparticle samples at different TPP volume addition was mixed with 100 µL 0.05% (v/v) TNBS solution in centrifuge tubes. All tubes were then incubated in a water bath for 3 h at 37 °C. Subsequently, 100 µL of the standard/sample solutions were transferred into a 96-well plate and mixed with 100 µL of 10% (w/v)

SDS and 75 µL of 1 M HCl respectively. The absorbance was then read at A_{335nm} and the utilized amine percentage was calculated using the following equation:

$$100 - (\text{Free amine percentage (\%)}) = \frac{A_{335} \text{ of CNP/pCNP}}{A_{335} \text{ of CS/pCS}} \times 100\%)$$
(at same concentration used)

Determination of PCA Encapsulation Efficiency (%EE) in Nanoparticle Samples:

The encapsulation efficiency (% EE) was analyzed by comparing the difference in absorbance at A_{296nm} between free PCA and the supernatant of encapsulated PCA. The nanoparticles samples were prepared as previously described. The samples were centrifuged at 18,000 rpm for 30 min. The supernatant of each sample was then collected, and the absorbance was read at A_{296nm} using an Implen NP80 UV/VIS spectrophotometer. The % EE was calculated using the following equation:

$$\% \text{ EE} = \frac{A_{296} \text{ of free PCA} - A_{296} \text{ of PCA in supernatant}}{A_{296} \text{ of free PCA}} \times 100\%$$

2.6. Assessment of In Vitro Vellular Efficacy of Nanoparticle Mediated PCA Uptake in A549 Lung Cancer Cell Line

The A549 lung cancer cell line was established and maintained by aseptic cell culture regimes in a T-25 flask with growth media consisting of 90% of 1X RPMI medium 1640 and 10% (FBS). The flask was maintained in incubator at 37 °C, supplied with 5% CO_2 and 90% humidity. About 100 µL of cells were seeded onto a 96-wells plate. The cells in 96-wells plate were treated with 100 µL of CNP, pCNP, PCA, CNP-PCA, and pCNP-PCA at different concentrations. At the end of each time point, the old media in each well were decanted and replaced with 170 µL fresh media solution and 30 µL of 5 mg/mL MTT solution. After 4 h incubation at 37 °C, all the solution in the wells was removed and replace with 100 µL DMSO. The absorbance was then read at A_{570nm} on a Bio-Rad iMark™ Microplate Absorbance Reader. Cell viability was then determined using the following equation:

$$\% \text{ Viability} = \frac{A_{570} \text{ of treated cells}}{A_{570} \text{ of untreated cells}} \times 100\%$$
(at same concentration used)

3. Results and Discussions

3.1. The Colume of Cross-Linker Governing the Size and PDI of Nanoparticles

The nanoparticles were spontaneously formed through the cross-linking of amine groups of chitosan polymer and phosphate groups of the cross-linker, TPP [45]. As shown in Figure 2A,B, nanoparticle size conferred a decreasing trend with increasing TPP volume until an optimum CS:TPP volume ratio was reached. Initially, when no TPP was added to the CS, the size of the polymer was 2720.33 ± 870.26 nm and slightly decreased to 2534.00 ± 1203.00 nm at 50 µL TPP volume addition, and significantly dropped to 241.57 ± 16.29 nm at 100 µL TPP volume addition. It was then gradually decreased to the smallest size upon 250 µL TPP volume addition. A similar trend was revealed by pCNP where the size of initial pCS at no TPP addition was 4560.33 ± 614.17 nm, dramatically dropped to 289.83 ± 8.92 nm at 50 µL TPP volume addition, and gradually decreased until it reached its smallest size of 90.23 ± 2.67 nm at 200 µL TPP addition. It showed that the minimum volumes of TPP required for a nano-sized particle to form were 100 µL and 50 µL for CNP and pCNP respectively. The initial size of CNP at 100 µL TPP volume addition was 241.57 ± 16.29 nm while pCNP at 50 µL TPP volume addition was 289.83 ± 8.92 nm. The smallest nanoparticle size of CNP obtained from this study was 82.24 ± 2.67 nm which by using 250 uL TPP while PCNP was 90.23 ± 2.67 nm by using 200 uL TPP. This result was congruent with the findings of Kavi Rajan et al. where the optimum chitosan to TPP ratio of about 3:1 [46]. Thereafter, particle size increased exponentially, indicating the formation of aggregates and nanoparticle clusters after this threshold resulting in the existence of excess TPP in the aqueous system, which may promote further interaction between the CNPs/pCNPs, thus initiating agglomerated nanoparticles with larger sizes [47,48]. Similarly, PDI values followed a similar decreasing trend with increased TPP volume. As shown in Figure 2A, the initial PDI value of CNP was 0.55 ± 0.25 and subsequently increased to 0.92 ± 0.08, and then decreased to 0.44 ± 0.03, 0.36 ± 0.03, 0.27 ± 0.01 until it reached the lowest point of 0.25 ± 0.01 at 250 µL of TPP addition. Next, the initial PDI value of pCNP was 0.85 ± 0.08 and decreased gradually to 0.49 ± 0.02, 0.37 ± 0.04, 0.37 ± 0.01 until it reached its lowest point of 0.25 ± 0.01 at 200 µL TPP volume addition. The previous study of Masarudin et al. revealed that the addition of 20 µL of TPP into 600 µL of CS has initiated the formation of nano-scale CNP and subsequently reached the smallest size with 200 µL of TPP, which is comparable to our current study [1]. Besides that, the PDI of 0.2–0.3 indicated the uniformity of the nanoparticles formed by the 3:1 CS/pCS to TPP volume ratio, which is supported by the findings of Koukarous [49]. This decreasing trend of size and PDI across both CNP and pCNP with increased TPP volumes addition was suggested due to the increased availability of the cross-linker to interact with the free amino groups existed in the fixed volume of chitosan polymer. Interestingly, despite the similar lowest PDI obtained by both CNP and pCNP, it was observed that the smallest size of pCNP was slightly larger than that of CNP by about 9.72%. This finding coincided with the previous study of Farhangi et al., where the conjugation of fatty acid chains into chitosan will result in an increased in size of the nanosystem [50]. This finding is comparable with our study, in which the conjugation of palmitic acid in the CS will correspondingly slightly increase in the size of nanoparticles due to the conjugation of extra component in CS polymer. Nonetheless, this insignificant particle size increment is expected and acceptable since the size of pCNP was still below 100 nm.

A

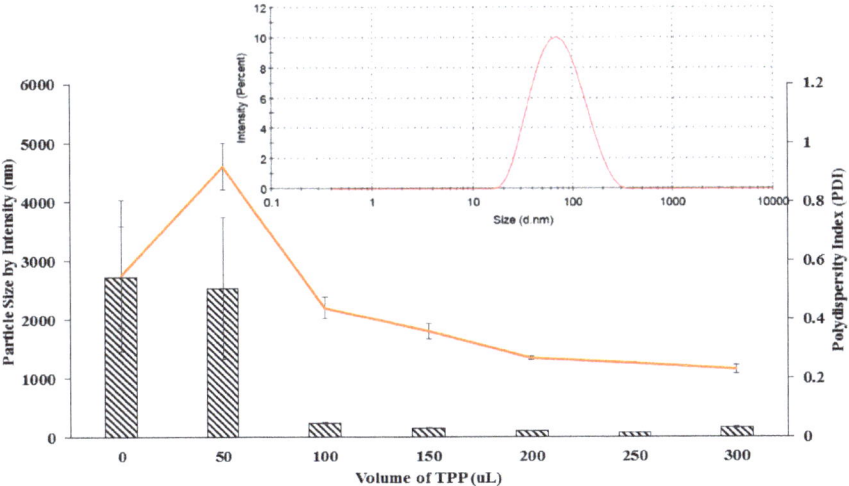

Figure 2. *Cont.*

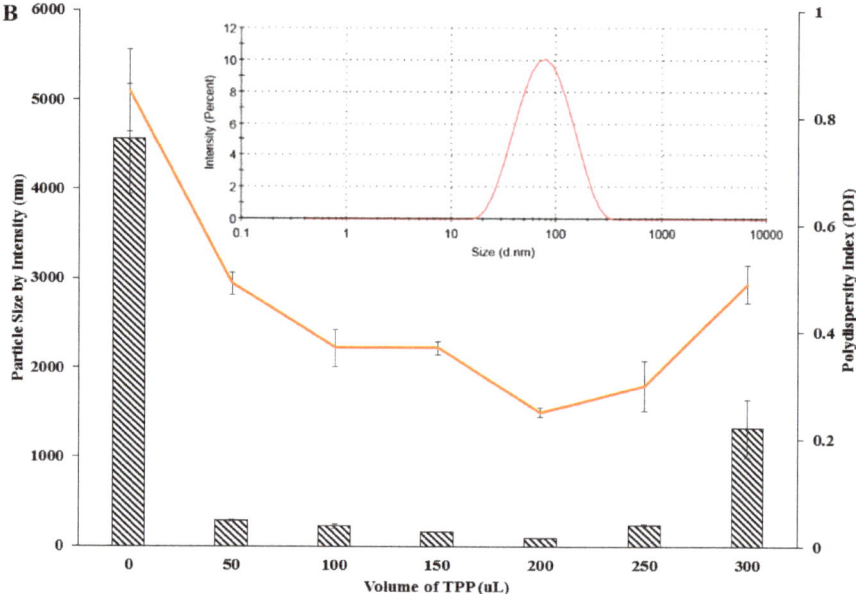

Figure 2. The PSD and PDI value of (**A**) CNP and (**B**) pCNP at different TPP volumes. Represented by bar (particle size) and line (PDI value) graphs. The smallest size of 82.24 ± 2.67 nm was obtained at 250 µL of TPP for CNP with PDI of 0.25 while 90.23 ± 2.67 nm at 200 µL for pCNP with PDI of 0.25. The DLS graph of CNP and pCNP synthesized by using 250 µL and 200 µL was shown inset in both Figure 2A,B, respectively. Error bars represent the SEM averaged from three independent experiment replicates. One-way ANOVA was performed with $p > 0.05$ for both particle size and PDI indicating no significant difference between the three experimental replicates.

3.2. The Formation of Nanoparticles Utilized Free Amine Group of CS/pCS Polymer

The TNBS assay is a well-known assay to quantify the free amine group as described earlier by Satake et al. [51]. As the formation of nanoparticles occurred through cross-linking between the cationic polymer and anionic cross-linker, amine group utilization was expected to show a decreasing value during conjugation and particle formation reactions. As shown in Figure 3, the utilization of amine groups increased following the increased TPP volumes used for both CNP and pCNP samples. In CNP samples, amine utilization of up to 26.75 ± 2.06% was achieved, while pCNP showed approximately 46.64 ± 0.94% amine utilization when a maximum of 300 µL TPP volume was used for cross-linking. Expectedly pCNP was shown to utilize a significantly higher percentage of amine groups compared with CNP. Considering that pCNP precedingly involved the conjugation of NHS-palmitoyl to CS prior to nanoparticle formation with TPP, an increase in its amine utilization suggested that the conjugation of palmitic acid was successfully obtained through the utilization of approximately 28.29% of amine groups. Data presented are similar to previous studies which postulate that the formation of CNP and pCNP will utilize the amine group of CS polymer and with greater utilization percentage in pCNP formation [52]. This result indicated a proportional increment in amine utilization with increased volume of the crosslinker regardless of their size and PDI value. It is because the formation of pCNP aggregates also happened through the cross-linking between the nanoparticles and excess TPP cross-linker which will further increased the amine utilization [53]. The pCNP has a greater utilization than CNP in all measured data, due to the utilization of amine groups through the conjugation of -NHS palmitic acid to the amine groups of chitosan. The study of Esquivel et al. reported a synthesis of thiol-modified chitosan with a utilization of 11% of amine groups in chitosan which is similar to our study that has utilized around 15% of free amine in chitosan following –NHS palmitic acid conjugation [54]. Besides that, the literature described by Mohammed et al. also proposed that chemical modifications of chitosan such as amphiphilic chitosan, carboxylated chitosan, and lactose-modified chitosan were achieved through the reaction between amine groups of chitosan and the modifying agents [41]. Thus, it can be deduced that the utilization of free amine percentage by pCNP will appear higher than CNP due to the utilization by the modification step. In addition, the excess percentage of amine utilization by pCNP than CNP was considered acceptable since there is still room for the cross-linking with TPP.

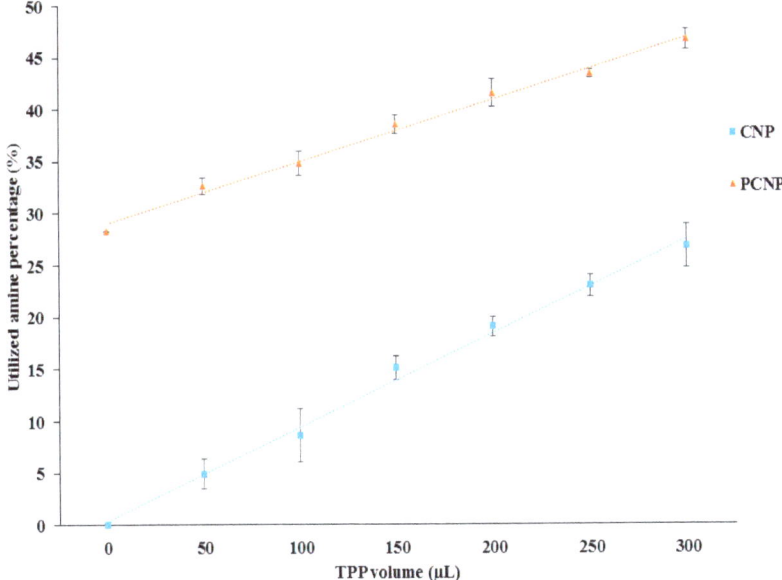

Figure 3. The utilization of amine percentage with different TPP volumes. The utilization of free amine percentage increased with increased TPP volume for CNP and pCNP. Data presented as mean ± SEM from three independent experiment replicates. Two tailed paired t-test was performed with p-value < 0.0001.

3.3. Formation of PCA-Encapsulated Nanoparticles

PCA was successfully encapsulated in both CNP and pCNP following spontaneous formation of nanoparticles after crosslinking with TPP. Encapsulation led to an increase in particle size to accommodate the phenolic compound within its internal structure. As shown in Figure 4, CNP-PCA particle size expanded 93.4% from 82.2 to 159.0 nm, whilst comparatively, a larger expansion was observed in its hydrophobically-modified counterpart. The particle size of pCNP-PCA increased from 90.2 to 196.3 nm, a 117.6% surge from empty pCNP. As this expansion correlated with previous study [46], the inclusion of a hydrophobic moiety within pCNP-PCA has affected its expansion compared to CNP-PCA at similar PCA concentrations used for encapsulation. This was ascribed to an increased amount of PCA encapsulated in pCNP-PCA, due in part towards a tighter hydrophobic-hydrophobic interaction forming between the palmitoyl groups in pCS with PCA prior to nanoparticle formation. Such interactions have also been similarly been reported by Wang et al., where they have conducted hydrophobic modification of chitosan using cholesterol conjugate through succinyl linkages and successfully increase the encapsulation efficiency of poorly water soluble epirubicin, an anthracycline topoisomerase inhibitor from 7.97% to 14.0% [55]. Since this palmitoyl anchor is absent in CNP, encapsulation reactions did not benefit from this extra interaction and were thus lower in terms of the amount of PCA within the nanoparticle core after formation. This correlated to a substantial difference in % *EE* values between CNP-PCA and pCNP-PCA as well, which further illustrates the enhanced compound loading properties following hydrophobic modifications of the nanoparticle.

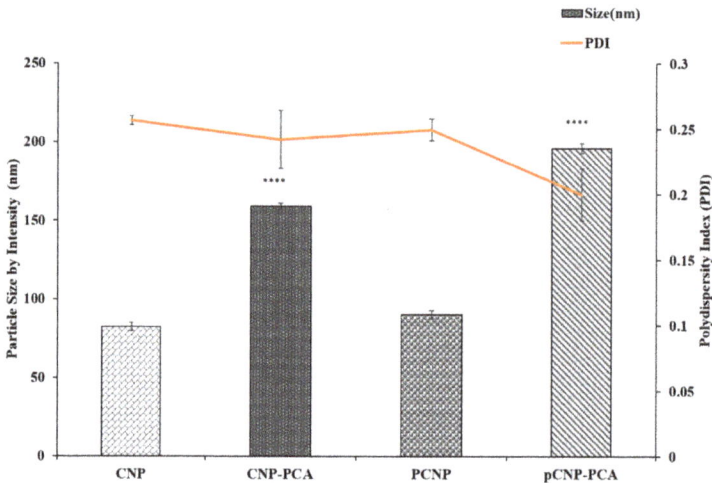

Figure 4. The expansion of nanoparticle size and PDI value after encapsulation. The results show that the size of empty nanoparticles expanded after encapsulation. Data are presented as mean ± SEM from three independent experiment replicates. One-way ANOVA was performed with **** $p < 0.0001$ indicating the significant difference in size between both CNP-PCA and pCNP-PCA with CNP and pCNP.

The % EE of PCA in CNP-PCA and pCNP-PCA has shown in the Table 1. The encapsulation efficiency of PCA was approximately 35.2 ± 1.7% in CNP-PCA, while pCNP-PCA had a significantly higher % EE of 54.4 ± 3.9%. A higher encapsulation efficiency was attained in pCNP-PCA as compared with CNP-PCA, which was mostly due to the hydrophobic anchor acquired by the presence of palmitoyl that associated with the modified pCS polymer, and consequently enable a higher encapsulation of PCA upon nanoparticles formation. This observation also correlated to a higher degree of expansion in the hydrophobically-modified nanoparticles, as described previously. Several studies have also reported enhanced encapsulation properties in hydrophobically-modified chitosan nanoparticles using other hydrophobic moieties such as deoxycholic acid, stearyl, phthaloyl, and N-acetyl histidine [56–59]. Zhang et al. also demonstrated that the hydrophobically modified chitosan nanoparticle was able to encapsulate the Doxorubicin to act as a carrier system for antitumor agents [60]. Previous literature studies suggested that the modification of chitosan by long alkyl chains (C6-C12) will gradually promote a more efficient hydrophobic interactions and intra-aggregation corresponding to the length of alkyl chains as compared with short alkyl chains (C5) [61]. In correlation with this study, palmitic acid with long alkyl chain (C15) was suggested to incur efficient hydrophobic interactions with PCA. Furthermore, Ways et al. have also highlighted that various chemical modifications of chitosan including trimethyl chitosan, thiolated chitosan, acrylated chitosan and acetylated chitosan will mainly occasioned in an enhancement in the loading, bioavailability and a substantial improvement of the therapeutic efficacy of some candidate drugs compared to unmodified chitosan [62]. The higher % *EE* attained by pCNP-PCA compared to CNP-PCA indicated a greater amount of PCA being encapsulated in pCNP, which may indirectly increase the therapeutic efficacy of pCNP-PCA. This assumption was supported by the previous study of Ong et al. where a greater % EE will enhance the bioavailability and consequently improved the absorption of encapsulated compounds as compared to their lower % EE counterpart [63]. Although the maximum % EE of pCNP-PCA was not assessed in this current study, it was suggested that % EE of pCNP-PCA may be improved by other strategies including dual or multiple loading of cargos [64]. However, these approaches may include complex and tedious procedures which may change the native structure of the therapeutic which may in turn affect its therapeutic efficacy [65].

Table 1. Encapsulation efficiency (% EE) of 500 µM PCA in CNP-PCA and pCNP-PCA nanoparticles. Efficiency of the phenolic compound in pCNP-PCA was higher than CNP-PCA due to its active hydrophobic-hydrophobic interaction with palmitoyl in the hydrophobically-modified nanoparticle. Data are presented as mean ± SEM from three independent experiment replicates.

Sample	Free PCA	CNP-PCA		pCNP-PCA	
	A_{296nm}	A_{296nm}	% EE	A_{296nm}	% EE
Replicate 1	0.79	0.53	32.91	0.42	46.84
Replicate 2	0.83	0.51	38.55	0.33	60.24
Replicate 3	0.82	0.54	34.15	0.36	56.10
Average			35.20 ± 1.71		54.39 ± 3.96

The PDI values of both CNP-PCA and pCNP-PCA were measured as of 0.20 ± 0.02 to 0.25 ± 0.01, as shown in Figure 4. This implied that the nanoparticle samples occurred at a high monodispersity and reproducibility [66]. Similarly, the hydrophobic modification in pCNP-PCA did not affects its dispersity. This suggested that the expansion of the size between both types of nanoparticles was almost similar. This postulation was parallel with the previous study of Maruyama et al. where the encapsulation of herbicides imazapic and imazapyr into CNP did not significantly alter the PDI of the system which indicating the homogeneity and stability of the nanoparticle system after encapsulation [67]. Additionally, the previous study of Othman et al. also signified that dual-loading of L-ascorbic acid and thymoquinone into CNP system has obtained similar PDI values of 0.19 ± 0.02 prior to, and 0.21 ± 0.01 after encapsulation has also suggested that encapsulation of therapeutics into CNP system not necessarily altered the PDI values [64]. In correlation with these findings, it was proposed that an equal or almost equal distribution of PCA occurred in both pCNP and CNP samples which resulted in an almost similar PDI being obtained in both systems.

3.4. Morphological Analysis of Nanoparticles

The morphological properties of CNP-PCA and pCNP-PCA was studied by using FESEM and TEM. Figure 5 showed the presence of single-spherical like particles with uniformed dispersity in both hydrophobically-modified and non-modified nanoparticles, with its approximate size correlating with DLS data. Particle size of CNP ranged from 73.0 to 91.5 nm for CNP (Figure 5A), while pCNP was in the range of 61.5 to 87.3 nm (Figure 5B). The morphology of pCNP was smooth and single-spherical like particle in shape and comparable with CNP, suggesting that the conjugation of palmitic acid in pCNP has no contributions to the surface morphology. This observation was expected since the palmitoyl group was likely to avoid from the surrounding water environment and resides in the interior of the pCNP [68]. Nevertheless, a population of pCNP with smaller sizes than CNP was observed in the figure. This qualitative morphology involved a randomly chosen site which was in contrast to the mean size of nanoparticles earlier, while the size reflected by the morphology was a relative approximation of nanoparticle size. The DLS measured the size average across all size populations while FESEM imaging observed at random spots which might be resulted in nanoparticles with sizes slightly deviated from the mean value obtained by DLS analysis [69]. Additionally, the difference in sizes may also due to the fundamental difference in the preparation of these two techniques, where the samples in DLS are hydrated, whereas in FESEM they are under vacuum, which will clearly have an important impact on the sizes measured. The morphology of CNP-PCA and pCNP-PCA was also observed as smooth and single-spherical like particles in shape in which similar to the blank nanoparticles. Meanwhile particle size upon PCA encapsulation showed an increased for both CNP-PCA and pCNP-PCA samples, which ranged from 121.3 to 191.6 nm for CNP-PCA (Figure 5C) and from 138.2 to 182.6 nm for pCNP-PCA (Figure 5D), which indicated interrelated measurements with data from DLS analysis. Moreover, a similar morphology was noticed in both CNP-PCA and pCNP-PCA which consisted a range of size populations. Although the % EE was higher in pCNP-PCA, the expansion of size between the CNP-PCA and the former is approximately not that significant. This observation suggested that the utilization of greater amine groups in pCS/pCNP due to conjugation of palmitic acid resulting in a lower net positive charge, together with the more specific hydrophobic-hydrophobic interactions between pCNP and PCA aid in developed pCNP-PCA of more compact nanoparticles and thus smaller in size, which explained the expansion of size of pCNP-PCA as comparable with CNP-PCA where an even higher % EE was attained [70].

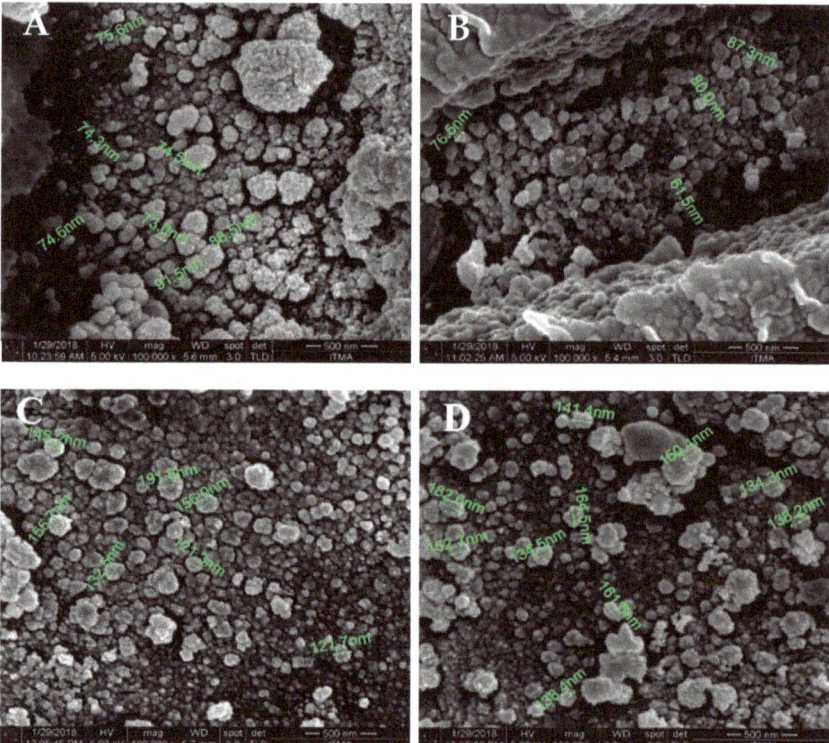

Figure 5. FESEM analysis of nanoparticle samples. (**A**) CNP, (**B**) pCNP, (**C**) CNP-PCA and (**D**) pCNP-PCA. The nanoparticle samples were revealed in single-spherical like particles with uniformed dispersity with different size range.

Conversely, Figure 6 showed the phase morphology of the nanoparticle samples. As mentioned by Mayeen et al., the electrons in TEM can penetrate through the samples and measures the changes of the electron beam to assess the internal structure of the samples; while FESEM works by scanning through the surface of samples through a raster scan pattern to assess the surface morphology of samples [71]. This feature was supported by Barhoum and García-Betancourt who further proposed FESEM and TEM to use in the morphology characterization analysis of nanoparticles to provide a detailed characterization of nanostructures [72]. TEM analysis has revealed single-spherical like CNP particles with sizes ranged from 77.9 to 138.1 nm (Figure 6A) while pCNP conferred a size from 68.7 to 144.5 nm (Figure 6B). Similarly, Figure 6C,D indicated an expansion in particle size following PCA encapsulation, where CNP-PCA expanded to a size range of 88.6 to 152.5 nm and pCNP-PCA from 110.6 and 184.1 nm. It was observed that a portion of CNP and pCNP was larger than 100.0 nm, which did not correlate to the DLS results. This was likely to be attributed towards an agglomeration of the nanoparticles. Such a phenomenon has been described as a consequence of the mechanical forces during synthesis. Dogan et al. reported that the aggregation of nanoparticles can occur during drying of samples on the TEM grid prior to observation [73]. Additionally, there exist some nanoparticles with sizes similar to CNP and pCNP observed in Figure 5C,D. These were probably due to the uneven distribution of PCA within CNP and pCNP, where not every CNP and pCNP nanoparticle was encapsulated with PCA resulting in samples that were comprised of empty nanoparticles and encapsulated nanoparticles. Interestingly, the TEM analysis of pCNP showed an additional layer contrast surrounding the inner surface of the outermost layer of pCNP which could be attributed to the palmitoyl groups (fatty acid chains) that was conjugated to the chitosan. This inference was made since this layer was not observed in the TEM analysis of unmodified CNP and it was because the only difference in structure between CNP and pCNP lies is the use of palmitic acid, which suggested this layer was contributed by the conjugation of palmitic acid. Comparing the results obtained between FESEM and TEM, both morphological analyses shown that the nanoparticles appeared single-spherical like in shape. Meanwhile, after encapsulation, the size of the nanoparticles expanded to become larger.

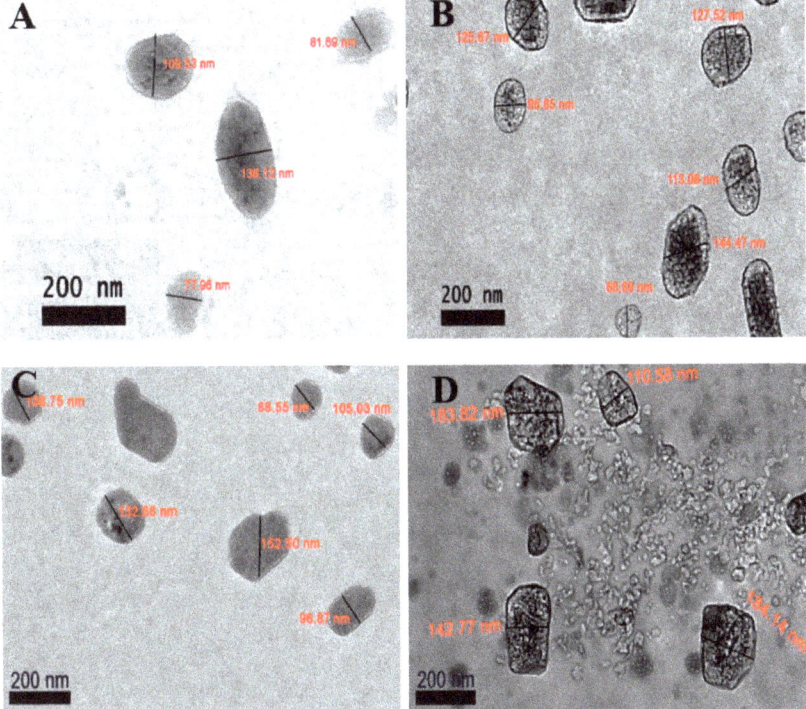

Figure 6. TEM analysis of nanoparticle samples. (**A**) CNP, (**B**) pCNP, (**C**) CNP-PCA and (**D**) pCNP-PCA. The nanoparticle samples were distributed with single-spherical like nanoparticles with uniformed dispersity with different size range.

3.5. Functional Group Annotation of Nanoparticle Samples Using FTIR Spectroscopy

FTIR analysis was used to annotate chemical functional groups and their occurrences in the nanoparticle samples. According to Coates et al., every single molecule has a unique infrared vibration spectrum which could be their specific "fingerprint" to be identified in a sample by comparing an "unknown" spectrum with the known spectra that had been recorded formerly [74]. The infrared spectra of CS, TPP, CNP and pCNP were listed in Figure 7 while the spectra of CNP, pCNP, CNP-PCA and pCNP-PCA were shown in Figure 8. The important functional groups corresponding to transmittance values of the samples were summarized in Table 2.

In Figure 7, a wide region around 3300 to 3500 nm^{-1} was detected in CS, CNP and PCNP at peaks of 3228, 3360 and 3383 cm^{-1} respectively, which corresponded to hydrogen-bonded O–H stretching and overlapped with primary amine stretching peaks [75]. The IR transmittance of amine group for CS was 49.26% and this increased to 71.24% and 55.82% upon CNP and pCNP formation, respectively. When the percentage of transmittance (% T) is high, the availability of active functional group is considered to be lower in the sample because fewer active function group is present to absorb the IR spectrum. This suggested that upon formation of nanoparticles, free amine groups in the CS were reduced leading a higher transmittance value in CNP and pCNP. This observation also corresponded well to TNBS assay data, showing a utilization of amine groups in chitosan. Additionally, the utilization of amine groups was also suggested by the differences shown by the characteristic peak of amine II group in the range of spectra between 1590 to 1650 cm^{-1}. It was observed that about 10% of transmittance at 1600 nm^{-1} for CS increased to 45.68% transmittance at 1629 cm^{-1} and 35.77% T at 1635 cm^{-1} for CNP and pCNP, respectively. The characteristic peak for phosphate groups (P=O) of TPP (1201 cm^{-1}) at 35.74% transmittance increased to 40.22% transmittance at 1155 cm^{-1} in CNP and 74.94% transmittance at 1281 cm^{-1} in pCNP; a similar observation was recorded previously by Martin et al. [76]. Next, the vibration of C–O–C stretching was found in CS at 1082 cm^{-1} with 32.20% transmittance, CNP at 1059 cm^{-1}, and pCNP at 1068 cm^{-1}, with 10% transmittance values [77].

Conversely, PCA exhibited numerous band peaks including functional groups at 3278 cm^{-1} (62.69% transmittance), 1658 cm^{-1} (15.56% transmittance), and 1300 cm^{-1} (69.14% transmittance), which was annotated for hydrogen bonding (O–H) stretching vibrations, C=C stretching, and carboxyl groups (C=O), respectively [78,79]. After PCA encapsulation, the transmittance of the O–H bond peaks increased to 75.50 and 71.13%, occurring at 3376 and 3227 cm^{-1} for CNP-PCA and pCNP-PCA, respectively. Transmittance of C=C bond peaks was also increased to 47.81% at 1625 cm^{-1} for CNP-PCA and 39.61% at 1625 cm^{-1} for pCNP-PCA. The C=O bond was found at 1389 cm^{-1} for 66.75% transmittance for CNP-PCA and at 1280 cm^{-1} for 73.83% transmittance for pCNP-PCA due to the presence of PCA. These results were comparable with the study of Usman et al. where the presence of several functional groups of PCA were found in the nanoparticles after encapsulation [80].

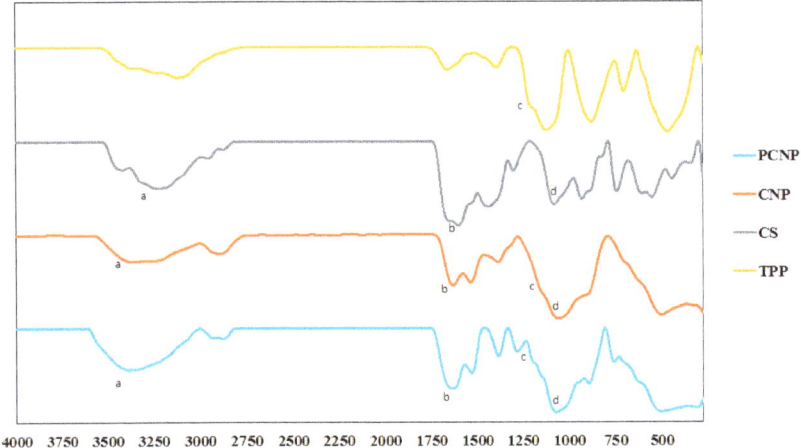

Figure 7. FTIR spectra of CS, TPP, CNP and PCNP have shown some of the important functional group in comparison. The functional groups are labeled as (a) amine group, (b) amine II group, (c) inorganic phosphate group and (d) C–O–C bond.

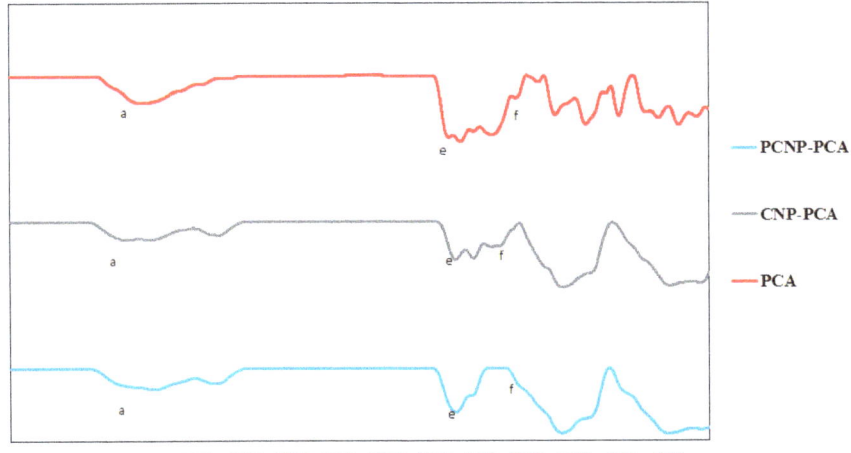

Figure 8. FTIR spectra of PCNP-PCA, PCNP, CNP-PCA, CNP and PCA have shown some of the important functional group in comparison. The functional groups are labeled as (a) hydrogen bond, (e) carbon double bond and (f) carbonyl group.

Table 2. The chemical functional groups present in CS, TPP, PCA, CNP, PCNP, CNP-PCA and PCNP-PCA. All the important functional groups detected were listed in the table with their respective wavenumber and percentage of transmittance.

Functional Group	Wavenumber (nm^{-1})	Percentage Transmittance (% T)	Sample
Hydrogen bond [a] (O—H)	3228	49.26	CS
	3360	71.24	CNP
	3383	55.82	PCNP
	3278	62.69	PCA
	3376	75.50	CNP-PCA
	3227	71.13	PCNP-PCA
Amine II group [b] (NH$_2$)	1600	10.00	CS
	1629	45.68	CNP
	1635	35.77	PCNP
Inorganic Phosphate [c] (P=O)	1201	35.74	TPP
	1155	40.22	CNP
	1281	74.94	pCNP
Ether group [d] (C-O-C)	1082	32.20	CS
	1059	10	CNP
	1068	10	pCNP
Carbon double bond [e] (C=C)	1658	15.56	PCA
	1625	47.81	CNP-PCA
	1625	39.61	pCNP-PCA
Carbonyl group [f] (C=O)	1300	69.14	PCA
	1389	66.75	CNP-PCA
	1280	73.83	pCNP-PCA

Annotations a–f reflects the assigned peaks as indicated in Figures 7 and 8.

3.6. Assessment of In Vitro Cytotoxicity of CNP and pCNP in A549 Lung Cancer Cells

CNP has been reported as a good biocompatible nanocarrier system [81,82]. In order to evaluate whether its hydrophobically-modified complement, pCNP is biocompatible as well, MTT cytotoxicity assay was performed against the A549 lung cancer cells. Figure 9 presents the cytotoxicity effects of CNP and pCNP in 24-h and 72-h treatments. The viability of A549 cells 24-h post-treatment after exposure to CNP and pCNP showed similar cytotoxic efficacies, which were 62.62% and 63.25% at the highest nanoparticle concentration of 0.25 mg/mL. At lower concentrations, the viability of cells was at least 80%. A similar cytotoxic effect was attained in 72-h treatments with a slightly lower viability recorded using pCNP compared to CNP at the highest concentration, which was 58.29% and 44.01%, respectively. This suggested that both nanoparticle systems may possess minimal cytotoxicity to the A549 cells. As CNP has been reported to be non-toxic to cells, cytotoxicity was possibly due to the number of nanoparticles that were formed by this parameter [83]. In higher concentrations of CNP/pCNP, the number of nanoparticles synthesized will be greater. This leads to cells being physically covered by the nanoparticles and subsequently cell death. The presence of high nanoparticle populations will easily agglomerate at the cell surface and consequently influence the absorption of nutrients and gaseous exchange and thus confer toxicity to cells [84] Additionally, previous studies have shown that CNP has no any significant cytotoxicity toward several cell lines such as HepG2 human liver cancer cell line, RAW 264.7 mouse macrophage, as well as the A549 cell lines [85–87].

The cytotoxicity of nanoparticles is always one of the major aspects to ascertain before being utilized as nanocarrier, especially for nanomedicine applications. Several aspects were taken into consideration when nanoparticles were employed as nanomedicine, such as the cell type of target, the properties of nanoparticle, and the dosage [87]. There are variations in cell physiology, proliferation state, membrane characteristics and phagocyte characteristics exist between different cell types which will have different reaction towards the nanomaterials [88]. Besides that, different sizes and shapes of nanoparticles demonstrated various biokinetic and biological impacts which consequently alter protein adsorption, cellular uptake, accumulation in organelles, and distribution of the body [89,90]. Hence, the optimization of the nanoparticle systems should be performed from time to time when encountering different cell lines.

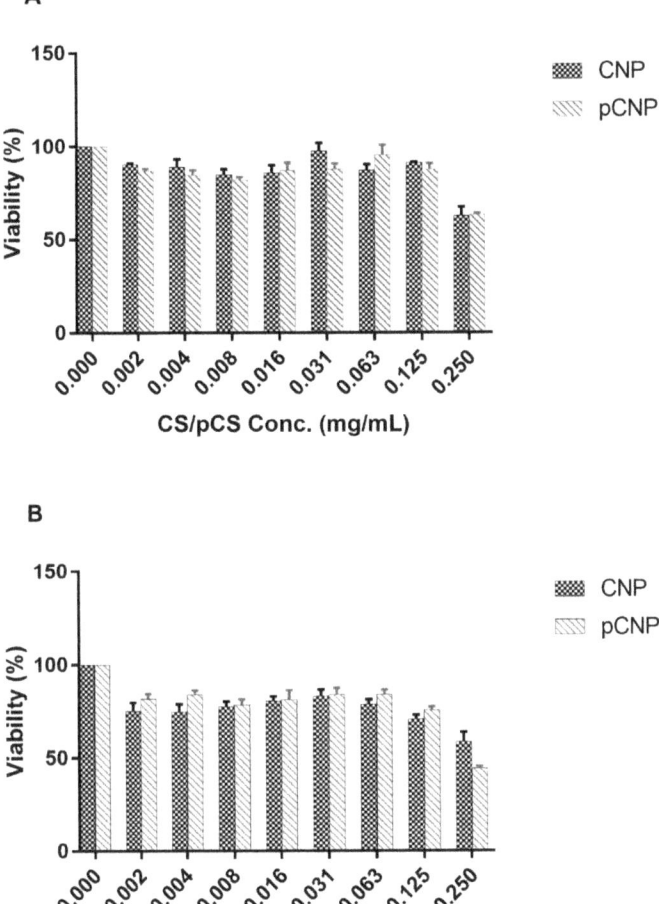

Figure 9. The cytotoxicity effect of CNP and pCNP at (**A**) 24 h post-treatment and (**B**) 72 h post-treatment. Data are presented as mean ± SEM from three independent experiment replicates. Two-tailed paired t-test was performed and $p > 0.05$ was obtained indicating no significant different between CNP and pCNP.

3.7. Assessment Of PCA Efficacy and Anticancer Properties Using Nanoparticle-Mediated In Vitro Cellular Delivery Systems

The cytotoxic efficacy of PCA, CNP-PCA and pCNP-PCA at different periods of time was tabulated in Figure 10. Approximately 500 µM of PCA was utilized for the encapsulation into CNP and pCNP, and halved after mixing with the media. As shown in Figure 10, the cytotoxic efficacy of the free PCA, CNP-PCA, and pCNP-PCA was shown according to the viability of A549 cell line against the concentration of PCA calculated according to the encapsulation efficiency (35.20 ± 1.71% and 54.39 ± 3.96% for CNP-PCA and pCNP-PCA respectively). In order to compare the efficacy among the three treatments, the IC_{50} values were calculated and are shown in Table 3. The results have shown that CNP-PCA and pCNP-PCA have significantly greater efficacy as compared with the non-encapsulated counterpart. Besides that, pCNP-PCA has achieved the lowest IC_{50} values in 24-h time point and 72-h time point of treatment but slightly lower than CNP-PCA in 48-h time point. At highest PCA concentration, the lowest % cell viability was revealed by pCNP-PCA with 34.25 ± 1.04% viability, followed by CNP-PCA (74.38 ± 1.05%) and free PCA (69.21 ± 1.70%) at 24-h time point. The % viability was subsequently dropped to 30.90 ± 2.37%, 46.76 ± 1.46% and 60.22 ± 1.43% for pCNP-PCA, CNP-PCA and PCA, respectively at 48-h time point. The % viability has further dropped to 12.54 ± 0.88%, 41.54 ± 0.65% and 53.83 ± 1.21% for pCNP-PCA, CNP-PCA, and PCA, respectively, at the 72-h time point. It was observed that PCA delivered by pCNP has the lowest % viability at all three time points and consistently dropped with increased time points. On the other hand, the IC_{50} values tabulated in Table 3 have showed that at 24-h time point, no IC_{50} value was detected for PCA alone and CNP-PCA, while pCNP-PCA was found at around 214.5 µM. Next, there is no IC_{50} found at 48 h post-treatment, as well for PCA alone, but IC_{50} of 448.4 and 412.1 µM were found for CNP-PCA and pCNP-PCA, respectively. After that, at 72 h post-treatment, IC_{50} was calculated at 407.3 and 130.7 µM for CNP-PCA and pCNP-PCA, and no IC_{50} was found for PCA alone.

Table 3. The IC_{50} values of different PCA treatments on A549 cell lines. The encapsulated PCA has greater efficacy than free PCA throughout all time points. Lower IC_{50} values indicate a higher cytotoxic efficiency.

Time Point	24 h	48 h	72 h
	* IC_{50} value (µM)		
PCA	N/A	N/A	N/A
CNP-PCA	191.50	75.78	63.27
pCNP-PCA	53.71	110.70	48.34

* the values were obtained through best-fit hypothetical calculation.

From the results above, we can deduce that the efficacy of nanoparticle-encapsulated PCA has an undeniably greater cytotoxic efficacy than non-encapsulated counterpart. These consequences may be due to the greater encapsulation efficiency of pCNP than CNP where a greater amount of PCA was encapsulated in pCNP than CNP. Moreover, the hydrophobic-hydrophobic interaction between pCNP and PCA maybe another factor that contributed to the slower release of PCA from pCNP than the conventional CNP that has no specific interaction with PCA [91]. The PCA encapsulation by chitosan nanoparticles has been previously characterized by Madureira and colleagues, and it was found that the bioavailability of PCA was enhanced by CNP encapsulation [92]. Besides that, Pham and coworkers have discovered that the PCA encapsulation by CNP has a greater effect in antifungal activity as compared with the non-encapsulated counterpart which further ascertained that encapsulation of PCA by nanoparticles can greatly enhanced its therapeutic efficacy [93]. The previous study conducted by Barahuie et al. performed nano-encapsulation of PCA by using zinc/aluminium-layered double hydroxide and assessed the cytotoxicity effect on human cervical, liver and colorectal cancer cell lines, and revealed that the anticancer efficacy of nano-encapsulated PCA was greater than the non-encapsulated PCA [79]. Fascinatingly, the IC_{50} of pCNP has no obvious orderly decreasing trend which could attributed by its controlled-release properties. The previous study of Hassan et al. has proposed that fluorescently labeled glutamic acids encapsulated CNP has revealed intracellular release and controlled accumulation properties that was coincided with our current study [94]. Anyhow, the study proposed that the in vitro cellular efficacy of the experimental samples on A549 cell line can be defined in this particular manner: PCA < CNP-PCA < pCNP-PCA.

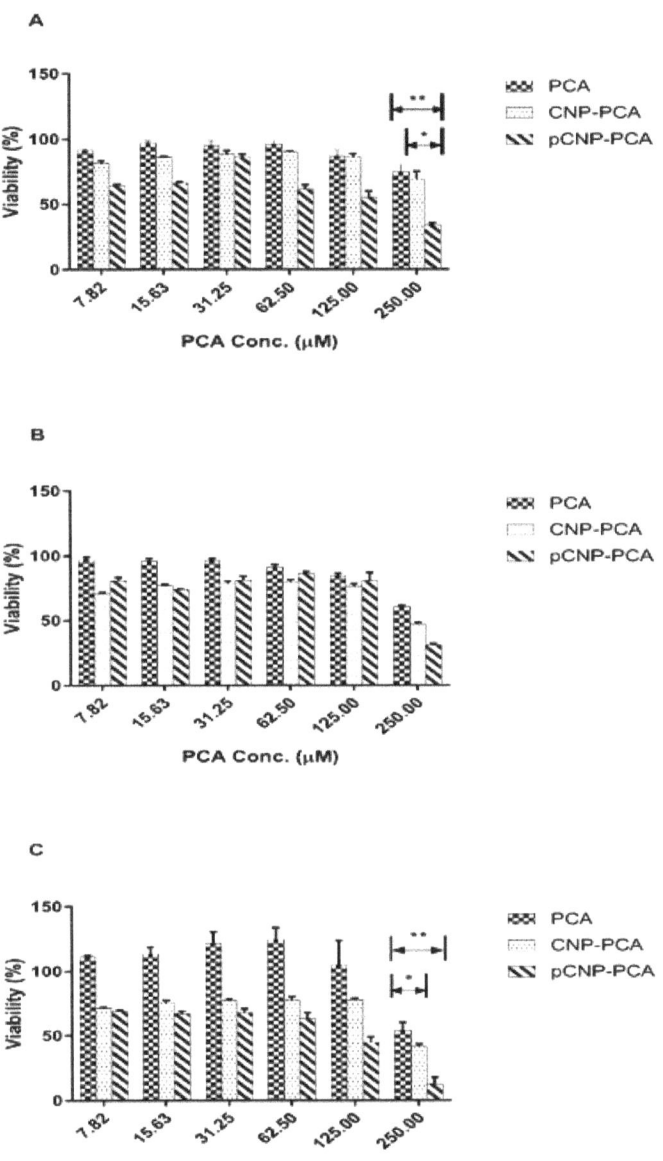

Figure 10. Cellular efficacy of PCA, CNP-PCA and pCNP-PCA at (**A**) 24 h post-treatment, (**B**) 48 h post-treatment and (**C**) 72 h post-treatment on A549 cell line. Data are presented as mean ± SEM from three independent experiment replicates. One-way ANOVA was performed and the significance was indicated by * = $p < 0.05$ and ** = $p < 0.01$ for all three graphs.

4. Conclusions

The utilization of the nanoparticulate delivery of poorly water soluble, naturally occurring phenolic compounds by using the hydrophobically modified chitosan nanoparticle system has revealed a greater cytotoxic efficacy towards the A549 human lung cancer cell line. In this current research, we found that the conjugation of NHS-palmitic acid to the chitosan has developed into a promising tool for the encapsulation of low water soluble phenolic compounds and PCA, which has a similar cytotoxic effect with the traditional chitosan counterpart but greater encapsulation efficiency and cytotoxic efficacy. Henceforth, this hydrophobic modification system perhaps presents a potential prominent delivery vector that could be customized for the delivery of low bioavailability cancer therapeutics.

Author Contributions: Conceptualization, C.Y.K., S.F., S.S.O., and M.J.M.; Data curation, C.Y.K.; Formal analysis, C.Y.K., T.G., S.F., S.S.O., and M.J.M.; Funding acquisition, M.J.M.; Investigation, C.Y.K., T.G., and M.J.M.; Methodology, C.Y.K., T.G., S.F., and M.J.M.; Project administration, M.J.M.; Resources, M.J.M.; Supervision, M.J.M.; Validation, S.F., S.S.O., and M.J.M.; Writing—original draft, C.Y.K. and M.J.M.; Writing—review & editing, C.Y.K., S.F., S.S.O., and M.J.M. All authors have read and agreed to the published version of the manuscript.

Funding: This research was funded by Universiti Putra Malaysia through the Inisiatif Geran Putra-Inisiatif Putra Berkumpulan grant scheme, grant number GP-IPB/2013/9425801. The APC was funded by Universiti Putra Malaysia.

Conflicts of Interest: The authors declare no conflict of interest.

References

1. Society, A.C. *American Cancer Society: Facts & Figures 2018*; American Cancer Society: Atlanta, GA, USA, 2018.
2. Nakano, T.; Shimizu, K.; Kawashima, O.; Kamiyoshihara, M.; Kakegawa, S.; Sugano, M.; Ibe, T.; Nagashima, T.; Kaira, K.; Sunaga, N.; et al. Establishment of a Human Lung Cancer Cell Line with High Metastatic Potential to Multiple Organs: Gene Expression Associated with Metastatic Potential in Human Lung Cancer. *Oncol. Rep.* **2012**, *28*, 1727–1735. [CrossRef] [PubMed]
3. Schirrmacher, V. From Chemotherapy to Biological Therapy: A Review of Novel Concepts to Reduce the Side Effects of Systemic Cancer Treatment (Review). *Int. J. Oncol.* **2019**, *54*, 407–419. [CrossRef]
4. Chanvorachote, P.; Chamni, S.; ninsontia, C.; Phiboonchaiyanan, P.P. Potential Anti-Metastasis Natural Compounds for Lung Cancer. *Anticancer Res.* **2016**, *36*, 5707–5718. [CrossRef]
5. Abotaleb, M.; Liskova, A.; Kubatka, P.; Büsselberg, D. Therapeutic Potential of Plant Phenolic Acids in the Treatment of Cancer. *Biomolecules* **2020**, *10*, 221. [CrossRef]
6. Mitra, S.; Dash, R. Natural Products for the Management and Prevention of Breast Cancer. *Evid.-Based Complement. Altern. Med.* **2018**, *2018*, 8324696. [CrossRef]
7. Demain, A.L.; Vaishnav, P. Natural Products for Cancer Chemotherapy. *Microb. Biotechnol.* **2011**, *4*, 687–699. [CrossRef] [PubMed]
8. Yuan, J.-M. Green Tea and Prevention of Esophageal and Lung Cancers. *Mol. Nutr. Food Res.* **2011**, *55*, 886–904. [CrossRef]
9. Fritz, H.; Seely, D.; Kennedy, D.A.; Fernandes, R.; Cooley, K.; Fergusson, D. Green Tea and Lung Cancer: A Systematic Review. *Integr. Cancer Ther.* **2013**, *12*, 7–24. [CrossRef]
10. Mehta, H.J.; Patel, V.; Sadikot, R.T. Curcumin and Lung Cancer—A Review. *Target. Oncol.* **2014**, *9*, 295–310. [CrossRef]
11. Tsai, J.-R.; Liu, P.-L.; Chen, Y.-H.; Chou, S.-H.; Cheng, Y.-J.; Hwang, J.-J.; Chong, I.-W. Curcumin Inhibits Non-Small Cell Lung Cancer Cells Metastasis through the Adiponectin/NF-κb/MMPs Signaling Pathway. *PLoS ONE* **2015**, *10*, e0144462. [CrossRef]
12. Ulasli, S.S.; Celik, S.; Gunay, E.; Ozdemir, M.; Ozyurek, A.; Koyuncu, T.; Unlu, M. Anticancer Effects of Thymoquinone, Caffeic Acid Phenethyl Ester and Resveratrol on A549 Non-Small Cell Lung Cancer Cells Exposed to Benzo (a) Pyrene. *Asian Pac. J. Cancer Prev.* **2013**, *14*, 6159–6164. [CrossRef]
13. Rajendra Prasad, N.; Karthikeyan, A.; Karthikeyan, S.; Venkata Reddy, B. Inhibitory Effect of Caffeic Acid on Cancer Cell Proliferation by Oxidative Mechanism in Human HT-1080 Fibrosarcoma Cell Line. *Mol. Cell. Biochem.* **2011**, *349*, 11–19. [CrossRef]
14. Liu, R.H. Potential Synergy of Phytochemicals in Cancer Prevention: Mechanism of Action. *J. Nutr.* **2004**, *134*, 3479S–3485S. [CrossRef]
15. Kakkar, S.; Bais, S. A Review on Protocatechuic Acid and Its Pharmacological Potential. *ISRN Pharmacol.* **2014**, *2014*, 1–9. [CrossRef] [PubMed]
16. Chao, C.-Y.; Yin, M.-C. Antibacterial Effects of Roselle Calyx Extracts and Protocatechuic Acid in Ground Beef and Apple Juice. *Foodborne Pathog. Dis.* **2009**, *6*, 201–206. [CrossRef] [PubMed]

17. Xican, L.; Wang, X.; Chen, D.; Chen, S. Antioxidant Activity and Mechanism of Protocatechuic Acid in Vitro. *Funct. Foods Health Dis.* **2011**, *1*, 232–244.
18. Tanaka, T.; Tanaka, T.; Tanaka, M. Potential Cancer Chemopreventive Activity of Protocatechuic Acid. *J. Exp. Clin. Med.* **2011**, *3*, 27–33. [CrossRef]
19. Scazzocchio, B.; Varì, R.; Filesi, C.; D'Archivio, M.; Santangelo, C.; Giovannini, C.; Iacovelli, A.; Silecchia, G.; Volti, G.L.; Galvano, F.; et al. Cyanidin-3-O-β-Glucoside and Protocatechuic Acid Exert Insulin-like Effects by Upregulating PPARγ Activity in Human Omental Adipocytes. *Diabetes* **2011**, *60*, 2234–2244. [CrossRef]
20. Shi, G.F.; An, L.J.; Jiang, B.; Guan, S.; Bao, Y.M. Alpinia Protocatechuic Acid Protects against Oxidative Damage in Vitro and Reduces Oxidative Stress in Vivo. *Neurosci. Lett.* **2006**, *403*, 206–210. [CrossRef]
21. Lende, A.B.; Kshirsagar, A.D.; Deshpande, A.D.; Muley, M.M.; Patil, R.R.; Bafna, P.A.; Naik, S.R. Anti-Inflammatory and Analgesic Activity of Protocatechuic Acid in Rats and Mice. *Inflammopharmacology* **2011**, *19*, 255–263. [CrossRef]
22. Yin, M.C.; Lin, C.C.; Wu, H.C.; Tsao, S.M.; Hsu, C.K. Apoptotic Effects of Protocatechuic Acid in Human Breast, Lung, Liver, Cervix, and Prostate Cancer Cells: Potential Mechanisms of Action. *J. Agric. Food Chem.* **2009**, *57*, 6468–6473. [CrossRef]
23. Hu, J.; Lin, S.; Huang, J.J.; Cheung, P.C.K. Mechanistic Study of the in Vitro and in Vivo Inhibitory Effects of Protocatechuic Acid and Syringic Acid on VEGF-Induced Angiogenesis. *J. Agric. Food Chem.* **2018**, *66*, 6742–6751. [CrossRef] [PubMed]
24. Tsao, S.M.; Hsia, T.C.; Yin, M.C. Protocatechuic Acid Inhibits Lung Cancer Cells by Modulating FAK, MAPK, and NF-B Pathways. *Nutr. Cancer* **2014**, *66*, 1331–1341. [CrossRef]
25. Savjani, K.T.; Gajjar, A.K.; Savjani, J.K. Drug Solubility: Importance and Enhancement Techniques. *ISRN Pharm.* **2012**, *2012*, 1–10. [CrossRef]
26. Zhu, L.; Lu, L.; Wang, S.; Wu, J.; Shi, J.; Yan, T.; Xie, C.; Li, Q.; Hu, M.; Liu, Z. Oral Absorption Basics. In *Developing Solid Oral Dosage Forms*; Elsevier: Amsterdam, The Netherlands, 2017; pp. 297–329. [CrossRef]
27. Sadhukha, T.; Prabha, S. Encapsulation in Nanoparticles Improves Anti-Cancer Efficacy of Carboplatin. *AAPS PharmSciTech* **2014**, *15*, 1029–1038. [CrossRef]
28. Tiwari, G.; Tiwari, R.; Bannerjee, S.; Bhati, L.; Pandey, S.; Pandey, P.; Sriwastawa, B. Drug Delivery Systems: An Updated Review. *Int. J. Pharm. Investig.* **2012**, *2*, 2. [CrossRef]
29. Benelmekki, M. An Introduction to Nanoparticles and Nanotechnology. *Des. Hybrid Nanopart.* **2014**, *10*, 5951–5959. [CrossRef]
30. Nagarajan, R. *Nanoparticles: Building Blocks for Nanotechnology*; ACS: Washington, DC, USA, 2008; pp. 2–14. [CrossRef]
31. Nasrollahzadeh, M.; Sajadi, S.M.; Sajjadi, M.; Issaabadi, Z. *Applications of Nanotechnology in Daily Life*, 1st ed.; Elsevier: Amsterdam, The Netherlands, 2019; Volume 28. [CrossRef]
32. Giustini, A.J.; Petryk, A.A.; Cassim, S.M.; Tate, J.A.; Baker, I.; Hoopes, P.J. Magnetic Nanoparticle Hyperthermia In Cancer Treatment. *Nano Life* **2010**, *1*, 17–32. [CrossRef]
33. Wu, C.-H.; Lan, C.H.; Wu, K.L.; Wu, Y.M.; Jane, W.N.; Hsiao, M.; Wu, H.C. Hepatocellular Carcinoma-Targeted Nanoparticles for Cancer Therapy. *Int. J. Oncol.* **2018**, *52*, 389–401. [CrossRef]
34. Khan, I.; Saeed, K.; Khan, I. Nanoparticles: Properties, Applications and Toxicities. *Arab. J. Chem.* **2017**, *12*, 908–931. [CrossRef]
35. Riccardi, C.; Musumeci, D.; Trifuoggi, M.; Irace, C.; Paduano, L.; Montesarchio, D. Anticancer ruthenium(III) Complexes and Ru(III)-Containing Nanoformulations: An Update on the Mechanism of Action and Biological Activity. *Pharmaceuticals* **2019**, *12*, 146. [CrossRef] [PubMed]
36. Schrand, A.M.; Rahman, M.F.; Hussain, S.M.; Schlager, J.J.; Smith, D.A.; Syed, A.F. Metal-Based Nanoparticles and Their Toxicity Assessment. *Wiley Interdiscip. Rev. Nanomed. Nanobiotechnol.* **2010**, *2*, 544–568. [CrossRef] [PubMed]
37. Wang, J.; Sun, P.; Bao, Y.; Liu, J.; An, L. Cytotoxicity of Single-Walled Carbon Nanotubes on PC12 Cells. *Toxicol. Vitr.* **2011**, *25*, 242–250. [CrossRef] [PubMed]
38. Pacurari, M.; Schwegler-Berry, D.; Friend, S.; Leonard, S.S.; Mercer, R.R.; Vallyathan, V.; Castranova, V. Raw Single-Walled Carbon Nanotube-Induced Cytotoxic Effects in Human Bronchial Epithelial Cells: Comparison to Asbestos. *Toxicol. Environ. Chem.* **2011**, *93*, 1045–1072. [CrossRef]
39. Schins, R.P.F. Mechanisms of Genotoxicity of Particles and Fibers. *Inhal. Toxicol.* **2002**, *14*, 57–78. [CrossRef]

40. Panyam, J.; Labhasetwar, V. Biodegradable Nanoparticles for Drug and Gene Delivery to Cells and Tissue. *Adv. Drug Deliv. Rev.* **2003**, *55*, 329–347. [CrossRef]
41. Mohammed, M.A.; Syeda, J.T.M.; Wasan, K.M.; Wasan, E.K. An Overview of Chitosan Nanoparticles and Its Application in Non-Parenteral Drug Delivery. *Pharmaceutics* **2017**, *9*, 53. [CrossRef]
42. El-Shabouri, M.H. Positively Charged Nanoparticles for Improving the Oral Bioavailability of Cyclosporin-A. *Int. J. Pharm.* **2002**, *249*, 101–108. [CrossRef]
43. Min, K.H.; Park, K.; Kim, Y.S.; Bae, S.M.; Lee, S.; Jo, H.G.; Park, R.W.; Kim, I.S.; Jeong, S.Y.; Kim, K.; et al. Hydrophobically Modified Glycol Chitosan Nanoparticles-Encapsulated Camptothecin Enhance the Drug Stability and Tumor Targeting in Cancer Therapy. *J. Control. Release* **2008**, *127*, 208–218. [CrossRef]
44. Masarudin, M.J.; Cutts, S.M.; Evison, B.J.; Pigram, P.J. Factors Determining the Stability, Size Distribution, and Cellular Accumulation of Small, Monodisperse Chitosan Nanoparticles as Candidate Vectors for Anticancer Drug Delivery: Application to the Passive Encapsulation of [14 C]—Doxorubicin. *Nanotechnol. Sci. Appl.* **2015**, *8*, 67–80. [CrossRef]
45. Shu, X.; Zhu, K. The Influence of Multivalent Phosphate Structure on the Properties of Ionically Cross-Linked Chitosan Films for Controlled Drug Release. *Eur. J. Pharm. Biopharm.* **2002**, *54*, 235–243. [CrossRef]
46. Kavi Rajan, R.; Hussein, M.Z.; Fakurazi, S.; Yusoff, K.; Masarudin, M.J. Increased ROS Scavenging and Antioxidant Efficiency of Chlorogenic Acid Compound Delivered via a Chitosan Nanoparticulate System for Efficient In Vitro Visualization and Accumulation in Human Renal Adenocarcinoma Cells. *Int. J. Mol. Sci.* **2019**, *20*, 4667. [CrossRef]
47. Naskar, S.; Koutsu, K.; Sharma, S. Chitosan-Based Nanoparticles as Drug Delivery Systems: A Review on Two Decades of Research. *J. Drug Target.* **2019**, *27*, 379–393. [CrossRef]
48. Bangun, H.; Tandiono, S.; Arianto, A. Preparation and Evaluation of Chitosan-Tripolyphosphate Nanoparticles Suspension as an Antibacterial Agent. *J. Appl. Pharm. Sci.* **2018**, *8*, 147–156. [CrossRef]
49. Koukaras, E.N.; Papadimitriou, S.A.; Bikiaris, D.N.; Froudakis, G.E. Insight on the Formation of Chitosan Nanoparticles through Ionotropic Gelation with Tripolyphosphate. *Mol. Pharm.* **2012**, *9*, 2856–2862. [CrossRef]
50. Farhangi, M.; Kobarfard, F.; Mahboubi, A.; Vatanara, A.; Mortazavi, S.A. Preparation of an Optimized Ciprofloxacin-Loaded Chitosan Nanomicelle with Enhanced Antibacterial Activity. *Drug Dev. Ind. Pharm.* **2018**, *44*, 1273–1284. [CrossRef] [PubMed]
51. Snyder, S.L.; Sobocinski, P.Z. An Improved 2,4,6-Trinitrobenzenesulfonic Acid Method for the Determination of Amines. *Anal. Biochem.* **1975**, *64*, 284–288. [CrossRef]
52. Kuen, C.; Fakurazi, S.; Othman, S.; Masarudin, M. Increased Loading, Efficacy and Sustained Release of Silibinin, a Poorly Soluble Drug Using Hydrophobically-Modified Chitosan Nanoparticles for Enhanced Delivery of Anticancer Drug Delivery Systems. *Nanomaterials* **2017**, *7*, 379. [CrossRef]
53. Huang, Y.; Lapitsky, Y. Monovalent Salt Enhances Colloidal Stability during the Formation of Chitosan/tripolyphosphate Microgels. *Langmuir* **2011**, *27*, 10392–10399. [CrossRef]
54. Esquivel, R.; Juárez, J.; Almada, M.; Ibarra, J.; Valdez, M.A. Synthesis and Characterization of New Thiolated Chitosan Nanoparticles Obtained by Ionic Gelation Method. *Int. J. Polym. Sci.* **2015**, *2015*, 1–18. [CrossRef]
55. Wang, Y.S.; Liu, L.R.; Jiang, Q.; Zhang, Q.Q. Self-Aggregated Nanoparticles of Cholesterol-Modified Chitosan Conjugate as a Novel Carrier of Epirubicin. *Eur. Polym. J.* **2007**, *43*, 43–51. [CrossRef]
56. Kim, Y.H.; Gihm, S.H.; Park, C.R.; Lee, K.Y.; Kim, T.W.; Kwon, I.C.; Chung, H.; Jeong, S.Y. Structural Characteristics of Size-Controlled Self-Aggregates of Deoxycholic Acid-Modified Chitosan and Their Application as a DNA Delivery Carrier. *Bioconjug. Chem.* **2001**, *12*, 932–938. [CrossRef] [PubMed]
57. Hu, F.Q.; Ren, G.F.; Yuan, H.; Du, Y.Z.; Zeng, S. Shell Cross-Linked Stearic Acid Grafted Chitosan Oligosaccharide Self-Aggregated Micelles for Controlled Release of Paclitaxel. *Colloids Surf. B Biointerfaces* **2006**, *50*, 97–103. [CrossRef] [PubMed]
58. Opanasopit, P.; Ngawhirunpat, T.; Rojanarata, T.; Choochottiros, C.; Chirachanchai, S. Camptothecin-Incorporating N-Phthaloylchitosan-G-mPEG Self-Assembly Micellar System: Effect of Degree of Deacetylation. *Colloids Surf. B Biointerfaces* **2007**, *60*, 117–124. [CrossRef] [PubMed]
59. Raja, M.A.; Arif, M.; Feng, C.; Zeenat, S.; Liu, C.G. Synthesis and Evaluation of pH-Sensitive, Self-Assembled Chitosan-Based Nanoparticles as Efficient Doxorubicin Carriers. *J. Biomater. Appl.* **2017**, *31*, 1182–1195. [CrossRef] [PubMed]

60. Zhang, J.; Chen, X.G.; Li, Y.Y.; Liu, C.S. Self-Assembled Nanoparticles Based on Hydrophobically Modified Chitosan as Carriers for Doxorubicin. *Nanomed. Nanotechnol. Biol. Med.* **2007**, *3*, 258–265. [CrossRef]
61. Ortona, O.; D'Errico, G.; Mangiapia, G.; Ciccarelli, D. The Aggregative Behavior of Hydrophobically Modified Chitosans with High Substitution Degree in Aqueous Solution. *Carbohydr. Polym.* **2008**, *74*, 16–22. [CrossRef]
62. Ways, T.M.M.; Lau, W.M.; Khutoryanskiy, V.V. Chitosan and Its Derivatives for Application in Mucoadhesive Drug Delivery Systems. *Polymers (Basel)* **2018**, *10*, 267. [CrossRef]
63. Ong, S.; Ming, L.; Lee, K.; Yuen, K. Influence of the Encapsulation Efficiency and Size of Liposome on the Oral Bioavailability of Griseofulvin-Loaded Liposomes. *Pharmaceutics* **2016**, *8*, 25. [CrossRef]
64. Othman, N.; Masarudin, M.; Kuen, C.; Dasuan, N.; Abdullah, L.; Md. Jamil, S. Synthesis and Optimization of Chitosan Nanoparticles Loaded with L-Ascorbic Acid and Thymoquinone. *Nanomaterials* **2018**, *8*, 920. [CrossRef]
65. Roberts, S.A.; Agrawal, N. Enhancing the Drug Encapsulation Efficiency of Liposomes for Therapeutic Delivery. 2017 IEEE Healthc. *Innov. Point Care Technol. HI-POCT* **2017**, *2017*, 136–139. [CrossRef]
66. Danaei, M.; Dehghankhold, M.; Ataei, S.; Hasanzadeh Davarani, F.; Javanmard, R.; Dokhani, A.; Khorasani, S.; Mozafari, M.R. Impact of Particle Size and Polydispersity Index on the Clinical Applications of Lipidic Nanocarrier Systems. *Pharmaceutics* **2018**, *10*, 57. [CrossRef] [PubMed]
67. Maruyama, C.R.; Guilger, M.; Pascoli, M.; Bileshy-José, N.; Abhilash, P.C.; Fraceto, L.F.; De Lima, R. Nanoparticles Based on Chitosan as Carriers for the Combined Herbicides Imazapic and Imazapyr. *Sci. Rep.* **2016**, *6*, 19768. [CrossRef] [PubMed]
68. Kruse, A.; Dinjus, E. Hot Compressed Water as Reaction Medium and Reactant. Properties and Synthesis Reactions. *J. Supercrit. Fluids* **2007**, *39*, 362–380. [CrossRef]
69. Malm, A.V.; Corbett, J.C.W. Improved Dynamic Light Scattering Using an Adaptive and Statistically Driven Time Resolved Treatment of Correlation Data. *Sci. Rep.* **2019**, *9*, 1–11. [CrossRef]
70. Jonassen, H.; Kjøniksen, A.L.; Hiorth, M. Effects of Ionic Strength on the Size and Compactness of Chitosan Nanoparticles. *Colloid Polym. Sci.* **2012**, *290*, 919–929. [CrossRef]
71. Mayeen, A.; Shaji, L.K.; Nair, A.K.; Kalarikkal, N. *Morphological Characterization of Nanomaterials*; Elsevier: Amsterdam, The Netherlands, 2018. [CrossRef]
72. Barhoum, A.; Luisa García-Betancourt, M. *Physicochemical Characterization of Nanomaterials: Size, Morphology, Optical, Magnetic, and Electrical Properties*; Elsevier: Amsterdam, The Netherlands, 2018. [CrossRef]
73. Dogan, Ü.; Çiftçi, H.; Cetin, D.; Suludere, Z.; Tamer, U. Nanoparticle Embedded Chitosan Film for Agglomeration Free TEM Images. *Microsc. Res. Tech.* **2017**, *80*, 163–166. [CrossRef]
74. Coates, J. *Interpretation of Infrared Spectra, A Practical Approach*; Meyer, R.A., Ed.; John Wiley & Sons: Chichester, UK, 2000.
75. N., M.D.; Eskandari, R.; Zolfagharian, H.; Mohammad, M. Preparation and in Vitro Characterization of Chitosan Nanoparticles Containing Mesobuthus Eupeus Scorpion Venom as an Antigen Delivery System. *J. Venom. Anim. Toxins Trop. Dis.* **2012**, *18*, 44–52.
76. Martins, A.F.; de Oliveira, D.M.; Pereira, A.G.B.; Rubira, A.F.; Muniz, E.C. Chitosan/TPP Microparticles Obtained by Microemulsion Method Applied in Controlled Release of Heparin. *Int. J. Biol. Macromol.* **2012**, *51*, 1127–1133. [CrossRef]
77. Queiroz, M.F.; Melo, K.R.T.; Sabry, D.A.; Sassaki, G.L.; Rocha, H.A.O. Does the Use of Chitosan Contribute to Oxalate Kidney Stone Formation? *Mar. Drugs* **2015**, *13*, 141–158. [CrossRef]
78. Bi, X.; Zhang, H.; Dou, L. Layered Double Hydroxide-Based Nanocarriers for Drug Delivery. *Pharmaceutics* **2014**, *6*, 298–332. [CrossRef] [PubMed]
79. Barahuie, F.; Hussein, M.Z.; Gani, S.A.; Fakurazi, S.; Zainal, Z. Synthesis of Protocatechuic Acid-Zinc/aluminium-Layered Double Hydroxide Nanocomposite as an Anticancer Nanodelivery System. *J. Solid State Chem.* **2015**, *221*, 21–31. [CrossRef]
80. Sani, M.; Mohd, U.; Hussein, Z.; Umar, A.; Sharida, K.; Mas, F.; Masarudin, J. Synthesis and Characterization of Protocatechuic Acid—Loaded Gadolinium—Layered Double Hydroxide and Gold Nanocomposite for Theranostic Application. *Appl. Nanosci.* **2018**, *8*, 5. [CrossRef]
81. Jiang, J.; Liu, Y.; Wu, C.; Qiu, Y.; Xu, X.; Lv, H.; Bai, A.; Liu, X. Development of Drug-Loaded Chitosan Hollow Nanoparticles for Delivery of Paclitaxel to Human Lung Cancer A549 Cells. *Drug Dev. Ind. Pharm.* **2017**, *43*, 1304–1313. [CrossRef]

82. Uppal, S.; Kaur, K.; Kumar, R.; Kaur, N.D.; Shukla, G.; Mehta, S.K. Chitosan Nanoparticles as a Biocompatible and Efficient Nanowagon for Benzyl Isothiocyanate. *Int. J. Biol. Macromol.* **2018**, *115*, 18–28. [CrossRef] [PubMed]
83. Cheung, R.; Ng, T.; Wong, J.; Chan, W. Chitosan: An Update on Potential Biomedical and Pharmaceutical Applications. *Mar. Drugs* **2015**, *13*, 5156–5186. [CrossRef]
84. Kong, B.; Seog, J.H.; Lee, S.B. Experimental Considerations on the Cytotoxicity of Nanoparticles Choice of Cell Types. *Nanomedicine* **2015**, *6*, 1–12. [CrossRef]
85. Loutfy, S.A.; El-din, H.M.A.; Elberry, M.; Allam, N.G.; Hasanin, M.T.M.; Abdellah, A.M. Synthesis, Characterization and Cytotoxic Evaluation of Chitosan Nanoparticles: In Vitro Liver Cancer Model. *Adv. Nat. Sci. Nanosci. Nanotechnol.* **2016**, *7*, 035008. [CrossRef]
86. Zhang, Y.; Xu, Y.; Xi, X.; Shrestha, S.; Jiang, P.; Zhang, W.; Gao, C. Amino Acid-Modified Chitosan Nanoparticles for Cu2+chelation to Suppress CuO Nanoparticle Cytotoxicity. *J. Mater. Chem. B* **2017**, *5*, 3521–3530. [CrossRef]
87. Grenha, A.; Grainger, C.I.; Dailey, L.A.; Seijo, B.; Martin, G.P.; Remuñán-López, C.; Forbes, B. Chitosan Nanoparticles Are Compatible with Respiratory Epithelial Cells in Vitro. *Eur. J. Pharm. Sci.* **2007**, *31*, 73–84. [CrossRef]
88. Díaz, B.; Sánchez-Espinel, C.; Arruebo, M.; Faro, J.; De Miguel, E.; Magadán, S.; Yagüe, C.; Fernández-Pacheco, R.; Ibarra, M.R.; Santamaría, J.; et al. Assessing Methods for Blood Cell Cytotoxic Responses to Inorganic Nanoparticles and Nanoparticle Aggregates. *Small* **2008**, *4*, 2025–2034. [CrossRef] [PubMed]
89. Powers, K.W.; Palazuelos, M.; Moudgil, B.M.; Roberts, S.M. Characterization of the Size, Shape, and State of Dispersion of Nanoparticles for Toxicological Studies. *Nanotoxicology* **2007**, *1*, 42–51. [CrossRef]
90. Chithrani, B.D.; Chan, W.C. Elucidating the Mechanism of Cellular Uptake and Removal of Protein-Coated Gold Nanoparticles of Different Sizes and Shapes. TL—7. *Nano Lett.* **2007**, *7*, 1542–1550. [CrossRef]
91. Chanphai, P.; Tajmir-Riahi, H.A. Encapsulation of Testosterone by Chitosan Nanoparticles. *Int. J. Biol. Macromol.* **2017**, *98*, 535–541. [CrossRef] [PubMed]
92. Madureira, A.R.; Pereira, A.; Pintado, M. Chitosan Nanoparticles Loaded with 2,5-Dihydroxybenzoic Acid and Protocatechuic Acid: Properties and Digestion. *J. Food Eng.* **2016**, *174*, 8–14. [CrossRef]
93. Pham, T.; Nguyen, T.; Thi, T.; Nguyen, T.-T.; Le, T.; Vo, D.; Nguyen, D.; Nguyen, C.; Nguyen, D.; Nguyen, T.; et al. Investigation of Chitosan Nanoparticles Loaded with Protocatechuic Acid (PCA) for the Resistance of Pyricularia Oryzae Fungus against Rice Blast. *Polymers (Basel)* **2019**, *11*, 177. [CrossRef]
94. Hassan, U.A.; Hussein, M.Z.; Alitheen, N.B.; Ariff, S.A.Y.; Masarudin, M.J. In Vitro Cellular Localization and Efficient Accumulation of Fluorescently Tagged Biomaterials from Monodispersed Chitosan Nanoparticles for Elucidation of Controlled Release Pathways for Drug Delivery Systems. *Int. J. Nanomed.* **2018**, *13*, 5075–5095. [CrossRef]

© 2020 by the authors. Licensee MDPI, Basel, Switzerland. This article is an open access article distributed under the terms and conditions of the Creative Commons Attribution (CC BY) license (http://creativecommons.org/licenses/by/4.0/).

Article

Chitosan-ZnO Nanocomposites Assessed by Dielectric, Mechanical, and Piezoelectric Properties

Evgen Prokhorov [1,*], Gabriel Luna-Bárcenas [1], José Martín Yáñez Limón [1], Alejandro Gómez Sánchez [1] and Yuriy Kovalenko [2]

[1] Cinvestav, Unidad Querétaro, Querétaro 76230, QRO, Mexico; gabriel.luna@cinvestav.mx (G.L.-B.); jmyanez@cinvestav.mx (J.M.Y.L.); alejandrogomez@cinvestav.mx (A.G.S.)
[2] Postgraduate Department, University of Aeronautics of Querétaro, Querétaro 76278, QRO, Mexico; kovalenko.yuriy@gmail.com
* Correspondence: prokhorov@cinvestav.mx

Received: 28 July 2020; Accepted: 28 August 2020; Published: 1 September 2020

Abstract: The aim of this work is to structurally characterize chitosan-zinc oxide nanoparticles (CS-ZnO NPs) films in a wide range of NPs concentration (0–20 wt.%). Dielectric, conductivity, mechanical, and piezoelectric properties are assessed by using thermogravimetry, FTIR, XRD, mechanical, and dielectric spectroscopy measurements. These analyses reveal that the dielectric constant, Young's modulus, and piezoelectric constant (d_{33}) exhibit a strong dependence on nanoparticle concentration such that maximum values of referred properties are obtained at 15 wt.% of ZnO NPs. The piezoelectric coefficient d_{33} in CS-ZnO nanocomposite films with 15 wt.% of NPs (d_{33} = 65.9 pC/N) is higher than most of polymer-ZnO nanocomposites because of the synergistic effect of piezoelectricity of NPs, elastic properties of CS, and optimum NPs concentration. A three-phase model is used to include the chitosan matrix, ZnO NPs, and interfacial layer with dielectric constant higher than that of neat chitosan and ZnO. This layer between nanoparticles and matrix is due to strong interactions between chitosan's side groups with ZnO NPs. The understanding of nanoscale properties of CS-ZnO nanocomposites is important in the development of biocompatible sensors, actuators, nanogenerators for flexible electronics and biomedical applications.

Keywords: chitosan; zinc oxide nanoparticles; interfacial layer; dielectric spectroscopy

1. Introduction

Zinc oxide nanoparticles (ZnO-NPs) are one of the most attractive materials due to their unique optical, piezoelectric, mechanical, and antibacterial properties. Nanocomposites based upon ZnO-NPs are widely used for the development of different optoelectronic, electronic, sensors, collar cells, etc., devices (see, for example [1–3]). Recently, there were published significant publications about the potential use of ZnO-NPs that include flexible devices such as supercapacitance [4], flexible piezoelectric nanogenerators with ZnO-polyvinylidene fluoride (PVDF) [5,6], piezoelectric vibration sensors based on polydimethylsiloxane (PDMS) and ZnO nanoparticle [7], soft thermoplastic material with polyurethane matrix [8], poly (ethylene oxide) and poly (vinyl pyrrolidone) blend matrix incorporated with zinc oxide (ZnO) nanoparticles for optoelectronic and microelectronic devices [9], gate transistors with ZnO and ethyl cellulose [10], chitosan-ZnO (CS-ZnO) nanocomposite for packing applications [11–13], CS-ZnO as antibacterial agent [13–15], and CS-ZnO nanocomposite for supercapacitor [16].

Based upon the above information, chitosan-based nanocomposites offer significant scientific and technological potential. In this regard, chitosan (CS), a polysaccharide obtained from the deacetylation of chitin, is a natural polymer with high absorption capacity, biodegradability, biocompatibility with antibacterial features. Additionally, chitosan is a hydrophilic polymer with NH_2 and OH side groups which can interact with ZnO nanoparticles via hydrogen bonding and form nanocomposites with new properties [17–19]. It is noteworthy that the literature reports publications that deal with different methods of CS-ZnO nanocomposite preparation and their antibacterial, optical, photocatalytic activity (see, for example [11–20]), and mechanical [15,21–23] properties. In the case of mechanical properties, it has been reported that the ZnO content improves mechanical properties not only in CS-ZnO composite [15,22,23] but also in CS-cellulose-ZnO [24] and in CS-PVA-ZnO [25] materials.

However, to the best our knowledge, the literature does not properly address the influence of ZnO content on the conductivity of CS-ZnO nanocomposite; in this regard there are two controversial articles related to the effect of ZnO additional on the dielectric constant of CS-ZnO membranes (in [23] with additional of ZnO NPs dielectric constant increase and in [26] decrease). The conductivity and dielectric properties play important role in applications of CS-ZnO nanocomposites in flexible organic electronics in a wide range of devices like transistors, sensors, flexible piezoelectric nanogenerators, ultraviolet photodetectors, photodiodes, etc., [4,23].

One of the most important questions related not only for application of CS-ZnO nanocomposites but also for all nanotechnology is how to find the best/optimum concentration of NPs with the best performance for different applications. CS-ZnO membrane composites consist of dielectric CS matrix and ZnO semiconductor fillers with wide bandgap with static dielectric constant ca. 8.5 [27] and low conductivity (ca. 10^{-4}–10^{-5} S/cm, which depends upon intrinsic defects created by oxygen vacancies [28–30]). Therefore, this material can be considered as a dielectric matrix with dielectric inclusions. It is noteworthy that for different polymer-ZnO NPs composites, the dielectric constant depends upon ZnO content where a maximum is observed; for instance, in PVDF-ZnO at 0.06 vol.% of ZnO [31]; in PVDF-ZnO 5.5 vol.% of ZnO [32]; in PVDF-ZnO at 15 wt.% [33]: in PVA/PVP-ZnO at 8 wt.% [34]; in PVA-ZnO at 10 mol% [35]. The explanation of the maximum in the dielectric constant proposed in these articles is based upon classical percolation theory; by increasing conductivity inclusions in dielectric matrix the conductivity of composites increases at the percolation threshold and upon higher concentration of fillers there appears a saturation. The dielectric constant also shows a maximum near the percolation threshold [36,37]; however, refs. [31–35] do not report a conductivity percolation effect. It is noteworthy that ZnO NPs exhibit low conductivity such that the PVA-ZnO composite's conductivity is ca. 10^{-7}–10^{-9} S/cm [35] and PVA/PVP-ZnO is less than 10^{-7} S/cm [34]. Therefore, such material cannot exhibit conductivity percolation phenomena and this model cannot be used to explain the maximum in dielectric constant.

Similarly, to the dielectric behavior, there is a maximum on the Young's modulus as a function of ZnO concentration. Ref. [38] reports a maximum on the Young's modulus in PHBV-ZnO (Poly(3-hydroxybutyrate-co-3-hydroxyvalerate-ZnO) at a composition of 4 wt.% of NPs; Ref. [39] reports that system PEEK-ZnO (poly(ether ether ketone)-ZnO) shows a maximum at 5 wt.% of ZnO; PMMA-ZnO at 1 wt.% of NPs [40]; PLA-ZnO at 2 wt.% [41]. In summary, such a maximum in Young's modulus has been related to the distribution of nanoparticles within the polymer matrix and strong interfacial adhesion that can enhance the mechanical properties of nanocomposites [42].

In general, in polymer-ZnO NPs composites both dielectric constant and Young's modulus share a common feature: by increasing NPs concentration both properties increase [43,44]. Consequently, it is conceivable that dielectric, mechanical, and piezoelectric properties can be optimized by varying the concentration of ZnO NPs in chitosan nanocomposites for different applications.

Based upon the above discussion, this work aims to investigate the structural properties of CS-ZnO films including their dielectric, conductivity, mechanical and piezoelectric properties by varying the concentration of ZnO nanoparticles. To assess this study, we take advantage of impedance spectroscopy, FTIR, XRD, thermogravimetry, and piezoelectric measurements.

2. Materials and Methods

Chitosan (CS, medium molecular weight, deacetylation ca. 72%), acetic acid (99.7%), and ZnO NPs dispersed in water (20 wt.% in water) with dimension ca. 40 nm were purchased from Sigma Aldrich (St. Louis, MO, USA) and used as received.

CS solution (1 wt.%) was prepared in acetic acid solution (1 vol.%) and stirred for 24 h. Different amounts of ZnO sonicate colloidal solutions with various weight percent of ZnO (5, 10, 15, and 20 wt.% with respect to CS dry-base) were dispersed in the CS solution by ultrasound for 30 min at 60 Hz. Finally, 18 mL of each nanocomposite solutions were placed in Petri dishes and dried during 20 h at 60 °C to obtain films with thickness ca. 40 µm. For impedance measurements, CS-ZnO films were gold-sputtered on both sides to serve as contacts.

The amount of free water was determined by thermogravimetric analysis (TGA) (TGA 4000—PerkinElmer, Walham, MA, USA). Measurements were made in the dry air with a heating rate of 10 °C/min. The interaction between CS functional groups with ZnO was analyzed by FTIR measurements on a Perkin Elmer Spectrum GX spectrophotometer using ATR (MIRacle™) sampling technique, with a diamond tip, in the range from 4000 to 650 cm^{-1} at room temperature. The crystalline structure of ZnO and CS-ZnO films were tested by an X-ray diffractometer (Rigaku Dmax 2100, The Woodlands, TX, USA) with Cu Kα radiation (λ = 0.154 nm).

Impedance measurements were carried out using Agilent 4249 A in the frequency range 40 Hz–100 MHz with an amplitude of AC voltage 100 mV at room temperature. DC resistance R and capacitance C at the limit of zero frequency were calculated from fitting impedance spectra using ZView program. Conductivity and static dielectric constant (at the limit of zero frequency) were calculated from the following relationship: $\sigma = d/(R \cdot S)$, $\varepsilon = (C \cdot d)/(\varepsilon_0 \cdot S)$, where d and S are the thickness and area of samples, respectively. Film thickness was measured in each sample using micrometer Mitutoyo with resolution 1 mkm. The mechanical test was performed on an Instron universal tensiometer material testing system (model TX2plus). Each composite film was cut with dimensions according to ASTM Standard D638-Epsilon. Each strip was held with a distance between clamps of 25 mm. The test was performed with the lower grip was fixed, and the upper grip rose at an extension rate of 1 mm/sec at room temperature. All the failures occurred in the middle region of the testing strips. This test was repeated six times for each specimen to confirm its repeatability.

The measurements of ferroelectric polarization loops (P versus E) and deformation curves as a function of the applied field (butterfly curves) were obtained simultaneously by placing the samples in a measuring cell with parallel electrodes immersed in silicone oil to avoid dielectric breakage of the surrounding medium with voltage step 100 V before sample breakdown. The polarization and deformation curves presented in this work correspond to the maximum applied voltage measured before film breakdown. The ferroelectric measurements were based on the principle of the Sawyer-Tower circuit using a Precision LC materials analyzer, Radiant Technologies Inc., coupled with a TREK Model 609E-6 voltage amplifier source. Results presented in this work correspond to the maximum voltage before the breakdown of the sample.

3. Results

XRD analysis can supply information about the crystalline structure of ZnO NPs and Cs-ZnO NPs films (Figure 1). XRD pattern of neat CS and CS-ZnO (with 20 wt.% of NPs) show the diffraction peak at $2\theta \approx 24°$ (hydrate crystalline phase, Form 1) and a weak peak at $2\theta \approx 16.9°$ (hydrate crystalline phase, Form 2) [45]. The peaks observed in ZnO NPs and CS-ZnO films were in good agreement with the database of hexagonal ZnO particles (JCPDS No. 36-1451) and the results are reported in refs. [15,19,23]. This means that structure of ZnO NPs was not modified by the presence of CS [18,20], but the intensity of broad CS peak at $2\theta \approx 24°$ decrease in CS-ZnO films indicated the increase in the degree of amorphous regions of the nanocomposites films due to the interaction between the CS matrix and ZnO NPs and decreasing of water content [25]. Measurements of CS-ZnO films were carried out on copper substrate; therefore, there are additional diffraction patterns associated to Cu substrate at 2θ ≈ 43.3° and 2θ ≈ 50.4°; this fact is indicated in Figure 1.

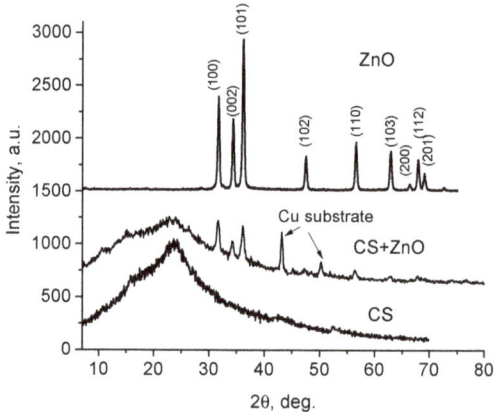

Figure 1. X-ray diffraction patterns of neat chitosan (CS) film, CS ZnO NPs composite, and ZnO nanoparticles (NPs).

The average size of particles (D) was calculated using the Debye–Scherrer equation [18,19]:

$$D = k\lambda / \beta \cos\theta \quad (1)$$

where the value of k is equal to 0.89, λ is the wavelength of X-ray (1.54∘A), β is the full width at half maximum, and θ is the half of the diffraction angle. The average value of crystallites (calculated using three diffraction peaks (100), (002), and (101)) were 45.3 nm which correlates well with the dimension of NPs (ca. 40 nm).

Figure 2 shows the FTIR spectra of neat CS films and CS-ZnO films with 10 and 20 wt.% of ZnO NPs. In the case of chitosan, the broadband characteristic peak centered at 3252 cm^{-1} corresponds to the overlap of stretching vibration of –NH and –OH groups shift in CS-ZnO films to lower wavenumber at 3227 cm^{-1}. The absorption peaks of CS at 1636 (amide I group), 1542 cm^{-1} (bending vibrations of NH$_3$), and 1065 (the stretching vibration of C–O–C of the glycosidic linkage) in CS-ZnO films shift to lower wavenumber (Figure 2) due to the interaction of these group with ZnO and formation of a hydrogen bond between ZnO and chitosan. This result is in good agreement with previous reports [15,18,19,23,46].

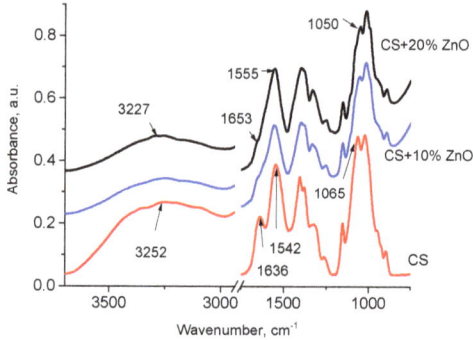

Figure 2. FTIR spectra of neat CS films and CS-ZnO films with 10 and 20 wt.% of ZnO.

Another confirmation of strong interaction between side groups of CS with ZnO NPs can be obtained from TGA measurements (Figure 3). It was previously reported [47] that neat CS exhibits a two-step weight loss. From room temperature to ca. 150 °C, the weight loss is related to the water evaporation and in the temperature range of 170–300 °C, the weight loss is due to the degradation of the CS [47]. Water absorption in CS is closely linked to the availability of amino and hydroxyl groups of CS that interact via hydrogen bonding with water molecules [47,48]. It has been observed that the water content depends upon ZnO NPs concentration (Figure 3) and it decreases with increasing weight% of NPs (11.7% in neat CS and 7.3% in CS-ZnO film with 20 wt.% of NPs, at the temperature 140 °C). As it was shown by FTIR analysis, these groups can bond with ZnO NPs; therefore, a decrease in the water absorption ability with increasing ZnO concentration is observed.

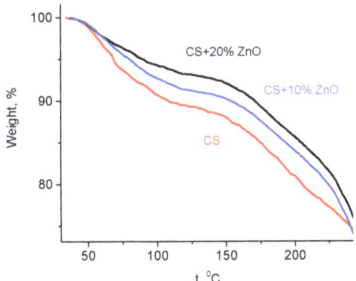

Figure 3. TGA measurements of pure CS and CS-ZnO NPs films with 10 and 20 wt.% of NPs.

The results obtained from XRD, FTIR, and TGA measurements have shown an interaction between CS matrix and ZnO NPs that play an important role in the explanation of electrical and mechanical properties of the nanocomposite.

Figure 4a shows the dependence of DC conductivity and Figure 4b shows the dependencies of the dielectric constant in the limit of zero frequency as a function of ZnO NPs wt.%. It is evident from Figure 4b that dependence of dielectric constant exhibits a maximum at a concentration of ZnO NPs of ca. 15 wt.% and it is higher than the static dielectric constant of neat ZnO (ca. 8.5 [27]). It is noteworthy that the conductivity of CS-ZnO nanocomposite (Figure 4a) is sufficiently lower than the ZnO conductivity (ca. 1×10^{-5} S/cm [28,29]) and it decreases with ZnO NPs wt.%.

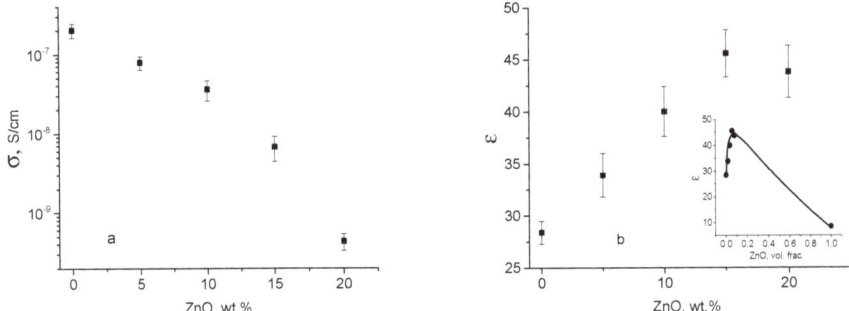

Figure 4. Dependences of (**a**) DC conductivity (σ) and (**b**) dielectric constant (ε) in the limit of zero frequency obtained in CS-ZnO films with different ZnO concentration at room temperature. Insert in Figure 4b shows dependence of ε on the volume fraction of ZnO NPs: points-experimental measurements and continuous line-results of the fitting.

Similarly, to the dielectric constant behavior, there is maximum in the dependence of Young's modulus on ZnO wt.% (Figure 5). Young's modulus increases from 1.7 GPa (in neat CS) to 9.09 GPa in films with 15 wt. = % of NPs. At concentration of ZnO NPs 20 wt.%, Young's modulus decreases to 4.5 GPa. The increasing of Young's modulus with ZnO concentration has been previously reported [12,15,20] and it has been interpreted by an additional energy-dissipating mechanism [12], weakness of intermolecular hydrogen bonds of CS formation of new hydrogen bonds between CS and ZnO [15]. CS-PVA-ZnO NPs membrane study [25] reported the maximum in Young's modulus at 10 wt.% of NPs which has been explained by the interaction of ZnO NPs with CS-PVA functional groups.

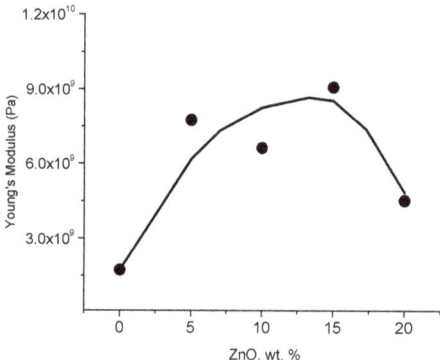

Figure 5. Dependence of Young's modulus of CS ZnO NPs films with different wt.% of NPs (points). The continuous line is a guide to the eye.

ZnO is a well-known material that exhibits both ferroelectric and piezoelectric behavior. Therefore, it is important to investigate these properties in CS-ZnO nanocomposites which can find applications in flexible electronics.

Figure 6 shows the ferroelectric hysteresis curve obtained in the CS-ZnO NPs film with 15 and 20 wt.% of NPs. The shape of the curve is typical for samples with electrical leakage, which prevents reaching saturation in the polarization [49]. This leakage current can be associated with the intrinsic proton conductivity of the CS matrix [50]. The corresponding deformation curves do not present symmetrical shape, probably because of the interaction between ZnO NPs and CS. The piezoelectric coefficient d_{33} was evaluated in the linear region of the deformation curve vs. applied voltages using the expression $d_{33} = \Delta l/\Delta V$ [51,52]. The piezoelectric coefficient d_{33} in CS-ZnO nanocomposite films with 15 wt.% of NPs (d_{33} = 65.9 pC/N) is higher than in neat ZnO NPs (between 0.4 and 12.4 pC/N [53,54]) and in poly(vinylidene fluoride) PVDF-ZnO flexible films (13.42 pC/N [55], 18.3 pC/N [56], 50 pC/N [52]) and compared with PVDF-PTTE-ZnO nanorods (70.3 pC/N with 15 wt.% of nanorods [57]). Note, that PVDF is a polymer with piezoelectric properties.

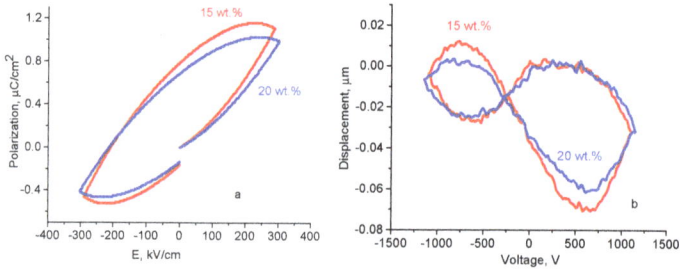

Figure 6. Ferroelectric hysteresis curve of (**a**) CS-ZnO NPs nanocomposites, (**b**) corresponding deformation curve Δl vs. V for CS-ZnO films with different wt.% of NPs indicate on the graph.

Such a high piezoelectric coefficient can be related to elastic properties of the CS matrix because the viscous and elastic properties play an important role in the piezoelectric performance of piezoelectric polymer composites [58].

Additionally, piezoelectric coefficient d_{33} demonstrate higher value (d_{33} = 65.9 pC/N) in CS-ZnO films with 15. wt.% of NPs than in films with 20 wt.% (d_{33} = 60.6 pC/N).

4. Discussion

Bulk conductivity of ZnO is ca. 10^{-4}–10^{-5} S/cm and it depends upon intrinsic defects created by oxygen vacancies [28–30]). Conductivity of nanoparticles depends upon grain size, morphology, and microstructure and it is ca. 1.5×10^{-7} S/cm [59]. Because of the low volume fraction of ZnO NPs in CS-ZnO films (from 0 to 0.15 wt.%; see below the volume fraction calculation), the effective conductivity of nanocomposites practically depends upon the conductivity of neat CS (ca. 10^{-7} S/cm). The conductivity of neat CS is related to the Grotthuss mechanism in which the protons are originated from the protonated amino groups that can move along the hydrated molecule in the hydrogen-bonding network via hopping process [50]. Because of the strong interaction of reactive CS side groups with ZnO NPs (as probed by FTIR measurements) the number of generated protons and the number of hydrated molecules decrease (TGA measurements); this plausible scenario is responsible for the decreasing of nanocomposite's conductivity.

As a rule, the effective dielectric constant ε and conductivity σ of a mixture of two materials with different ε and σ can be calculated using models as the Maxwell, Bruggeman, Lichtenecker, or different percolation model [60–62]. However, all these models produce a monotonic decreasing of the effective dielectric constant with increasing concentration of nanoparticles in the polymer (because ε of CS films is ca. 25 and ε of ZnO is ca. 8.5 [27]); it also can show a monotonic increasing of effective conductivity with increasing concentration of (because σ of CS is ca. 2×10^{-7} S/cm and σ of ZnO is ca. 10^{-4}–10^{-5} S/cm).

In contrast to those models, refs. [63,64] proposed a three-phase model to describe the dielectric properties of polymer–ceramic composites. Here, the effective dielectric constant of such composite materials depends upon the ε of the polymer matrix, the ε of fillers, and the ε of interfacial layer between filler and the dielectric matrix. To describe such three-phase system, refs. [63,64] introduce a parameter K, termed the interfacial volume constant, which accounted for the matrix–filler interaction strength as:

$$\Phi_{int} = K\Phi_{NPs}\Phi_{pol} \qquad (2)$$

where Φ_{int}, Φ_{NPs}, and Φ_{pol} are the volume fractions of interfacial phase, dielectric particles, and polymer, respectively. K depends upon the degree of particle clustering.

In the case if dielectric constant of interfacial layer is higher than dielectric constant of polymer matrix and fillers,

$$K > 0, \; \varepsilon_{interfacial} > \varepsilon_{polymer}, \; \varepsilon_{interfacial} > \varepsilon_{filler}, \qquad (3)$$

with increasing of NPs concentration effective dielectric constant of nanocomposite increases, as observed in Figure 4b.

This model has demonstrated that the dependence of the dielectric constant on NPs concentration is nonmonotonic and can exhibit a maximum as a function of NPs concentration. This maximum appears when there is an overlap of interfacial layers due to NPs agglomeration, thus reducing the interfacial volume fraction that effectively decreases the effective value of dielectric constant of nanocomposite.

In this work, we experimentally fit obtained values of dielectric constant ε in CS-ZnO NPs films by equations proposed in refs. [63,64] using the Scilab program. The least-squares fitting was performed using standard genetic algorithm optimization functions in the Scilab [65] numerical computational package. The dielectric constant of CS was obtained from measurements on neat CS and ε of ZnO was taken 8.5 [27]. Only the values of K and ε interfacial parameters are the adjustable parameters.

To convert weight fraction (Wt) to volume fraction (V) of the ZnO NPs, the next equation can be used [66]:

$$V = \frac{W_t}{W_t + (\rho_{ZnO}/\rho_{CS})(1 - W_t)} \quad (4)$$

where, ρ_{ZnO} and ρ_{CS} denote the ZnO and CS density.

The density of CS films is ca. 1.5 g cm^{-3} [67,68], and the true density of ZnO is 5.6 g cm^{-3} [3].

As a result of optimization, the fitted values are K = 18.3, and interface dielectric constant equals 69.9. The results of the referred fittings are shown on inset of Figure 4b as a continuous line. One can see that this three-phase model fits well the experimental results by predicting a maximum in the dielectric constant. Positive value of K means that there are significant interfacial interactions between CS and ZnO NPs; these observations were confirmed by FTIR and TGA measurements. Additionally, an interfacial layer dielectric constant value of 69.9 is higher than that of CS and ZnO NPs which is responsible for the observed maximum in Figure 4b. In summary, the three-phase model is able to capture the correct physics of the nanocomposite by corroborating the behaviors and trends of the experimental measurements.

Similarly, a maximum at 15 wt.% of ZnO is observed as in the dependency of Young's modulus on ZnO concentration (Figure 5). According to refs. [69,70], the Young's modulus in CS films increases with decreasing of water content due to plasticizing effect of water and change in the glass transition temperature. In the TGA measurements water content decreases ca. 5% in CS-ZnO films with 20 wt.% of NPs when compared with neat CS. Based upon the results reported in ref. [69,70], this decreasing of water content corresponds to the increasing of Young's modulus approximately to 0.5–1 GPa. Therefore, all contributions in elastic module can be related to the change of NPs concentration. Furthermore, a maximum at 15 wt.% of ZnO is observed as in the dependency of Young's modulus and dielectric constant on ZnO concentration (Figures 4b and 5). The explanations proposed in the literature on the change of Young's modulus in ZnO nanocomposites are the following: An additional energy-dissipating mechanism [12], the weakness of intermolecular hydrogen bonding of CS, or the formation of new hydrogen bonding between CS and ZnO [15]; these explanations cannot be directly applied to properly address the existence of a maximum in both dielectric and mechanical properties. According to refs. [71–73], the mechanical properties exhibit a similar dependency on the formation of an interfacial layer. Above the percolation threshold, the interfacial regions surrounding the ZnO NPs overlap indicating percolation in the clusters which dominates in the mechanical properties [72]. Because of the agglomeration of NPs, there is a maximum in both dielectric constant and Young's modulus because further agglomeration of NPs tends to destroy the interfacial regions.

Now let us discuss the dependence of the piezoelectric coefficient on the concentration of NPs. There have been several reports that proposed different models, for instance Refs. [58,74–76] reported an increase of d_{33} with the volume fraction of the piezoelectric NPs. Moreover, in polymer nanocomposites with NPs with high dielectric constant (such as PZT, BaTiO$_3$, BZT) there is an increase of d_{33} with an increase of the dielectric constant with NPs content [74–76]. However, in BaTiO$_3$-epoxy-ZnO (with a fixed concentration of BaTiO$_3$) [77] and PVDF-ZnO [56], the dependences of the piezoelectric coefficient and the dielectric constant on the concentration of ZnO exhibit a maximum (similarly to the maximum presented in Figure 4b). This maximum cannot be explained by classical conductivity percolation effect because the conductivity of CS-ZnO films decreases with ZnO NPs wt.% (Figure 4a). However, the proposed three-phase model describes well the maximum in dielectric constant (an overlap of interfacial layers due to NPs agglomeration) that effectively decreases the value of dielectric constant and most likely the piezoelectric coefficient. A similar hypothesis on the decreasing of d_{33} due to NPs agglomeration (based on the SEM measurements in BaTiO$_3$-epoxy-ZnO) is proposed in ref. [75].

The piezoelectric coefficient d_{33} in CS-ZnO nanocomposite films with 15 wt.% of NPs (d_{33} = 65.9 pC/N) is higher than most of polymer-ZnO nanocomposites (see Table 1) and it compares with PVDF-PTTE-ZnO nanorods (70.3 pC/N).

Table 1. Comparison of piezoelectric d_{33} values in polymer nanocomposites with ZnO NPs.

Polymer Nanocomposite	NPs Dimension (nm)	d_{33} (pC/N)	Refs
ZnO, bulk	-	0.4–12.4	[53,54]
photo-epoxy/ZnO films	Less 100	15–23	[78]
PVDF/ZnO films	50–80	13.42	[55]
PVDF/ZnO nanoporous films	35–45	18.3	[56]
PVDF β-phase/ZnO	50–150	50	[54]
PVDF-PTTE/ZnO nanorods	-	70.3	[57]
PHB/ZnO scaffolds	80–100	13.7	[79]
PVDF/ZnO nanorods	-	−1.17	[80]
CS/ZnO films	40	65.9	This work

Note that PVDF is a polymer with piezoelectric properties.

It is well-known that in classic ferroelectric materials such as BaTiO$_3$, a significant increase in the piezoelectric coefficient in materials with nanodomain structure was observed. The value of d_{33} in samples with grains 50 nm increases more than twice to 416 pC/N [81] compared with grains of 500 nm (200 pC/N). In the case of the ZnO NPs with dimension ca. 40 nm the size of the nanodomains must be less than 40 nm that can increase d_{33} in CS-ZnO nanocomposite. Furthermore, high value of d_{33} in CS-ZnO nanocomposite films can be related to their elastic properties that play an important role in the piezoelectric performance of piezoelectric polymer composites [58]. Additionally, refs. [82,83] reported that CS films with 91.2% of deacetylation degree exhibit a piezoelectric coefficient d_{33} between 7 and 18.4 *pC/N*. These piezoelectric properties are observed because of the fact that crystalline part of CS has non-centrosymmetry orthorhombic structure with piezoelectric properties [82]. However, in our work we did not observe piezoelectric properties of neat CS films; this observation may be traceable to the lower deacetylation degree of CS (*ca.* 72%). Nevertheless, in the presence of an electrical field (at which we measured d_{33}) an alignment of CS chains can be observed. The alignment of CS chains can be correlated with the increase in dielectric and piezoelectric properties of the composites [84]. Therefore, the high d_{33} in CS/ZnO films can be traced to the synergistic effect of nanodomain structure of NPs, piezoelectricity of CS, elastic properties of films, and optimum NPs concentration.

5. Conclusions

The investigation of CS-ZnO films with different wt.% of NPs shows that a maximum value is observed in both dielectric constant and Young's modulus. Similarly, at the same concentration of NPs there appears the highest value of the piezoelectric coefficient. This maximum can be related to cluster agglomeration of ZnO NPs above the dielectric and mechanical percolation threshold (15 wt.% of ZnO NPs). These properties of nanocomposite's films are interpreted by using a three-phase model which includes: (1) CS matrix, (2) ZnO NPs, and (3) interfacial layer between ZnO and CS matrix. This interface layer is responsible for the higher dielectric constant when compared with the ε of neat CS and ZnO, a higher Young's modulus, and a higher d_{33}. The piezoelectric coefficient d_{33} in CS-ZnO nanocomposite films with 15 wt. % of NPs (d_{33} = 65.9 *pC/N*) is higher than in the most of polymer-ZnO nanocomposites because of the synergistic effect of nanodomain structure of NPs, piezoelectricity of CS, elastic properties of CS, and optimum NPs concentration.

Based upon the presented methodology, one can try to fine-tune the desired properties by manipulating the concentration and agglomeration of NPs that ultimately control the molecular interactions. It is noteworthy that these variables not only depend upon the dimension of NPs but also on the method of preparation and the chemistry of constituents. However, the methodology presented here allows to determine the direct relationship between dielectric, mechanical, piezoelectrical properties, and the concentration of nanoparticles that may prove useful in the design and optimization of polymer-based nanocomposites for different applications.

In summary, the molecular understanding of nanoscale properties of CS-ZnO nanocomposites is relevant in the development of biocompatible sensors, actuators, nanogenerators, etc. for flexible electronics, and biomedical applications.

Author Contributions: Conceptualization, methodology, investigation, writing—original draft preparation E.P.; writing—investigation, discussion, review, and editing G.L.-B.; investigation J.M.Y.L.; investigation A.G.S.; software Y.K. All authors have read and agreed to the published version of the manuscript.

Funding: This research was funded and supported by CONACYT, Mexico (grant A1-S-9557).

Acknowledgments: The authors are grateful to J.A. Muñoz Salas for technical assistance in electrical measurements, R.A. Mauricio-Sánchez for assistance in FTIR measurements, M.A. Hernandez Landaverde for assistance in XRD measurements, and R. Flores Farias for assistance in piezoelectric measurements.

Conflicts of Interest: The authors declare no conflict of interest.

References

1. Kalpana, V.N.; Rajeswari, V.D. A Review on Green Synthesis, Biomedical Applications, and Toxicity Studies of ZnO NPs. *Bioinorg. Chem. Appl.* **2018**, *2018*, 3569758. [CrossRef] [PubMed]
2. Mishra, Y.K.; Adelung, R. ZnO tetrapod materials for functional applications. *Mater. Today* **2018**, *21*, 631–651. [CrossRef]
3. Klingshirn, C. ZnO: Material, Physics and Applications. *Chem. Phys. Chem.* **2007**, *8*, 782–803. [CrossRef] [PubMed]
4. Selvaraj, T.; Perumal, V.; Khor, S.F.; Anthony, L.S.; Gopinath, S.C.B.; Mohamed, N.M. The recent development of polysaccharides biomaterials and their performance for supercapacitor applications. *Mater. Res. Bull.* **2020**, *126*, 110839. [CrossRef]
5. Jin, C.; Hao, N.; Xu, Z.; Trase, I.; Nie, Y.; Dong, L.; Closson, A.; Chen, Z.; Zhang, J.X.J. Flexible piezoelectric nanogenerators using metal-doped ZnO-PVDF films. *Sens. Actuators A Phys.* **2020**, *305*, 111912. [CrossRef]
6. Fakhri, P.; Amini, B.; Bagherzadeh, R.; Kashfi, M.; Latifi, M.; Yavari, N.; Kani, S.A.; Kong, L. Flexible hybrid structure piezoelectric nanogenerator based on ZnO nanorod/PVDF nanofibers with improved output. *RSC Adv.* **2019**, *9*, 10117. [CrossRef]
7. Sinar, D.; Knopf, G.K. Disposable piezoelectric vibration sensors with PDMS/ZnO transducers on printed graphene-cellulose electrodes. *Sens. Actuators A Phys.* **2020**, *302*, 111800. [CrossRef]
8. Buzarovska, A.; Dinescu, S.; Lazar, A.D.; Serban, M.; Pircalabioru, G.G.; Costache, M.; Gualandi, C.; Averous, L. Nanocomposite foams based on flexible biobased thermoplastic polyurethane and ZnO nanoparticles as potential wound dressing materials. *Mater. Sci. Eng. C* **2019**, *104*, 109893. [CrossRef]
9. Choudhary, S. Structural, optical, dielectric and electrical properties of (PEO–PVP)–ZnO nanocomposites. *J. Phys. Chem. Solids* **2018**, *121*, 196–209. [CrossRef]
10. Carvalho, J.T.; Dubceac, V.; Grey, P.; Cunha, I.; Fortunato, E.; Martins, R.; Clausner, A.; Zschech, E.; Pereira, L. Fully Printed Zinc Oxide Electrolyte-Gated Transistors on Paper. *Nanomaterials* **2019**, *9*, 169. [CrossRef]
11. Souza, V.G.L.; Rodrigues, C.; Valente, S.; Pimenta, C.; Pires, J.R.A.; Alves, M.M.; Santos, C.F.I.; Coelhoso, M.; Fernando, A.L. Eco-Friendly ZnO/Chitosan Bionanocomposites Films for Packaging of Fresh Poultry Meat. *Coatings* **2020**, *10*, 110. [CrossRef]
12. Ridwan, R.; Rihayat, T.; Suryani, S.; Ismi, A.S.; Nurhanifa, N.; Riskina, S. Combination of poly lactid acid zinc oxide nanocomposite for antimicrobial packaging application. *IOP Conf. Ser. Mater. Sci. Eng.* **2020**, *830*, 042018. [CrossRef]
13. Rodrigues, C.; De Mello, J.M.M.; Dalcanton, F.; Macuvele, D.L.P.; Padoin, N.; Fiori, M.A.; Soares, C.; Riella, H.G. Mechanical, Thermal and Antimicrobial Properties of Chitosan-Based-Nanocomposite with Potential Applications for Food Packaging. *J. Polym. Environ.* **2020**, *28*, 1216–1236. [CrossRef]
14. Vaseeharan, B.; Sivakamavalli, J.; Thaya, R. Synthesis and characterization of chitosan-ZnO composite and its antibiofilm activity against aquatic bacteria. *J. Compos. Mater.* **2013**, *49*, 177–184. [CrossRef]
15. Li, L.-H.; Deng, J.-C.; Deng, H.-R.; Liu, Z.-L.; Xin, L. Synthesis and characterization of chitosan/ZnO nanoparticle composite membranes. *Carbohydr. Res.* **2010**, *345*, 994–998. [CrossRef] [PubMed]

16. Anandhavelu, S.; Dhanasekaran, V.; Sethuraman, V.; Park, H.J. Chitin and Chitosan Based Hybrid Nanocomposites for Super Capacitor Applications. *J. Nanosci. Nanotechnol.* **2017**, *17*, 1321–1328. [CrossRef]
17. Qiu, B.; Xu, X.-F.; Deng, R.-H.; Xia, G.-Q.; Shang, X.-F.; Zhou, P.-H. Construction of chitosan/ZnO nanocomposite film by in situ precipitation. *Int. J. Biol. Macromol.* **2019**, *122*, 82–87. [CrossRef]
18. Abarna, B.; Preethi, T.; Rajarajeswari, G.R. Single-pot solid-state synthesis of ZnO/chitosan composite for photocatalytic and antitumour applications. *J. Mater. Sci. Mater. Electron.* **2019**, *30*, 21355–21368. [CrossRef]
19. Rahman, M.; Muraleedaran, P.K.; Mujeeb, V.M.A. Applications of chitosan powder with in situ synthesized nano ZnO particles as an antimicrobial agent. *Int. J. Biol. Macromol.* **2015**, *77*, 266–272. [CrossRef]
20. Jayasuriya, A.C.; Aryaei, A.; Jayatissa, A.H. ZnO nanoparticles induced effects on nanomechanical behavior and cell viability of chitosan films. *Mater. Sci. Eng. C* **2013**, *33*, 3688–3696. [CrossRef]
21. Das, K.; Maiti, S.; Liu, D. Morphological, Mechanical and Thermal Study of ZnO Nanoparticle Reinforced Chitosan Based Transparent Biocomposite Films. *J. Inst. Eng. (India) Ser. D* **2014**, *95*, 35–41. [CrossRef]
22. Ummartyotin, S.; Pechyen, C. Physico-Chemical Properties of ZnO and Chitosan Composite for Packaging Material. *J. Biobased Mater. Bioenergy* **2017**, *11*, 183–192. [CrossRef]
23. Rahman, P.M.; Mujeeb, V.M.A.; Muraleedharan, K.; Thomas, S.K. Chitosan/nano ZnO composite films: Enhanced mechanical, antimicrobial and dielectric properties. *Arab. J. Chem.* **2018**, *11*, 120–127. [CrossRef]
24. Indumathi, M.P.; Sarojini, K.S.; Rajarajeswari, G.R. Antimicrobial and biodegradable chitosan/cellulose acetate phthalate/ZnO nano composite films with optimal oxygen permeability and hydrophobicity for extending the shelf life of black grape fruits. *Int. J. Biol. Macromol.* **2019**, *132*, 1112–1120. [CrossRef]
25. Hezma, A.M.; Rajeh, A.; Mannaa, M.A. An insight into the effect of zinc oxide nanoparticles on the structural, thermal, mechanical properties and antimicrobial activity of Cs/PVA composite. *Colloids Surf. A Physicochem. Eng. Asp.* **2019**, *581*, 123821. [CrossRef]
26. Alturki, A.M. Effect of Preparation Method on the Particles Size, Dielectric Constant and Antibacterial Properties of ZnO Nanoparticles and Thin Film of ZnO/Chitosan. *Orient. J. Chem.* **2018**, *34*, 548–554. [CrossRef]
27. Warlimont, H.; Martienssen, W. (Eds.) *Springer Handbook of Materials Data*; Springer Nature: Cham, Switzerland, 2018.
28. Swaroop, K.; Naveen, C.S.; Jayanna, H.S.; Somashekarappa, H.M. Effect of gamma irradiation on DC electrical conductivity of ZnO nanoparticles. In *AIP Conference Proceedings*; AIP Publishing LLC: Melville, NY, USA, 2015; Volume 1665, p. 050100. [CrossRef]
29. Naveen, C.S.; Jayanna, H.S.; Lamani, A.R.; Rajeeva, M.P. Temperature dependent DC electrical conductivity studies of ZnO nanoparticle thick films prepared by simple solution combustion method. In *AIP Conference Proceedings*; AIP Publishing LLC: Melville, NY, USA, 2014.
30. Wang, R.S.; An, J.; Ong, H. *Studies of Interfacial Optical and Electrical Properties on Transparent Dielectrics/ZnO Systems*; Cambridge University Press (CUP): Cambridge, UK, 2006; Volume 928, pp. 35–78.
31. Wang, G.-S.; Wu, Y.-Y.; Zhang, X.; Li, Y.; Guo, L.; Cao, M. Controllable synthesis of uniform ZnO nanorods and their enhanced dielectric and absorption properties. *J. Mater. Chem. A* **2014**, *2*, 8644–8651. [CrossRef]
32. Wang, G.; Deng, Y.; Xiang, Y.; Guo, L. Fabrication of radial ZnO nanowire clusters and radial ZnO/PVDF composites with enhanced dielectric properties. *Adv. Funct. Mater.* **2008**, *18*, 2584–2592. [CrossRef]
33. Rajesh, K.; Crasta, V.; Kumar, N.B.R.; Shetty, G. Effect of ZnO nanofiller on dielectric and mechanical properties of PVA/PVP blend. In *AIP Conference Proceedings*; AIP Publishing LLC: Melville, NY, USA, 2019; Volume 2162, p. 020096.
34. Wang, G.; Deng, Y.; Guo, L. Single-Crystalline ZnO Nanowire Bundles: Synthesis, Mechanism and Their Application in Dielectric Composites. *Chem. A Eur. J.* **2010**, *16*, 10220–10225. [CrossRef]
35. Hemalatha, K.S.; Rukmani, K. Concentration dependent dielectric, AC conductivity and sensing study of ZnO-polyvinyl alcohol nanocomposite films. *Int. J. Nanotechnol.* **2017**, *14*, 961–974. [CrossRef]
36. Efros, A.L.; Shklovskii, B.I. Critical Behaviour of Conductivity and Dielectric Constant near the Metal-Non-Metal Transition Threshold. *Phys. Status Solidi (B)* **1976**, *76*, 475–485. [CrossRef]
37. Bergman, D.J.; Stroud, D. Physical Properties of Macroscopically Inhomogeneous Media. *Methods Exp. Phys.* **1992**, *46*, 147–269. [CrossRef]
38. Díez-Pascual, A.M.; Díez-Vicente, A.L. ZnO-Reinforced Poly(3-hydroxybutyrate-co-3-hydroxyvalerate) Bionanocomposites with Antimicrobial Function for Food Packaging. *ACS Appl. Mater. Interfaces* **2014**, *6*, 9822–9834. [CrossRef] [PubMed]

39. Díez-Pascual, A.M.; Xu, C.; Luque, R. Development and characterization of novel poly(ether ether ketone)/ZnO bionanocomposites. *J. Mater. Chem. B* **2014**, *2*, 3065. [CrossRef]
40. Poddar, M.K.; Sharma, S.; Moholkar, V.S. Investigations in two-step ultrasonic synthesis of PMMA/ZnO nanocomposites by in–situ emulsion polymerization. *Polymer* **2016**, *99*, 453–469. [CrossRef]
41. Pantani, R.; Gorrasi, G.; Vigliotta, G.; Murariu, M.; Dubois, P. PLA-ZnO nanocomposite films: Water vapor barrier properties and specific end-use characteristics. *Eur. Polym. J.* **2013**, *49*, 3471–3482. [CrossRef]
42. Abbas, M.; Buntinx, M.; Deferme, W.; Peeters, R. (Bio)polymer/ZnO Nanocomposites for Packaging Applications: A Review of Gas Barrier and Mechanical Properties. *Nanomaterials* **2019**, *9*, 1494. [CrossRef]
43. Díez-Pascual, A.M.; Díez-Vicente, A.L. High-Performance Aminated Poly(phenylene sulfide)/ZnO Nanocomposites for Medical Applications. *ACS Appl. Mater. Interfaces* **2014**, *6*, 10132–10145. [CrossRef]
44. AboMostafa, H.M.; El Komy, G.M. Enhancement of Structural, Dielectric and Mechanical Properties of Ps: Fe Doped ZnO Based Polymer Nanocomposites. *J. Inorg. Organomet. Polym. Mater.* **2019**, *29*, 908–916. [CrossRef]
45. Kumar-Krishnan, S.; Prokhorov, E.; Ramirez-Cardona, M.; Hernández-Landaverde, M.A.; Zárate-Triviño, D.G.; Kovalenko, Y.; Sanchez, I.C.; Mendez-Nonell, J.; Bárcenas, G.L. Novel gigahertz frequency dielectric relaxations in chitosan films. *Soft Matter* **2014**, *10*, 8673–8684. [CrossRef]
46. Abdelhady, M.M. Preparation and Characterization of Chitosan/Zinc Oxide Nanoparticles for Imparting Antimicrobial and UV Protection to Cotton Fabric. *Int. J. Carbohydr. Chem.* **2012**, *2012*, 1–6. [CrossRef]
47. González-Campos, J.; Prokhorov, E.; Bárcenas, G.; Sanchez, I.C.; Lara-Romero, J.; Mendoza-Duarte, M.E.; Villaseñor, F.; Guevara-Olvera, L. Chitosan/silver nanoparticles composite: Molecular relaxations investigation by dynamic mechanical analysis and impedance spectroscopy. *J. Polym. Sci. Part B Polym. Phys.* **2010**, *48*, 739–748. [CrossRef]
48. Corazzari, I.; Nisticò, R.; Turci, F.; Faga, M.G.; Franzoso, F.; Tabasso, S.; Magnacca, G. Advanced physico-chemical characterization of chitosan by means of TGA coupled on-line with FTIR and GCMS: Thermal degradation and water adsorption capacity. *Polym. Degrad. Stab.* **2015**, *112*, 1–9. [CrossRef]
49. Panomsuwan, G.; Manuspiya, H. A comparative study of dielectric and ferroelectric properties of sol–gel-derived BaTiO3 bulk ceramics with fine and coarse grains. *Appl. Phys. A* **2018**, *124*, 713. [CrossRef]
50. Prokhorov, E.; Luna-Bárcenas, G.; González-Campos, J.B.; Kovalenko, Y.; García-Carvajal, Z.Y.; Mota-Morales, J. Proton conductivity and relaxation properties of chitosan-acetate films. *Electrochim. Acta* **2016**, *215*, 600–608. [CrossRef]
51. Vyshatko, N.P.; Brioso, P.M.; De La Cruz, J.P.; Vilarinho, P.M.; Kholkin, A.L. Fiber-optic based method for the measurements of electric-field induced displacements in ferroelectric materials. *Rev. Sci. Instrum.* **2005**, *76*, 85101. [CrossRef]
52. Fialka, J. Determination of the Piezoelectric Charge Constant D33 Measured by the Laser Interferometer and Frequency Method. In *Annals of DAAAM for 2010 & Proceedings of the 21st International DAAAM Symposium*; DAAAM International: Vienna, Austria, 2010; Volume 21.
53. Sinha, N.; Ray, G.; Bhandari, S.; Godara, S.; Kumar, B. Synthesis and enhanced properties of cerium doped ZnO nanorods. *Ceram. Int.* **2014**, *40*, 12337–12342. [CrossRef]
54. Thakur, P.; Kool, A.; Hoque, N.A.; Bagchi, B.; Khatun, F.; Biswas, P.; Brahma, D.; Roy, S.; Banerjee, S.; Das, S. Superior performances of in situ synthesized ZnO/PVDF thin film based self-poled piezoelectric nanogenerator and self-charged photo-power bank with high durability. *Nano Energy* **2018**, *44*, 456–467. [CrossRef]
55. Satthiyaraju, M.; Ramesh, T. Nanomechanical, Mechanical Responses and Characterization of Piezoelectric Nanoparticle-Modified Electrospun PVDF Nanofibrous Films. *Arab. J. Sci. Eng.* **2019**, *44*, 5697–5709. [CrossRef]
56. Zhao, P.; Wang, S.; Kadlec, A. Piezoelectric and dielectric properties of nanoporous polyvinylidence fluoride (PVDF) films. *Behav. Mech. Multifunct. Mater. Compos. 2016* **2016**, *9800*, 98000. [CrossRef]
57. Singh, H.H.; Khare, N. Flexible ZnO-PVDF/PTFE based piezo-tribo hybrid nanogenerator. *Nano Energy* **2018**, *51*, 216–222. [CrossRef]
58. Li, J.; Zhu, Z.; Fang, L.; Guo, S.; Erturun, U.; Zhu, Z.; West, J.E.; Ghosh, S.; Kang, S.H. Analytical, numerical, and experimental studies of viscoelastic effects on the performance of soft piezoelectric nanocomposites. *Nanoscale* **2017**, *9*, 14215–14228. [CrossRef]

59. Godavarti, U.; Mote, V.; Dasari, M. Role of cobalt doping on the electrical conductivity of ZnO nanoparticles. *J. Asian Ceram. Soc.* **2017**, *5*, 391–396. [CrossRef]
60. Barber, P.; Balasubramanian, S.; Anguchamy, Y.; Gong, S.; Wibowo, A.; Gao, H.; Ploehn, H.J.; Loye, H.-C. Polymer Composite and Nanocomposite Dielectric Materials for Pulse Power Energy Storage. *Materials* **2009**, *2*, 1697–1733. [CrossRef]
61. Kirkpatrick, S. Percolation and Conduction. *Rev. Mod. Phys.* **1973**, *45*, 574–588. [CrossRef]
62. Cai, W.-Z.; Tu, S.-T.; Gong, J.-M. A Physically Based Percolation Model of the Effective Electrical Conductivity of Particle Filled Composites. *J. Compos. Mater.* **2006**, *40*, 2131–2142. [CrossRef]
63. Vo, H.T.; Shi, F.G. Towards model-based engineering of optoelectronic packaging materials: Dielectric constant modeling. *Microelectron. J.* **2002**, *33*, 409–415. [CrossRef]
64. Todd, M.G.; Shi, F.G. Characterizing the interphase dielectric constant of polymer composite materials: Effect of chemical coupling agents. *J. Appl. Phys.* **2003**, *94*, 4551. [CrossRef]
65. Campbell, S.L.; Chancelier, J.P.; Nikoukhah, R. *Modeling and Simulation in Scilab/Scicos with ScicosLab 4.4*; Springer: New York, NY, USA, 2010.
66. Zakaria, A.Z.; Shelesh-Nezhad, K. The effects of interphase and interface haracteristics on the tensile behaviour of POM/CaCO3 nanocomposites. *Nanomater. Nanotechnol.* **2004**, *17*, 1–10.
67. Nunthanid, J.; Laungtana-Anan, M.; Sriamornsak, P.; Limmatvapirat, S.; Puttipipatkhachorn, S.; Lim, L.-Y.; Khor, E. Characterization of chitosan acetate as a binder for sustained release tablets. *J. Control. Release* **2004**, *99*, 15–26. [CrossRef] [PubMed]
68. Mezina, E.A.; Lipatova, I.M.; Losev, N.V. Effect of mechanical activation on rheological and film-forming properties of suspensions of barium sulfate in chitosan solutions. *Russ. J. Appl. Chem.* **2011**, *84*, 486–490. [CrossRef]
69. Aguirre-Loredo, R.Y.; Hernández, A.I.R.; Morales-Sánchez, E.; Gómez-Aldapa, C.A.; Velazquez, G. Effect of equilibrium moisture content on barrier, mechanical and thermal properties of chitosan films. *Food Chem.* **2016**, *196*, 560–566. [CrossRef] [PubMed]
70. Cazón, P.; Vázquez, M.; Velazquez, G. Environmentally Friendly Films Combining Bacterial Cellulose, Chitosan, and Polyvinyl Alcohol: Effect of Water Activity on Barrier, Mechanical, and Optical Properties. *Biomacromolecules* **2019**, *21*, 753–760. [CrossRef] [PubMed]
71. Baxter, S.C.; Burrows, B.J.; Fralick, B.S. Mechanical percolation in nanocomposites: Microstructure and micromechanics. *Probabilistic Eng. Mech.* **2016**, *44*, 35–42. [CrossRef]
72. Padmanabhan, V. Percolation of high-density polymer regions in nanocomposites: The underlying property for mechanical reinforcement. *J. Chem. Phys.* **2013**, *139*, 144904. [CrossRef]
73. Fralick, B.S.; Gatzke, E.P.; Baxter, S.C. Three-dimensional evolution of mechanical percolation in nanocomposites with random microstructures. *Probabilistic Eng. Mech.* **2012**, *30*, 1–8. [CrossRef]
74. Hua, Z.; Shi, X.; Chen, Y. Preparation, Structure, and Property of Highly Filled Polyamide 11/BaTiO3 Piezoelectric Composites Prepared Through Solid-State Mechanochemical Method. *Polym. Compos.* **2017**, *40*, E177–E185. [CrossRef]
75. Choudhury, A. Dielectric and piezoelectric properties of polyetherimide/BaTiO3 nanocomposites. *Mater. Chem. Phys.* **2010**, *121*, 280–285. [CrossRef]
76. Hemeda, O.M.; Tawfik, A.; El-Shahawy, M.M.; Darwish, K.A. Enhancement of piezoelectric properties for [poly(vinylidene fluoride)/barium zirconate titanate] nanocomposites. *Eur. Phys. J. Plus* **2017**, *132*, 333. [CrossRef]
77. Tuff, W.; Manghera, P.; Tilghman, J.; Van Fossen, E.; Chowdhury, S.; Ahmed, S.; Banerjee, S. BaTiO3–Epoxy–ZnO-Based Multifunctional Composites: Variation in Electron Transport Properties due to the Interaction of ZnO Nanoparticles with the Composite Microstructure. *J. Electron. Mater.* **2019**, *48*, 4987–4996. [CrossRef]
78. Kandpal, M.M.; Sharan, C.; Poddar, P.; Prashanthi, K.; Apte, R.; Rao, V.R. Photopatternable nano-composite (SU-8/ZnO) thin films for piezo-electric applications. *Appl. Phys. Lett.* **2012**, *101*, 104102. [CrossRef]
79. Zviagin, A.S.; Chernozem, R.V.; Surmeneva, M.A.; Pyeon, M.; Frank, M.; Ludwig, T.; Tutacz, P.; Ivanov, Y.F.; Mathur, S.; Surmeneva, M.A. Enhanced piezoelectric response of hybrid biodegradable 3D poly(3-hydroxybutyrate) scaffolds coated with hydrothermally deposited ZnO for biomedical applications. *Eur. Polym. J.* **2019**, *117*, 272–279. [CrossRef]

80. Singh, H.H.; Singh, S.; Khare, N. Enhanced β-phase in PVDF polymer nanocomposite and its application for nanogenerator. *Polym. Adv. Technol.* **2017**, *29*, 143–150. [CrossRef]
81. Shen, Z.-Y.; Li, J.-F. Enhancement of piezoelectric constant d33 in BaTiO3 ceramics due to nano-domain structure. *J. Ceram. Soc. Jpn.* **2010**, *118*, 940–943. [CrossRef]
82. Praveen, E.; Murugan, S.; Jayakumar, K. Investigations on the existence of piezoelectric property of a bio-polymer—Chitosan and its application in vibration sensors. *RSC Adv.* **2017**, *7*, 35490–35495. [CrossRef]
83. Hänninen, A.; Sarlin, E.; Lyyra, I.; Salpavaara, T.; Kellomäki, M.; Tuukkanen, S. Nanocellulose and chitosan based films as low cost, green piezoelectric materials. *Carbohydr. Polym.* **2018**, *202*, 418–424. [CrossRef]
84. Van den Ende, D.A.; Bory, B.F.; Groen, W.A.; Van der Zwaag, S. Improving the d33 and g33 properties of 0-3 piezoelectric composites by Dielectrophoresis. *J. Appl. Phys.* **2010**, *107*, 024107. [CrossRef]

© 2020 by the authors. Licensee MDPI, Basel, Switzerland. This article is an open access article distributed under the terms and conditions of the Creative Commons Attribution (CC BY) license (http://creativecommons.org/licenses/by/4.0/).

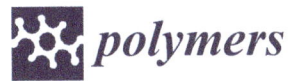

Article

Chitosan-Sulfated Titania Composite Membranes with Potential Applications in Fuel Cell: Influence of Cross-Linker Nature

Andra-Cristina Humelnicu [1], Petrisor Samoila [1,*], Mihai Asandulesa [1], Corneliu Cojocaru [1], Adrian Bele [1], Adriana T. Marinoiu [2], Ada Sacca [3] and Valeria Harabagiu [1,*]

[1] "Petru Poni" Institute of Macromolecular Chemistry, Aleea Grigore Ghica Voda 41A, 700487 Iasi, Romania; humelnicu.andra@icmpp.ro (A.-C.H.); asandulesa.mihai@icmpp.ro (M.A.); cojocaru.corneliu@icmpp.ro (C.C.); bele.adrian@icmpp.ro (A.B.)
[2] National Research and Development Institute for Cryogenics and Isotopic Technologies – ICSI Rm. Valcea, 240050 Ramnicu Valcea, Romania; adriana.marinoiu@icsi.ro
[3] National Research Council of Italy, Institute for Advanced Energy Technologies "Nicola Giordano" (CNR-ITAE), via S. Lucia sopra Contesse 5, 98126 Messina, Italy; ada.sacca@itae.cnr.it
* Correspondence: samoila.petrisor@icmpp.ro (P.S.); hvaleria@icmpp.ro (V.H.); Tel.: +40-232-217454 (P.S. & V.H.)

Received: 30 March 2020; Accepted: 9 May 2020; Published: 14 May 2020

Abstract: Chitosan-sulfated titania composite membranes were prepared, characterized, and evaluated for potential application as polymer electrolyte membranes. To improve the chemical stability, the membranes were cross-linked using sulfuric acid, pentasodium triphosphate, and epoxy-terminated polydimethylsiloxane. Differences in membranes' structure, thickness, morphology, mechanical, and thermal properties prior and after cross-linking reactions were evaluated. Membranes' water uptake capacities and their chemical stability in Fenton reagent were also studied. As proved by dielectric spectroscopy, the conductivity strongly depends on cross-linker nature and on hydration state of membranes. The most encouraging results were obtained for the chitosan-sulfated titania membrane cross-linked with sulfuric acid. This hydrated membrane attained values of proton conductivity of 1.1×10^{-3} S/cm and 6.2×10^{-3} S/cm, as determined at 60 °C by dielectric spectroscopy and the four-probes method, respectively.

Keywords: chitosan; sulfated titania; cross-linking; polyelectrolyte composite membranes

1. Introduction

Polymer electrolyte membranes (PEMs) play a key role in fuel cells as they transport charges between electrodes and prevent fuel leaks [1,2]. Nafion™ PEMs are recognized for their high proton conductivity, good mechanical properties, and chemical and electrochemical stability [3]. However, Nafion membranes are extremely expensive, their preparation and use are hazardous for the environment, and they are also characterized by fuel crossover and low durability [4–6]. Other shortcomings of these membranes consist in water management problems, need for a full humidification since their high proton conductivity is strictly linked to the H_3O^+ hydration, and, as a consequence, the optimal operative temperature is limited below 100 °C [7].

In this respect, biopolymer-based membranes are more and more considered as strong candidates to develop cheaper and more environmentally friendly PEMs. Recently, chitosan membranes were proposed as substitutes Nafion PEMs [8–10]. In spite of promising results, there is still research effort required to improve the mechanical strength and the conductivity of chitosan membranes [5]. The solutions often proposed were related to their doping with inorganic fillers (e.g., solid superacids known for their high proton conductivities and hygroscopic properties) [11] and/or chemical modification of the polysaccharide by chemical cross-linking [12]. The cross-linking is the one of the most efficient ways to improve the characteristics of pristine chitosan membranes, by ameliorating their thermal and mechanical properties, as well as their water uptake capacities, with direct impact on the conductivity increasing [12]. However, to the best of our knowledge, little attention was paid on the influence of the cross-linker nature on the properties of chitosan-solid superacid hybrid PEMs.

The objective of this study was to develop novel chitosan-based composite membranes and to follow their structure, morphology, mechanical, and thermal properties, as a function of the cross-linker nature. To this end, we selected sulfated titanium dioxide as inorganic filler and sulfuric acid, pentasodium triphosphate and epoxy-terminated polydimethylsiloxane as cross-linkers, the last one not reported for the preparation of PEMs considered for fuel cell application. Typical tests used to describe PEM performances—such as water uptake, chemical stability and conductivity behavior—were also carried out.

2. Materials and Methods

2.1. Materials

Chitosan (CS) with molecular weight of 290 kDa (determined by a viscometric study and according to Equation (S1) on Supplementary Materials) and 82% degree of deacetylation (determined by ^{1}H NMR (Figure S1 on Supplementary Materials), TiO_2 (anatase, particle size < 25 nm, 99.7% purity), bis(glycidyloxypropyl)-terminated polydimethylsiloxane of M_n = 980 Da (PDMS), acetic acid, sulfuric acid 98%, pentasodium triphosphate (TPP) and sodium hydroxide were purchased from Sigma Aldrich (Taufkirchen, Germany). Methanol (purris p.a.) was supplied by Chemical Company (Iasi, Romania). All reagents were of analytical grade and used without further purification.

Sulfated TiO_2 (TS) was obtained by adapting the methods described by Li et al. [13] and Ayyaru et al. [14]. For this purpose, 22.5 mg of TiO_2 nanoparticles were dispersed in a solution containing 10 mL of methanol and 5 mL of 1 M H_2SO_4 through sonication for 30 min in a Emmi 12HC bath. The suspension was then centrifuged, washed with water, and the solid was dried in oven at 105 °C for 6 h to obtain the sulfated TiO_2 sample (TS) with a content of 0.354 mmol sulfate groups/g (evaluated according to Equation (S2) on Supplementary Materials).

2.2. Preparation of Membranes

2.2.1. Preparation of Chitosan-Sulfated Titania Composite Membranes (CS-TS)

CS-TS composite membranes were prepared using 5 wt % of sulfated TiO_2 relative to the amount of chitosan. In the first step, a stock solution of chitosan (3% w/v) was prepared by dissolving the polysaccharide in 2% acetic acid solution. Subsequently, precisely determined amount of TS was dispersed in 5 mL water, sonicated for 15 min and added over 15 mL of chitosan solution, under continuous stirring at 600 rpm for 30 min. The final mixture was poured into 9.6 cm diameter Petri dish and dried at 30 °C for 48 h (up to constant weight) to obtain CS-TS membrane.

2.2.2. Membrane Cross-Linking by Sulfuric Acid (HS) and by Pentasodium Tripolyphosphate (TPP)

The cross-linking was carried out by adapting already published procedures. Thus, dried CS-TS membranes were immersed into 1 M H_2SO_4 solution (pH = 0.38) for 15 min [15] or in 2% TPP for 2 h solution (pH = 8.64) [16]. Subsequently, the membranes were washed with distilled water and dried at room temperature for 48 h until constant weight was achieved. The resulted composite membranes, with average thicknesses of about 70 μm (from SEM images of the cross-sections), were labeled as CS-TS-HS and CS-TS-TPP, respectively.

2.2.3. Cross-Linking by Bis(glycidyloxypropyl)-Terminated Polydimethylsiloxane (PDMS)

The PDMS cross-linked composite membrane (CS-TS-PDMS) was obtained adapting a procedure previously described for the preparation of PDMS modified chitosan [17]. Shortly, PDMS (NH_2/epoxy = 1/1 molar ratio) was added in situ to the CS/TS dispersion prepared as described in the previous paragraph and the mixture was stirred at 40 °C until complete homogenization. After casting into Petri dishes and oven drying at 30 °C, the obtained membrane was washed with distilled water and dried at room temperature for 48 h until constant weight was attained. The average thickness of the dried membrane was of around 130 µm, as determined from SEM image of the membrane cross-section.

2.3. Materials Characterization

Structural characterization of TS and pristine chitosan (CS) intermediates as well as of the composite membranes was performed by FTIR spectroscopy using a Bruker Vertex 70 spectrometer (Bruker Optics, Ettlingen, Germany) and KBr pellet method. The surface and cross-section morphology and elemental composition of the composite membranes were investigated by scanning electron microscopy using an (ESCM) Quanta 200 device (SEM, FEI Company, Brno, Czech Republic)) coupled with energy dispersive X-ray (EDX) system (EDAX, Mahwah, NJ, USA). The cross-sections were obtained by tearing the liquid nitrogen frozen membranes.

The thermal properties of the samples were studied under nitrogen atmosphere in the temperature range of 20–700 °C, with a heating rate of 10 °C/min on a Jupiter thermal analysis system TG-DSC Model STA449F1 (NETZSCH, Selb, Germany). The strain-stress curves were obtained by using a two-column Instron Model 3365 device equipped with a 500 N cell force (Instron, Norwood, MA, USA). In this respect, dumb-bell shaped samples (L = 5 cm; l = 4 mm; active length = 3.5 cm) were cut using a press and were tested for uniaxial stress–strain curves with a 50 mm/min elongation speed.

The kinetics of water uptake was evaluated at 25, 60 and 80 °C. In this respect, 1 cm^2 of membrane samples were dried until constant weight (W_{dry}). Subsequently, the samples were immersed in 30 mL distilled water at different temperatures. At regular time intervals, the membranes were extracted, the water excess was removed by buffering the samples on filter paper and weighting (W_{wet}). The water uptake ($WU\%$) was calculated for each sample using the formula

$$WU(\%) = \left((W_{wet} - W_{dry})/W_{dry}\right) \times 100 \tag{1}$$

The oxidative stability of the membranes was studied by immersing 1 cm^2 dry samples in 10 mL of freshly prepared Fenton reagent (3 vol % H_2O_2 solution containing 4 ppm $Fe(SO_4)_2 \cdot 7H_2O$) at room temperature. The membranes were extracted from the solution after 1 h or after 24 h, dried and weighed in order to determine the weight loss of the sample as a function of time.

Dielectric spectroscopy measurements were performed with a broadband dielectric spectrometer (Novocontrol, Montabaur, Germany) equipped with a high-resolution Alpha-A analyzer and a Quatro Cryosystem temperature controller. Complex dielectric permittivity spectra were recorded under isothermal conditions by applying an alternating electrical field of 1 V in a broad range of frequency (0.1–10^7 Hz). The composite membranes were sandwiched between two gold-plated flat electrodes and measurements were carried out under pure nitrogen, preventing the moisture from environment. The measurements were performed on dry (samples kept at 80 °C into a vacuum oven for 12 h) and hydrated membranes (obtained by immersing the samples in distilled water for 1 h at room temperature prior to dielectric spectroscopy measurements). The dielectric spectra were collected in steps of 5 °C with 0.1 °C stability and high reproducibility, at temperatures from 0 to 160 °C and from 0 to 80 °C for dry and hydrated sample, respectively.

The proton conductivity (PC) measurement on CS-TS-HS membrane was carried out in the longitudinal direction (in-plane) by the four-probes method at two different temperatures (30 and 60 °C), at fully humidification level (100% RH), P = 1 atm, using a hydrogen flux of 1000 sccm, as suggested by the supplier, and DC current by using a PTFE commercial BT-112 Bekktech conductivity cell (Bekktech,

LLC acquired by Scribner Associates Inc. in 2011, Southern Pines, NC, USA) with a 5 cm² fixture hardware by Fuel Cell Technologies, Inc. (BekkTech product no. ACC-920). The cell was connected to a test station and a potentiostat-galvanostat (AMEL mod.551) [18,19]. A membrane sample of about 2.5 × 0.52 cm² was cut by a sample punch (BekkTech product no. ACC-960) and its size was measured through a width measurement tool (BekkTech product no. ACC-940) with a magnification of 11× and a reticule with 0.1 mm gradients. The thickness was measured by a Mitutoyo electronic gauge. The membrane was assembled in the cell and placed in contact with the two fixed platinum electrodes. By an indirect imposition of the current, a voltage drop between the two fixed electrodes was measured. The electrical resistance values were obtained by extrapolating the data from the plot of current as a potential function. At the end, the PC (σ, S·cm^{-1}) was calculated using the formula

$$\sigma = L/R \cdot W \cdot T \tag{2}$$

where L = 0.425 cm, fixed distance between the two Pt electrodes; R = resistance in Ω; W = sample width in cm; T = sample thickness in cm. The measurement and cell set-up are detailed in Figure S2.

3. Results and Discussions

3.1. Membrane Preparation

CS-TS membrane of a content of 5 wt % TS filler was prepared by simple mixing the components and drying. Ionic interactions between Lewis acid sites on filler surface and amino basic groups of CS were proposed to stabilize the inorganic particles into CS matrix [13] (Scheme 1).

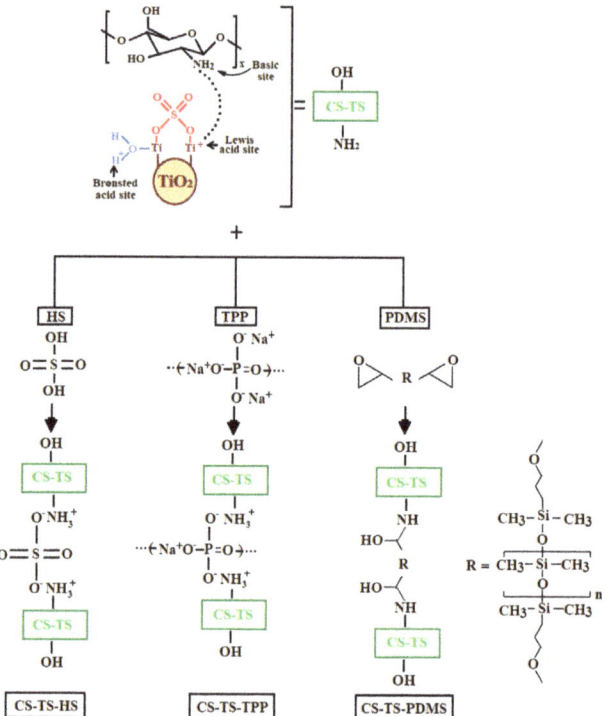

Scheme 1. Preparation of CS-based composite membranes.

Three different agents, such as sulfuric acid (HS), pentasodium tripolyphosphate (TPP), and bis(glycidoxypropyl)-terminated polydimethylsiloxane (PDMS) were further used for cross-linking and to study their influence on the structure and the properties of the composite membranes. The first two cross-linkers provide a well-known electrostatic interactions between protonated amino groups of CS and cross-linker anions [15], while PDMS undergoes covalent cross-linking through the well-known reaction between CS amino groups and epoxy units attached to the siloxane chains [17] (Scheme 1). Based on its intrinsic properties (high hydrophobicity, very low T_g, thermal stability up to 300 °C, relatively low variation of its properties with temperature [20], PDMS cross-linker is expected to provide to the membrane a higher flexibility and lower shrinkage during drying, as well as a controllable hydrophilic–hydrophobic balance.

3.2. FTIR Characterization

The structure of the prepared materials was first assessed through FTIR spectroscopy. Figure 1 compares the normalized FTIR spectra of TS and CS intermediates with those of the composite membranes. Apart from the absorption band at 1633 cm^{-1} (deformation vibration, adsorbed water) and of a broad band in the range of 879–409 cm^{-1} (Ti–O), in the spectrum of TS, the presence of four other absorption bands between 1243 and 967 cm^{-1} (1243 and 1141 cm^{-1}, asymmetric and symmetric stretching vibrations of S=O groups; 1055 cm^{-1} and 967 bands, asymmetric and symmetric stretching vibrations of S–O units) confirms the sulfating process and indicates a bidentate coordination of sulfate groups to Ti atoms, as previously stated [13,21].

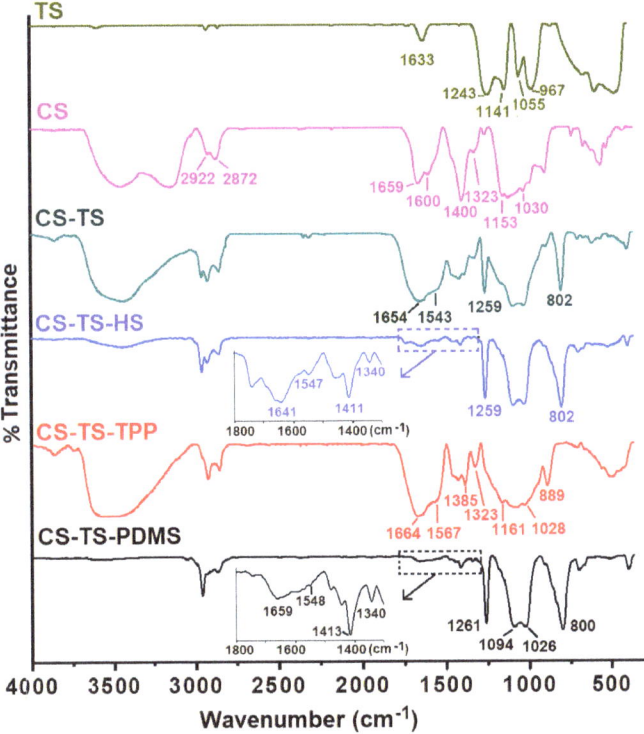

Figure 1. FTIR spectra of sulfated TiO$_2$ (TS), pristine chitosan (CS) and CS-TS, CS-TS-HS, CS-TS-TPP, and CS-TS-PDMS composite membranes.

For CS sample the characteristic absorption bands appear in the range 4000–800 cm^{-1} [22]. Thus, broad bands between 3500–3100 cm^{-1} (O–H and N–H stretching vibration) are visible in the CS spectrum, while asymmetric and symmetric stretching vibrations of C–H groups are located at 2922 and 2872 cm^{-1}, respectively. Chitosan shows also characteristic bands at 1659 cm^{-1} (C=O, amide I), 1600 cm^{-1} (N–H in plane deformation, primary amine), 1400 cm^{-1} (N–H deformation, amide II) and 1323 cm^{-1} (C–N stretching vibrations, amide III). In the range 1153–1030 cm^{-1} the bands were attributed to C–O–C groups [23].

Modified FTIR spectra as compared to those of the pristine TS and CS components are observed for the composite membranes confirming the presence of these components in the membrane structure and the interaction of the cross-linker with chitosan matrix.

Thus, the presence of acidic groups determines the protonation of amino groups of chitosan and the shifting of the primary amine absorbance from 1600 cm^{-1} in CS spectrum to 1543 and 1547 cm^{-1}, respectively in CS-TS and CS-TS-HS spectra (see insert) [15]. Moreover, the interaction of sulfate groups with chitosan determines shifting of amide I and amide III bands to lower (decreased bond strength), respectively higher wavenumbers (increased bond strength). Both composite membranes also show strong bands attributed to the sulfate (1259 cm^{-1}) and to Ti–O–Ti vibrations (superposed on the NH$_3^+$ rocking vibrations [24,25] and located at 802 cm^{-1}).

The successful cross-linking of CS-TS membrane with TPP is confirmed by the shifting of the primary amine absorbance to 1567 cm^{-1} due to its protonation and interaction with the phosphate groups as well as the presence of the P–O–P bridge asymmetric stretching vibration at 889 cm^{-1} [26]. The other phosphate characteristic bands are overlapped on C–O–C vibrations of CS in the range of 1161–1028 cm^{-1} and different effect of phosphate groups on amide bands as compared to that of sulfate groups is evidenced, i.e., the increase and decrease of amide I, amide II bonds strength respectively, while amide III absorption remain unchanged.

In the FTIR spectrum of the CS-TS-PDMS membrane, the covalent linking through the reaction of amino groups of CS and epoxy units attached to the siloxane chains, as well as the increasing of membrane hydrophobic character induced by siloxane component determines a notable diminishing of the absorbance bands between 3500–3100 cm^{-1}. Moreover, strong bands characteristic to siloxane moiety (1261 and 800 cm^{-1} Si–CH$_3$; 1094–1026 cm^{-1}, Si–O–Si) [17], partially covering the bands of CS are also visible in the spectrum.

One should also mention the diminishing of the characteristic bands of CS in the region 1670–1300 cm^{-1} for both CS-TS-HS and CS-TS-PDMS samples due to the effect of the strong Si–O–Si (1100–1000 cm^{-1}) and sulfate (1259 cm^{-1}) bands on shorter bands in normalized spectra. (see inserts).

3.3. SEM Characterization

Representative SEM images of membranes recorded for membrane cross-sections are shown in Figure 2. From the analysis of SEM micrographs, one may observe that all prepared membranes are dense, with no detectable interconnected pores. TiO$_2$ nanoparticles are quite well dispersed throughout the cross-section of membranes for all studied materials. In addition, CS-TS-PDMS membrane shows a phase separated morphology, with distinct domains at micro scale level, as a consequence of strong incompatibility between hydrophilic chitosan and hydrophobic polysiloxanes sequences [20]. Otherwise, according to the literature, siloxanes polymers tend to migrate at the surface exposed to air [27]. In this respect, elemental analysis was performed on CS-TS-PDMS membrane surfaces and in the cross-section (Figure 3). The EDX data confirm the presence of Si in higher concentrations at air exposed interface compared to polymer/glass interface. Nevertheless, one may notice that most of Si atoms are mainly concentrated in cross-section.

As one may also see from Figure 2, the thicknesses of the cross-linked membranes (about 70 μm for CS-TS-HS and CS-TS-TPP samples and about 130 μm for CS-TS-PDMS membrane) are larger as compared to CS-TS pristine membrane (about 30 μm), due to the incorporation of the cross-linking agents. Similar results were obtained on poly(styrene–2-vinylpyrridine) films crosslinked by quaternization of pyridine units with diiodobutane [28]. The SEM micrographs recorded for the membrane surfaces (Figure S3), confirm that all the obtained membranes are dense, with uniform dispersion of TiO_2.

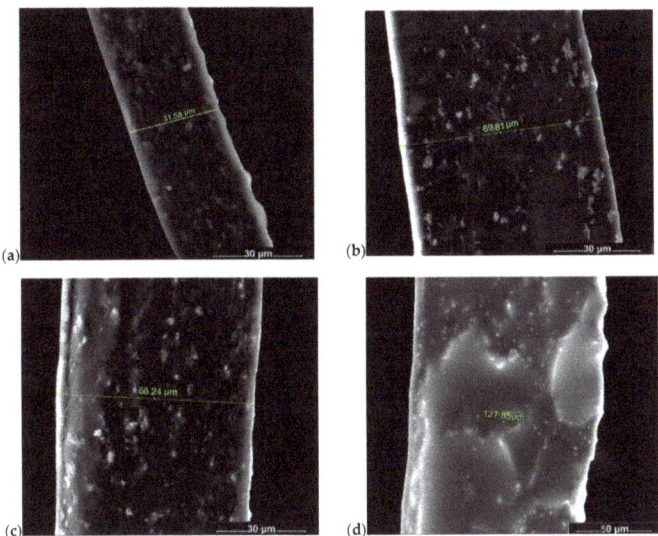

Figure 2. Representative cross-section SEM images of (**a**) CS-TS, (**b**) CS-TS-HS, (**c**) CS-TS-TPP, and (**d**) CS-TS-PDMS composite membranes.

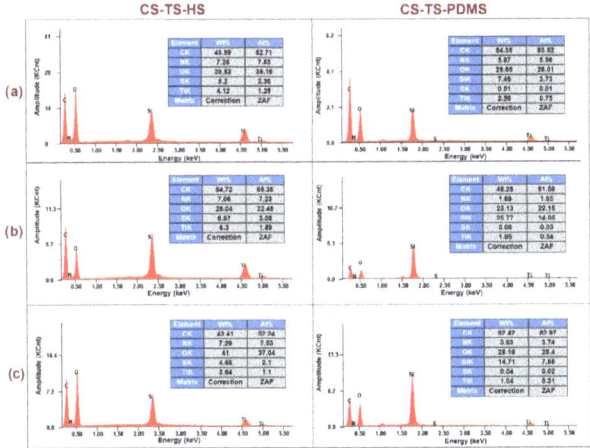

Figure 3. EDX spectra and elemental compositions of CS-TS-HS (**left**) and CS-TS-PDMS (**right**) composite membranes (**a**) polymer/glass interface, (**b**) cross-section, and (**c**) polymer/air interface.

EDX analysis for CS-TS-HS and CS-TS-PDMS Figure 3) also showed an average content of nitrogen of 7.2%, respectively 3.7%, confirming the presence of CS in these samples that showed small amide absorptions in their FTIR spectra.

3.4. Thermogravimetric Analysis

Figure 4 presents the thermogravimetric (TG) and thermogravimetric derivative (DTG) curves for pristine CS and composite membranes registered in nitrogen atmosphere, between 20 and 700 °C. CS degrades through a complex mechanism involving deacetylation followed by dehydration, deamination, and depolymerization processes. At temperatures lower than 300 °C, two main weight losses are noticed, centered at T_{max} = 73 °C and 300 °C, respectively, as also reported previously by others [29]. The first stage (T_{max} = 73 °C) is attributed to the evaporation of water and residual acetic acid solvent and the second stage (T_{max} = 300 °C) is concerned with degradation of chitosan chains. At temperatures higher than 300 °C, the sample continuously degrades up to a residual percentage weight of 34% at 700 °C.

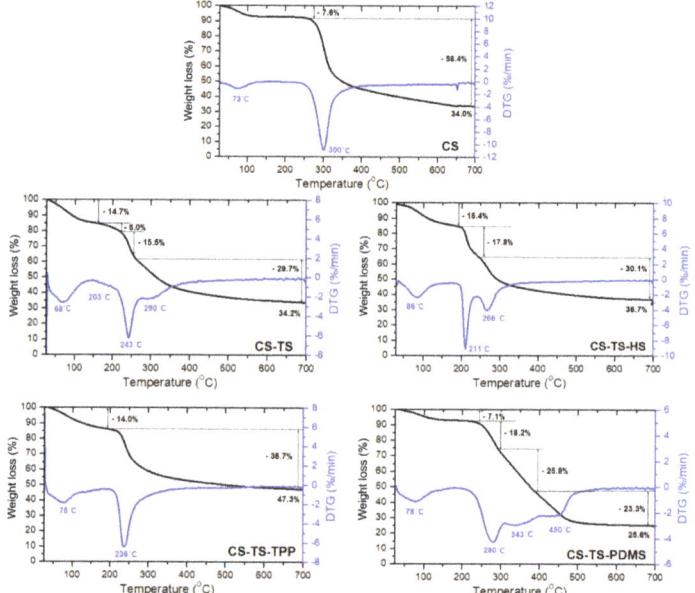

Figure 4. TG/DTG curves of CS-TS, CS-TS-HS, CS-TS-TPP, and CS-TS-PDMS composite membranes compared with pristine chitosan (CS).

The thermal decomposition of composite membranes strongly depends on the nature of cross-linking agent. Similarly to CS, all of them showed water solvent and acetic acid traces evaporation phenomena with T_{max} between 56 and 86 °C. The presence of acidic groups either linked to TS or coming from HS and TPP cross-linkers reduces the stability of all composite membranes, the second decomposition T_{max} being about 20–90 °C lower than the value corresponding to pristine CS. However, comparable weight losses at 700 °C are observed for CS, CS-TS and CS-TS-HS samples, while lower and higher weight losses are registered for CS-TS-TPP and CS-TS-PDMS membranes, respectively. Moreover, multi-stage decomposition behavior is observed for composite membranes.

CS-TS membrane shows the second major weight loss with two maxima at 203 °C, attributed to the loss of sulfate groups, and 243 °C, attributed to the decomposition of amorphous part of CS matrix. The CS more crystalline domains are decomposing at slightly lower T_{max} (290 °C) as compared to pristine CS. CS-TS-HS membrane presents a second stage of decomposition at T_{max} = 211 °C (sulfate groups) and a third one (CS chains) at T_{max} = 266 °C due to the action of higher amounts of sulfate groups on the polysaccharide structure. CS-TS-TPP membrane showed only a second stage degradation with T_{max} of 237 °C. One may notice that this step occurs at a temperature close to the loss of the sulfate groups. On the other hand, literature data indicate that TPP cross-linked chitosan degrades at lower temperatures than pure chitosan (around 230 °C) and this behavior is explained by a decrease in crystallinity of the polysaccharide following crosslinking with TPP [30]. Thus, at this stage, the degradation of both sulfate groups and polymer occurs. Apart the first step of water evaporation, the thermal degradation of the CS-TS-PDMS sample is more complex and involves simultaneous decomposition of the functional sulfate, hydroxyl and alkyl amines units (at 280 °C) and of CS and PDMS chains at 342 and 450 °C, respectively Note that all prepared membranes are thermally stable to relatively high temperatures (above 200 °C).

3.5. Mechanical Properties

To study the mechanical properties of the produced membranes, typical mechanical tests were performed by recording the stress–strain profiles (Figure 5). The mechanical properties were ascertained in terms of fracture strain, tensile stress, and Young's modulus (Table 1).

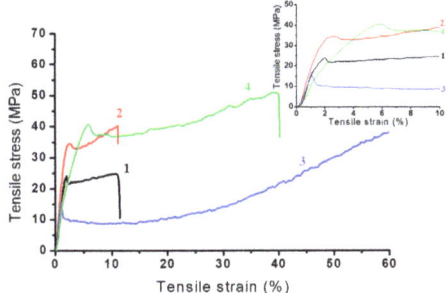

Figure 5. Mechanical stress–strain profiles at high and low (insert) tensile strain values of CS-TS (**1**), CS-TS-HS (**2**), CS-TS-TPP (**3**), and CS-TS-PDMS (**4**) composite membranes.

Table 1. Mechanical properties, water uptake, and stability in oxidative environment and of chitosan-sulfonated titania composite membranes.

Membrane	Mechanical Properties			Water Uptake (%) (24 h)			Weight Loss (%) in Fenton Reagent	
	Tensile Strain (%)	Tensile Stress (MPa)	Young's Modulus [2] (GPa)	25 °C	60 °C	80 °C	1 h	24 h
CS-TS	10.6	24.9	1.54	dissolution			dissolution	
CS-TS-HS	11.8	40.2	1.87	184	172	163	10	15
CS-TS-TPP	-[1]	15.3	1.46	170	121	135	8	22
CS-TS-PDMS	39.7	50.9	1.01	118	88	92	9	30

[1] Sample slips in gripping, no break occurred. [2] Calculated at 1% strain.

As expected, membrane mechanical properties were influenced by the cross-linker nature. All samples present stress–strain curves specific to plastic materials with a clear elastic domain at low strains, followed by a yield strength and plastic deformation. Due to the nature of the siloxane bond (highly flexible), the yield strength of CS-TS-PDMS shifted to larger strains. Moreover, the cross-linking of CS-TS membrane with TPP resulted in a tougher material, since no strain break occurred. These observations were corroborated by the data presented in Table 1.

The cross-linking process of composite membranes affects the calculated values for fracture strain, tensile stress, and Young's modulus. Thus, compared to CS-TS sample, one may observe that the elongation at break is higher for the cross-linked membranes. It should be noted that the same trends were found for the tensile stress values, except for CS-TS-TPP sample. One may also notice that the cross-linking processes decrease Young's modulus values whatever the cross-linking agent, except the membrane cross-linked with sulfuric acid, but its values are higher than 1 GPa for all samples.

3.6. Water Uptake Capacity of Composite Membranes

Sufficient water content is essential in the operation of fuel cells determining the membrane performance, stability, and durability [31]. As expected, CS-TS sample was completely dissolved in water after less than 5 min. Water uptake kinetics curves of cross-linked membranes are given in Supplementary Materials (Figure S4) and the values observed after 24 h are listed in Table 1. All cross-linked membranes revealed very high water uptake capacities for all temperatures considered, with values ranging from 88% (for amphiphilic CS-TS-PDMS sample at 60 °C) to 184% (for CS-TS-HS sample at 80 °C). Moreover, water uptake mainly occurred in the first minutes of experiment. Generally, the cross-linked membranes exhibit superior capabilities at 25 °C compared to 60 and 80 °C. These findings are in good agreement with the literature and confirm that the moisture adsorbed by chitosan films decrease with the increase of temperature [32].

3.7. Chemical Stability of Composite Membranes

The oxidative stability of membranes are often used in the evaluation of PEMs [33]. In this respect, the membranes were challenged with freshly prepared Fenton's reagent (Table 1). The unmodified membrane (CS-TS) was completely dissolved in Fenton's reagent, while the cross-linked membranes were relatively stable. The resistance to oxidation of the membranes decreased in 24 h from weight loss values of up to 10 wt % in 1 h, to 15, 22, and 30 wt % for the samples cross-linked with HS, TPP, and PDMS, respectively, the CS-TS-HS membrane showing the best oxidation resistance.

3.8. Broadband Dielectric Spectroscopy

3.8.1. Overall Dielectric Behavior of Dry Membranes

Figure 6 displays the evolution of dielectric constant (ε') and dielectric loss (ε'') with frequency for dry CS-TS-HS membrane, as a representative example. The spectra corresponding to dry CS-TS, CS-TS-TPP, and CS-TS-PDMS samples are found on Supplementary Materials (Figure S5).

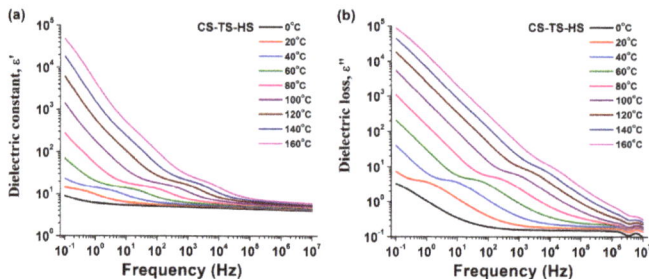

Figure 6. Evolution of the dielectric constant (**a**) and of the dielectric loss with frequency (**b**) for dry CS-TS-HS composite membrane.

As is generally known, the dielectric constant is related to the orientation of chemical dipoles in the direction of an alternating electrical field. ε' diminishes gradually with increasing frequency, since the dipoles can no longer follow the oscillations of alternative field [34]. According to Figure 6a, the CS-TS-HS membrane provided a high dielectric constant in the considered integral frequency range, revealing an intense dipolar activity. The strong decrease of the dielectric constant, especially at low frequencies is an effect of ionic polarization induced by the sulfate groups [34]. Moreover, the magnitude of ε' increased with temperature due to increased mobility of polymer segments having a dipole moment.

The dielectric loss parameter comprises the dissipated energy for the dipole alignment motions and the energy required to move ions in response to the alternating electrical field. As a consequence, both the polarization and the electrical conductivity signals are observed. In Figure 6b, the dielectric loss curves revealed a linear evolution with frequency, especially at low frequencies and high temperatures, with a slope close to −1, which is characteristic for the ionic conductivity-type signal. In fact, the high dielectric signals and their gradual decline with increasing frequency are a result of ionic polarization in the polymer membrane matrix [34].

The electrical conductivity, σ (S/cm), is related to the dielectric loss and was further estimated with the relation (3) [35]

$$\sigma = 2\pi \, \varepsilon_0 f \, \varepsilon'' \tag{3}$$

where ε_0 is the permittivity of the free space and f is the applied electric field frequency.

Figure 7 displays the behavior of conductivity with frequency at selected temperatures from 0 to 160 °C for dry membranes. For CS-TS-HS sample (Figure 7a), at low temperature (0 °C), the conductivity exhibits an approximately linear evolution with log frequency and is generally attributed to electronic-type conduction of the bulk membrane [36]. At higher temperatures, deviations from linearity are observed, especially at low frequencies. This region appears in the same frequency range with the linear-type behavior of dielectric constant and dielectric loss. The observed correspondences were previously reported by Pochard et al. [37]. According to literature, the low frequency-independent conductivity plateau could be attributed to the transport of protons through the polymer membrane [36]. Additionally, one may observe an increasing step in $\sigma(f)$ spectra that shifts progressively to higher frequencies with temperature. This particular signal called as Maxwell–Wagner–Sillars (MWS) polarization generally appears in heterogeneous systems, at the interface between the components with different dielectric constant (polar sulfate groups within the chitosan matrix) [34,38,39]. As shown in Figure 7a, one may notice that the conductivity values at a frequency of 1 Hz are ranging from 5.9× 10^{-13} S/cm at $T = 0$ °C to 8.5 × 10^{-9} S/cm at $T = 160$ °C. The relatively low values correspond to the conductivity of the bulk material and reveal the dielectric-type of the dry CS-TS-HS sample [4]. The $\sigma(f)$ spectra of dry CS-TS, CS-TS-TPP, and CS-TS-PDMS systems are presented in Supplementary Materials, Figure S6 and reveal similar dielectric behavior.

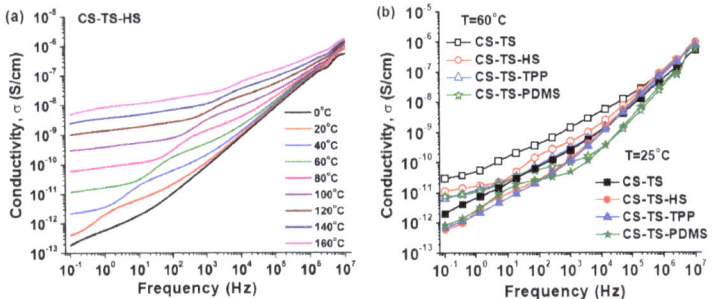

Figure 7. Evolution of the measured conductivity with frequency and temperature for dry membranes: (a) CS-TS-HS membrane (b) comparison of conductivity behavior of all prepared membranes.

The high values of dielectric constant and dielectric loss, especially at low frequencies, suggests the possibility of the use of these materials as PEMs suitable for fuel cell application in a considerable temperature range [36]. Moreover, the frequency evolution of dielectric loss with the slope of −1 could suggest that the segmental dynamics controls the conductivity signal [40]. Since the glass transition of polymer membranes appears above 210 °C, the segmental relaxation should corresponds to the side chain movements from chitosan together with the attached acid groups. In this respect, a comparative evolution of conductivity as function of frequency for all considered dry membranes is displayed in Figure 7b. At lower temperatures (25 °C), the conductivity is mostly electronic, with a small contribution of ionic conductivity localized in the low frequency spectral region (more evident for CS-TS-TTP sample) and generally assigned with proton transfer through different sulfate acid sites.

Furthermore, the σ(f) profiles of cross-linked chitosan membranes furnishes similar behavior probably due to reduced mobility of active sites. The CS-TS-HS, CS-TS-TPP, and CS-TS-PDMS composite membranes presented lower proton conductivity than the CS-TS membrane since the cross-linking process restrict the mobility of polymer segments, hindering the transport of charges [34].

As seen in Figure 7b, at higher temperature (60 °C), the plateau region of proton conductivity is enlarged to higher frequencies, revealing an increased mobility of charge carriers. In this temperature region, the magnitude of σ(f) spectra for CS-TS is still higher than those of cross-linked membranes, due to restrictions imposed by the membrane network.

The effects of cross-linking with different agents are shown in Table 2, where the proton conductivity values are extracted from σ(f) spectra of the prepared membranes, choosing the frequency of 1 Hz.

Table 2. Values of conductivity at various frequencies for dry and hydrated chitosan-sulfated titania membranes.

Membrane	Conductivity, σ (S/cm) at a Frequency of 1 Hz for Dry Membranes			Conductivity, σ (S/cm) at Low and High Frequencies for Hydrated Membranes			
				$f = 1$ Hz		$f = 10^6$ Hz	
	25 °C	60 °C	100 °C	25 °C	60 °C	25 °C	60 °C
CS-TS	6.6×10^{-12}	5.2×10^{-11}	3.3×10^{-9}	-	-	-	-
CS-TS-HS	2.2×10^{-12}	1.7×10^{-11}	3.9×10^{-10}	2.5×10^{-6}	8.1×10^{-6}	2.1×10^{-3}	1.1×10^{-3}
CS-TS-TPP	1.8×10^{-12}	1.2×10^{-11}	7.8×10^{-11}	5.7×10^{-8}	5.7×10^{-8}	5.7×10^{-5}	4.5×10^{-5}
CS-TS-PDMS	2.6×10^{-12}	1.1×10^{-11}	7.4×10^{-11}	7.0×10^{-8}	1.9×10^{-7}	1.4×10^{-5}	3.1×10^{-5}

As seen from Table 2, at 25 °C, σ values of all dry membranes are of the order of 10^{-12} S/cm and are increasing to 10^{-11} S/cm at 60 °C. At 100 °C (see also Figure S7), the values of the conductivity for dry CS-TS-TPP and CS-TS-PDMS are surprisingly similar, revealing that the TPP and PDMS cross-linking agents have comparable effects on the transport of protons through the chitosan membrane. This finding is somewhat surprising since the acid sites of pentasodium tripolyphosphate were expected to enhance the proton conductivity of the membrane. By contrast, the conductivity of CS-TS-HS is at least one order higher than those of CS-TS-TPP and CS-TS-PDMS indicating that the protonation of the amine groups of chitosan by sulfuric acid promotes the protonic conductivity. Thus, among various types of cross-linking agents, the sulfuric acid conducts to superior protonic conductivity at low humidity and high temperatures.

3.8.2. Influence of Water Absorption on the Protonic Conductivity of Membranes

The hydrated membranes were obtained by immersing the samples in distilled water for 1 h at room temperature prior to dielectric spectroscopy measurements. The hydrated CS-TS membrane was not examined, because the water incorporation completely damaged the sample. The dielectric spectra of membranes were collected at different temperatures from 0 to 60 °C, in steps of 5 °C. No reliable spectra were obtained at temperatures higher than 60 °C due to quick evaporation of water.

Previous studies have concluded that the hydrophilic sulfonic acid groups from the membranes is primarily responsible for water uptake [2]. As stated above, the water uptake of the studied membranes varies with the cross-linker nature. The resulting conductivity dependences as function of frequency and temperature are shown in Figure 8. σ(f) spectra similar to that corresponding to hydrated CS-TS-HS membrane were obtained for hydrated CS-TS-TPP and CS-TS-PDMS samples (Figure S8). For CS-TS-HS hydrated membrane (Figure 8a), one may observe dramatically increased σ(f) magnitudes in the integral frequency range as compared to the dry sample (Figure 7a). Therefore, the membrane saturated with water exhibits conductivity values with about 4 orders of magnitude higher than the corresponding dry membrane, thus suggesting that the conductivity is strongly enhanced by proton migration between polar water molecules. Similar differences between dry and hydrated membranes were previous reported for the standard Nafion 117 [4]. Likewise, the σ(f) profiles reveal an additional frequency independent conductivity plateau located at high frequencies.

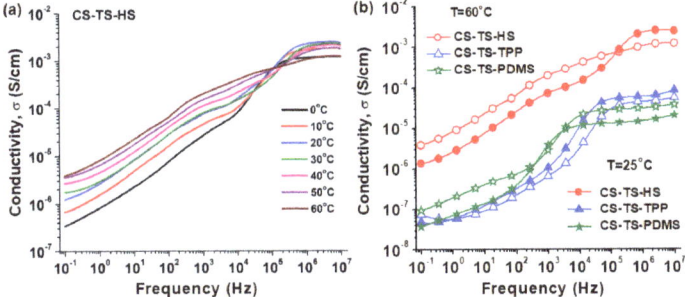

Figure 8. Evolution of conductivity with frequency at different temperatures for hydrated samples: (a) CS-TS-HS membrane, (b) comparison of conductivity behavior of all cross-linked membranes.

The comparative σ(f) spectra of hydrated membranes from Figure 8b and the conductivity values collected in Table 2 reveal that both the low frequency and the high frequency independent frequency plateaus are different in magnitude and their enlargement is depending on membrane composition, water content and temperature. The conductivity of hydrated CS-TS-HS membrane is two orders higher than those of CS-TS-TPP and CS-TS-PDMS because of superior content of sulfuric acid and water uptake capacity (see Section 3.6). Moreover, the conductivities of CS-TS-TPP and CS-TS-PDMS membranes are almost similar, revealing no particular influence of the cross-linking agent.

According to Figure S9, the conductivity of dry membranes increases linearly with temperature, suggesting that the relaxation dynamics influences the overall conductivity. By contrast, the increase of conductivity with temperature for hydrated samples is noticeably reduced, indicating that the conductivity is primarily highlighted by polar water molecules.

The activation energy for proton transport in chitosan membranes, E_σ, was determined with the Arrhenius-type relation

$$\sigma = \sigma_0 exp(-E_\sigma/kT) \qquad (4)$$

where σ_0 is a pre-exponential factor, k is the Boltzmann constant, and T is the absolute temperature. The activation energy is related to the energy required for proton transport between different polar sites [35].

The activation energy values of the dry membranes are ranging from 53 to 69 kJ/mol. The calculated values are comparable with other systems previously reported [36,41]. Besides, the activation energy for hydrated membranes is much lower than that of dry membranes (ranging from 23 to 31 kJ/mol). According to literature, the values for hydrated membranes suggest that the proton transport occurs primarily via the Grotthuss mechanism, i.e., proton migration through hydrogen bond of water molecules by jumping [34,42,43].

The measurement of in-plane proton conductivity by four-probes [44,45] method, as described in Section 2.3, was carried out as a confirmation only on the most promising membrane CS-TS-HS that supplied the highest proton conductivity value by dielectric spectroscopy. Such measurement was repeated twice for each temperature in order to have a statistically valid value and hence the result provided is the average between them. Thus, the test through four-probes method (in-plane) at 30 and 60 °C indicated proton conductivity values of 3.0×10^{-3} and 6.2×10^{-3}, respectively.

The results obtained supply a complete overlapping at 30 °C and a good coherence at 60 °C, if compared to the results on the same membrane arising from the dielectric spectroscopy taking into account the substantial differences of the techniques. In Figure S2, the technique used together to the test station and cell for the measurement is described. Such a result confirms the promising capacity of the membrane CS-TS-HS, cross-linked by sulfuric acid.

4. Conclusions

Chitosan-sulfated titania composite membranes with appropriate properties for fuel cell applications were produced and the influence of three different cross-linkers—sulfuric acid, pentasodium tripolyphosphate, and polydimethylsiloxane-diglycidyl ether terminated—were studied to obtain their properties. The chemical interaction between chitosan and sulfated titania, as well as the success of the cross-linking reactions, was proved by FTIR structural analysis. The morphological analysis by SEM showed the formation of dense membranes with thicknesses ranging from 31 to 130 μm and uniform dispersion of inorganic filler. The mechanical and thermal measurements indicated that cross-linking processes conducted to tougher materials with thermal stabilities values up to 200 °C. Typical tests usually applied for PEM evaluation, such as water uptake and chemical stability, indicated that the cross-linked membranes developed in the present study can be recommended for fuel cell application. The proton conductivity performances evaluated by dielectric spectroscopy were proven to strongly depend on cross-linker nature and on hydration state of membranes. The most promising membrane was achieved by using sulfuric acid as cross-linker. In addition, according to calculated values of activation energy, the proton transport can occur mainly via the Grothuss mechanism.

Supplementary Materials: The following are available online at http://www.mdpi.com/2073-4360/12/5/1125/s1, Figure S1: 1H NMR spectrum of pristine chitosan (CS); Figure S2: (a) Test station for proton conductivity measurement connected to cell and potentiat-galvanostat; (b) conductivity cell and formula used; Figure S3: Representative surface SEM images of (a) composite chitosan–sulfonated titania membrane (CS-TS) and composite chitosan–sulfonated titania membranes cross-linked with (b) sulfuric acid, (c) pentasodium tripolyphosphate, and (d) polydimethylsiloxane (CS-TS-HS, CS-TS-TPP, and CS-TS-PDMS, respectively); Figure S4: Water uptake kinetics of composite chitosan–sulfonated titania membranes cross-linked with sulfuric acid, pentasodium tripolyphosphate and polydimethylsiloxane (CS-TS-HS, CS-TS-TPP, and CS-TS-PDMS, respectively) at (a) 25 °C, (b) 60 °C and (c) 80 °C; Figure S5: Dielectric constant and dielectric loss evolution with frequency for dry CS-TS, CS-TS-TPP, and CS-TS-PDMS composite membranes; Figure S6 Evolutions of the measured conductivity with frequency for dry (a) CS-TS, (b) CS-TS-TPP, and (c) CS-TS-PDMS composite membranes; Figure S7: The evolution of conductivity with frequency at 100 °C for dry membranes; Figure S8: Evolutions of the measured conductivity with frequency for hydrated CS-TS-TPP and CS-TS-PDMS composite membranes; Figure S9: The evolution of conductivity with temperature at 0.1 Hz for dry and hydrated membranes; and Equation (S1) for Determination of sulfate groups content by back-titration method.

Author Contributions: Conceptualization: P.S. and C.C.; Methodology: P.S. and C.C.; Validation—experiments and analysis: A.-C.H., M.A., A.B., A.S., and A.T.M.; Writing—original draft preparation: A.-C.H. and P.S.; Writing—review and editing: P.S., V.H., C.C., and A.S.; Supervision: V.H. All authors have read and agreed to the published version of the manuscript.

Funding: This work was supported by a grant of the Romanian Ministry of Research and Innovation, CCCDI-UEFISCDI, project number PN-III-P1-1.2-PCCDI-2017-0194/25PCCDI/2018, within PNCDI III.

Conflicts of Interest: The authors declare no conflict of interest.

References

1. Zhu, M.; Song, Y.; Hu, W.; Li, X.; Jiang, Z.; Guiver, M.D.; Liu, B. SPAEK-based binary blends and ternary composites as proton exchange membranes for DMFCs. *J. Membr. Sci.* **2012**, *415*, 520–526. [CrossRef]
2. Devrim, Y.; Erkan, S.; Bac, N.; Eroglu, I. Preparation and characterization of sulfonated polysulfone/titanium dioxide composite membranes for proton exchange membrane fuel cells. *Int. J. Hydrog. Energy* **2009**, *34*, 3467–3475. [CrossRef]
3. Mazzapioda, L.; Panero, S.; Navarra, M.A. Polymer Electrolyte Membranes Based on Nafion and a Superacidic Inorganic Additive for Fuel Cell Applications. *Polymers* **2019**, *11*, 914. [CrossRef]
4. Di Noto, V.; Piga, M.; Pace, G.; Negro, E.; Lavina, S. Dielectric Relaxations and Conductivity Mechanism of Nafion: Studies Based on Broadband Dielectric Spectroscopy. *ECS Trans.* **2008**, *16*, 1183–1193. [CrossRef]

5. Shaari, N.; Kamarudin, S.K. Chitosan and alginate types of bio-membrane in fuel cell application: An overview. *J. Power Sources* **2015**, *289*, 71–80. [CrossRef]
6. Saccà, A.; Carbone, A.; Gatto, I.; Pedicini, R.; Freni, A.; Patti, A.; Passalacqua, E. Composites Nafion-titania membranes for Polymer Electrolyte Fuel Cell (PEFC) applications at low relative humidity levels: Chemical physical properties and electrochemical performance. *Polym. Test.* **2016**, *56*, 10–18. [CrossRef]
7. Kreuer, K.D. Hydrocarbon membranes. In *Handbook of Fuel Cells: Fundamentals, Technology and Applications*. In *Fuel Cell Technology and Applications*; Vielstich, W., Lamm, A., Gasteiger, H., Eds.; John Wiley & Sons Ltd.: London, UK, 2003; Volume 3, pp. 420–435.
8. Divya, K.; Rana, D.; Alwarappan, S.; Saraswathi, M.S.S.A.; Nagendran, A. Investigating the usefulness of chitosan based proton exchange membranes tailored with exfoliated molybdenum disulfide nanosheets for clean energy applications. *Carbohydr. Polym.* **2019**, *208*, 504–512. [CrossRef]
9. Wang, W.; Shan, B.; Zhu, L.; Xie, C.; Liu, C.; Cui, F. Anatase titania coated CNTs and sodium lignin sulfonate doped chitosan proton exchange membrane for DMFC application. *Carbohydr. Polym.* **2018**, *187*, 35–42. [CrossRef]
10. Santamaria, M.; Pecoraro, C.; Di Franco, F.; Di Quarto, F.; Gatto, I.; Saccà, A. Improvement in the performance of low temperature H2–O2 fuel cell with chitosan–phosphotungstic acid composite membranes. *Int. J. Hydrog. Energy* **2016**, *41*, 5389–5395. [CrossRef]
11. Wang, J.; Zhang, Y.; Wu, H.; Xiao, L.; Jiang, Z. Fabrication and performances of solid superacid embedded chitosan hybrid membranes for direct methanol fuel cell. *J. Power Sources* **2010**, *195*, 2526–2533. [CrossRef]
12. Wang, J.; Gong, C.; Wen, S.; Liu, H.; Qin, C.; Xiong, C.; Dong, L. Proton exchange membrane based on chitosan and solvent-free carbon nanotube fluids for fuel cells applications. *Carbohydr. Polym.* **2018**, *186*, 200–207. [CrossRef] [PubMed]
13. Li, C.C.; Zheng, Y.P.; Wang, T.H. Sulfated mesoporous Au/TiO2 spheres as a highly active and stable solid acid catalyst. *J. Mater. Chem.* **2012**, *22*, 13216. [CrossRef]
14. Ayyaru, S.; Dharmalingam, S. Improved performance of microbial fuel cells using sulfonated polyether ether ketone (SPEEK) TiO2-SO3H nanocomposite membrane. *RSC Adv.* **2013**, *3*, 25243–25251. [CrossRef]
15. Cui, Z.; Xiang, Y.; Si, J.; Yang, M.; Zhang, Q.; Zhang, T. Ionic interactions between sulfuric acid and chitosan membranes. *Carbohydr. Polym.* **2008**, *73*, 111–116. [CrossRef]
16. Gierszewska-Drużyńska, M.; Ostrowska-Czubenko, J. Influence of Crosslinking Process Conditions on Molecular and Supermolecular Structure of Chitosan Hydrogel Membrane. *Prog. Chem. Appl. Chitin Deriv.* **2011**, *16*, 15–22.
17. Enescu, D.; Hamciuc, V.; Pricop, L.; Hamaide, T.; Harabagiu, V.; Simionescu, B.C. Polydimethylsiloxane-modified chitosan I. Synthesis and structural characterisation of graft and crosslinked copolymers. *J. Polym. Res.* **2008**, *16*, 73–80. [CrossRef]
18. Saccà, A.; Carbone, A.; Gatto, I.; Pedicini, R.; Passalacqua, E. Synthesized Yttria Stabilised Zirconia as filler in Proton Exchange Membranes (PEMs) with enhanced stability. *Polym. Test.* **2018**, *65*, 322–330. [CrossRef]
19. Angjeli, K.; Nicotera, I.; Baikousi, M.; Enotiadis, A.; Gournis, D.; Saccà, A.; Passalacqua, E.; Carbone, A. Investigation of layered double hydroxide (LDH) Nafion-based nanocomposite membranes for high temperature PEFCs. *Energy Convers. Manag.* **2015**, *96*, 39–46. [CrossRef]
20. Enescu, D.; Hamciuc, V.; Ardeleanu, R.; Cristea, M.; Ioanid, A.; Harabagiu, V.; Simionescu, B.C. Polydimethylsiloxane modified chitosan. Part III: Preparation and characterization of hybrid membranes. *Carbohydr. Polym.* **2009**, *76*, 268–278. [CrossRef]
21. Kumar, K.S.; Rajendran, S.; Prabhu, M.R. A Study of influence on sulfonated TiO2-Poly (Vinylidene fluoride-co-hexafluoropropylene) nano composite membranes for PEM Fuel cell application. *Appl. Surf. Sci.* **2017**, *418*, 64–71. [CrossRef]
22. Humelnicu, A.-C.; Cojocaru, C.; Dorneanu, P.P.; Samoila, P.; Harabagiu, V. Novel chitosan-functionalized samarium-doped cobalt ferrite for adsorptive removal of anionic dye from aqueous solutions. *Comptes Rendus Chim.* **2017**, *20*, 1026–1036. [CrossRef]
23. Das, G.; Kim, C.Y.; Kang, D.H.; Kim, B.H.; Yoon, H.H. Quaternized Polysulfone Cross-Linked N,N-Dimethyl Chitosan-Based Anion-Conducting Membranes. *Polymers* **2019**, *11*, 512. [CrossRef]

24. Yang, D.; Li, J.; Jiang, Z.; Lu, L.; Chen, X. Chitosan/TiO 2 nanocomposite pervaporation membranes for ethanol dehydration. *Chem. Eng. Sci.* **2009**, *64*, 3130–3137. [CrossRef]
25. Ngah, W.W.; Fatinathan, S.; Yosop, N. Isotherm and kinetic studies on the adsorption of humic acid onto chitosan-H2SO4 beads. *Desalination* **2011**, *272*, 293–300. [CrossRef]
26. Loutfy, S.A.; Alam El-Din, H.M.; Elberry, M.H.; Allam, N.G.; Hasanin, M.T.M.; Abdellah, A.M. Synthesis, characterization and cytotoxic evaluation of chitosan nanoparticles: In vitro liver cancer model. *Adv. Nat. Sci. Nanosci. Nanotechnol.* **2016**, *7*, 35008. [CrossRef]
27. Simionescu, C.I.; Rusa, M.; David, G.; Pinteala, M.; Harabagiu, V.; Simionescu, B.C. Block and graft copolymers with polysiloxane and poly(N-acyliminoethylene) sequences. *Angew. Chemie Makromol.* **1997**, *253*, 139–149. [CrossRef]
28. Hayward, R.C.; Chmelka, B.F.; Kramer, E.J. Template Cross-Linking Effects on Morphologies of Swellable Block Copolymer and Mesostructured Silica Thin Films. *Macromolecules* **2005**, *38*, 7768–7783. [CrossRef]
29. Ziegler-Borowska, M.; Chelminiak-Dudkiewicz, D.; Kaczmarek, H. Thermal stability of magnetic nanoparticles coated by blends of modified chitosan and poly(quaternary ammonium) salt. *J. Therm. Anal. Calorim.* **2014**, *119*, 499–506. [CrossRef]
30. Pati, F.; Adhikari, B.; Dhara, S. Development of chitosan – tripolyphosphate fibers through pH dependent ionotropic gelation. *Carbohydr. Res.* **2011**, *346*, 2582–2588. [CrossRef]
31. Kandlikar, S.G.; Garofalo, M.L.; Lu, Z. Water Management in A PEMFC: Water Transport Mechanism and Material Degradation in Gas Diffusion Layers. *Fuel Cells* **2011**, *11*, 814–823. [CrossRef]
32. Aguirre-Loredo, R.Y.; Rodriguez-Hernandez, A.; Velázquez, G. Modelling the effect of temperature on the water sorption isotherms of chitosan films. *Food Sci. Technol.* **2016**, *37*, 112–118. [CrossRef]
33. Escorihuela, J.; Garcia-Bernabe, A.; Montero, A.; Sahuquillo, Ó.; Giménez, E.; Compañ, V. Ionic Liquid Composite Polybenzimidazol Membranes for High Temperature PEMFC Applications. *Polymers* **2019**, *11*, 732. [CrossRef]
34. Ramly, N.N.; Aini, N.A.; Sahli, N.; Aminuddin, S.F.; Yahya, M.Z.A.; Ali, A.M.M. Dielectric behavior of UV-crosslinked sulfonated poly (ether ether ketone) with methyl cellulose (SPEEK-MC) as proton exchange membrane. *Int. J. Hydrog. Energy* **2017**, *42*, 9284–9292. [CrossRef]
35. Bronnikov, S.; Podshivalov, A.; Kostromin, S.; Asandulesa, M.; Cozan, V. Electrical conductivity of polyazomethine/fullerene C60 nanocomposites. *Phys. Lett. A* **2017**, *381*, 796–800. [CrossRef]
36. Gu, H.; England, D.; Yan, F.; Texter, J. New high charge density polymers for printable electronics, sensors, batteries, and fuel cells. In Proceedings of the 2008 2nd IEEE International Nanoelectronics Conference, Shanghai, China, 24–27 March 2008; pp. 863–868.
37. Pochard, I.; Vall, M.; Eriksson, J.; Farineau, C.; Cheung, O.; Frykstrand, S.; Welch, K.; Strømme, M. Amine-functionalised mesoporous magnesium carbonate: Dielectric spectroscopy studies of interactions with water and stability. *Mater. Chem. Phys.* **2018**, *216*, 332–338. [CrossRef]
38. Samet, M.; Levchenko, V.; Boiteux, G.; Seytre, G.; Kallel, A.; Serghei, A. Electrode polarization vs. Maxwell-Wagner-Sillars interfacial polarization in dielectric spectra of materials: Characteristic frequencies and scaling laws. *J. Chem. Phys.* **2015**, *142*, 194703. [CrossRef]
39. Asandulesa, M.; Musteata, V.E.; Bele, A.; Dascalu, M.; Bronnikov, S.; Racles, C. Molecular dynamics of polysiloxane polar-nonpolar co-networks and blends studied by dielectric relaxation spectroscopy. *Polym.* **2018**, *149*, 73–84. [CrossRef]
40. Zhang, S.; Runt, J. Segmental Dynamics and Ionic Conduction in Poly(vinyl methyl ether)−Lithium Perchlorate Complexes. *J. Phys. Chem. B* **2004**, *108*, 6295–6302. [CrossRef]
41. Ali, A.; Mohamed, N.; Arof, A.K. Polyethylene oxide (PEO)–ammonium sulfate ((NH4)2SO4) complexes and electrochemical cell performance. *J. Power Sources* **1998**, *74*, 135–141. [CrossRef]
42. Wang, J.; Bai, H.; Zhang, H.; Zhao, L.; Chen, H.; Li, Y. Anhydrous proton exchange membrane of sulfonated poly(ether ether ketone) enabled by polydopamine-modified silica nanoparticles. *Electrochim. Acta* **2015**, *152*, 443–455. [CrossRef]
43. Fischer, S.A.; Dunlap, B.; Gunlycke, D. Proton transport through hydrated chitosan-based polymer membranes under electric fields. *J. Polym. Sci. Part B Polym. Phys.* **2017**, *50*, 9–1109. [CrossRef]

44. Saccà, A.; Carbone, A.; Pedicini, R.; Portale, G.; D'Ilario, L.; Longo, A.; Martorana, A.; Passalacqua, E. Structural and electrochemical investigation on re-cast Nafion membranes for polymer electrolyte fuel cells (PEFCs) application. *J. Membr. Sci.* **2006**, *278*, 105–113. [CrossRef]
45. Saccà, A.; Gatto, I.; Carbone, A.; Pedicini, R.; Maisano, S.; Stassi, A.; Passalacqua, E. Influence of doping level in Yttria-Stabilised-Zirconia (YSZ) based-fillers as degradation inhibitors for proton exchange membranes fuel cells (PEMFCs) in drastic conditions. *Int. J. Hydrog. Energy* **2019**, *44*, 31445–31457. [CrossRef]

© 2020 by the authors. Licensee MDPI, Basel, Switzerland. This article is an open access article distributed under the terms and conditions of the Creative Commons Attribution (CC BY) license (http://creativecommons.org/licenses/by/4.0/).

Article

Nanostructured Chitosan/Maghemite Composites Thin Film for Potential Optical Detection of Mercury Ion by Surface Plasmon Resonance Investigation

Nurul Illya Muhamad Fauzi [1], Yap Wing Fen [1,2,*], Nur Alia Sheh Omar [2], Silvan Saleviter [2], Wan Mohd Ebtisyam Mustaqim Mohd Daniyal [2], Hazwani Suhaila Hashim [1] and Mohd Nasrullah [3]

[1] Department of Physics, Faculty of Science, Universiti Putra Malaysia, UPM Serdang 43400, Selangor, Malaysia; illyafauzi97@gmail.com (N.I.M.F.); hazwanisuhaila@gmail.com (H.S.H.)
[2] Functional Devices Laboratory, Institute of Advanced Technology, Universiti Putra Malaysia, UPM Serdang 43400, Selangor, Malaysia; nuralia.upm@gmail.com (N.A.S.O.); silvansaleviter94@gmail.com (S.S.); wanmdsyam@gmail.com (W.M.E.M.M.D.)
[3] Faculty of Civil Engineering Technology, Universiti Malaysia Pahang (UMP), Gambang 26300, Kuantan, Pahang, Malaysia; nasrul.ump@gmail.com
* Correspondence: yapwingfen@gmail.com or yapwingfen@upm.edu.my

Received: 26 May 2020; Accepted: 12 June 2020; Published: 4 July 2020

Abstract: In this study, synthesis and characterization of chitosan/maghemite (Cs/Fe$_2$O$_3$) composites thin film has been described. Its properties were characterized using Fourier transform infrared spectroscopy (FTIR), atomic force microscopy (AFM) and ultraviolet-visible spectroscopy (UV-Vis). FTIR confirmed the existence of Fe–O bond, C–N bond, C–C bond, C–O bond, O=C=O bond and O–H bond in Cs/Fe$_2$O$_3$ thin film. The surface morphology of the thin film indicated the relatively smooth and homogenous thin film, and also confirmed the interaction of Fe$_2$O$_3$ with the chitosan. Next, the UV-Vis result showed high absorbance value with an optical band gap of 4.013 eV. The incorporation of this Cs/Fe$_2$O$_3$ thin film with an optical-based method, i.e., surface plasmon resonance spectroscopy showed positive response where mercury ion (Hg^{2+}) can be detected down to 0.01 ppm (49.9 nM). These results validate the potential of Cs/Fe$_2$O$_3$ thin film for optical sensing applications in Hg^{2+} detection.

Keywords: chitosan; maghemite; optical; mercury ion; surface plasmon resonance

1. Introduction

Organic polymeric materials made up of many repeating monomer units have made a significant impact on biological and biomedical research activities because of the flexibility and the ease of fabrication [1]. One of the well-known organic polymeric materials is chitosan, easily derived from partial deacetylation of chitin with a degree of 50% or greater [2–4]. To be more specific, chitosan is a family of linear polysaccharide as a part of glucosamine and N-acetyl glucosamine units linked via β-1,4 glucosidic bonds [5,6]. Chitosan contains three types of reactive functional groups, primary amine groups and primary and secondary hydroxyl groups, respectively, at positions C-2, C-3 and C-6. Among the three types of functional groups, the primary amine groups at C-2 positions are the most favorable sites interacting with the biological molecules, metal ions and organic halogen substances. Taking the advantages of chitosan with high absorption capacity and high biocompatibility, chitosan is known as an ideal substrate for enzyme immobilization [7]. Other excellent advantages of chitosan including non-toxicity, great film-forming ability, powerful adhesion property and high mechanical strength, offers great room for sensor applications [8–10]. However, the problem of poor stableness of chitosan because of the hydrophilic character and pH sensitivity restricts its application [11,12]. Previous reports showed that the stability of chitosan could be improved by combining with oxide or metal oxides and the product can be effectively used as recognition elements for chemical sensors and biosensors [13–15].

Iron (III) oxide or ferric oxide is the inorganic compound with the Fe_2O_3 formula, which varies in color depending on its phase [16]. Fe_2O_3 materials have four polymorphs phases such as α-Fe_2O_3 (hematite), β-Fe_2O_3, γ-Fe_2O_3 (maghemite) and ε-Fe_2O_3 [17,18]. The differences of the phases are known from their originality, for examples, hematite and maghemite are naturally obtained and the other two of phases are synthesized in laboratory [19,20]. Among the phases, γ-Fe_2O_3 is one of the chief interests. It is the second most common sustainable form of Fe_2O_3, known as completely oxidized magnetite. Maghemite has a high curie temperature, but has a lower saturation magnetization at room temperature and a supermagnetism property that makes it quite efficient in removing heavy metal pollutants from water [21,22]. Moreover, it is believed that Fe_2O_3 can improve and provide better mechanical properties to chitosan [23].

Accumulation of heavy metals in water and food production, primarily mercury (Hg) is the most hazardous heavy-metal pollutants even at a very low concentration. The most toxic chemical forms of Hg are ionic Hg (Hg^{2+}), causes serious damage to human health such as brain damage, immune dysfunction and paralysis [24–26]. Therefore, the removal and detection of Hg^{2+} in the aqueous environment are of great significance [27–31]. Among the existing optical techniques to detect Hg^{2+} are colorimetric, fluorescent, chemosensor, electrochemiluminescence (ECL) and photoluminescent (PL) [32–34]. Though these techniques are widely used, they encounter from many drawbacks, such as high instrument operating costs, repetitive pretreatment procedures and long initiation times [35].

Corresponding to the previous methods, surface plasmon resonance (SPR) proposed a cost-effective, label-free detection method for convenient usage, rapid detection and excellent sensitivity and selectivity to heavy metal ions [36–40]. Since enormous efforts devoted to creating sensors with high sensitivity to Hg^{2+} are greatly needed currently, selection of the metallic layer such as the gold layer is an important aid in producing higher sensor sensitivity in SPR [41]. Over the last decade, the surface SPR technique has emerged as an effective optical technique for various applications including detection of heavy metal ions [42–51]. Unfortunately, the main problem to detect optically the heavy metal ions solution is the similar refractive indices of heavy metal ions for lowest concentration, which eventually becomes the goal of researchers. Hence, many researchers have dedicated their time to develop chitosan-based materials onto SPR interfaces in lowering the detection limit of Hg^{2+}, specifically [52–54]. A recent study documented the utilization of polypyrrole-chitosan/nickel-ferrite nanoparticles as an active layer to a prism-based on SPR technique for Hg^{2+} sensing, which reached a limit of detection (LOD) as low as 1.94 µM [54]. Other recent studies using chitosan-based materials as sensing layers for the detection of Hg^{2+} by SPR are summarized in Table 1. It is of interest to further improve the LOD using chitosan-based SPR sensor.

Table 1. Chitosan based material by surface plasmon resonance (SPR) for the detection of Hg^{2+}.

Ref.	Sensing Layer	LOD
[38]	MMW chitosan (glutaraldehyde-crosslinked)	2.49 µM
[52]	Polypyrrole-chitosan conducting polymer composite	2.50 µM
[53]	Chitosan/graphene oxide	0.50 µM
[54]	Polypyrrole-chitosan/nickel-ferrite nanoparticles	1.94 µM

Ref.: reference. LOD: limit of detection.

To the best of our knowledge, the study for Cs/γ-Fe_2O_3 composite to detect Hg^{2+} using the SPR technique is not reported yet. There is also a lack of studies on the structural and optical properties of these composites. Therefore, an effort was made to apply the chitosan/γ-Fe_2O_3 thin film onto a thin gold surface, as a novel active layer for the SPR technique in sensing Hg^{2+} as low as nanomolar.

Besides, the studies of structural and optical properties of Cs/γ-Fe$_2$O$_3$ thin film on the gold surface are also reported and explored.

2. Materials and Methods

2.1. Reagent and Materials

The Fe$_2$O$_3$ was purchased from R&M Marketing, Essex, U.K. The medium molecular weight chitosan and acetic acid were purchased from Aldrich (Saint louis, MO, USA). Standard solution of Hg^{2+} with concentration of 1000 ppm was purchased from Merck (Darmstatd, Germany).

2.2. Preparation of Chemical

Firstly, 50 mL distilled water was added into Fe$_2$O$_3$ (4 mg/mL). Then 10 mL of NH$_3$ (25%) and 0.615 mg of ethylenediaminetetra acetic acid (EDTA) was added as precipitation agent and as capping agent to the solution with stirring respectively. The reaction was allowed to proceed for 1 h at 50 °C with constant stirring. Finally, the black precipitate of nano-Fe$_2$O$_3$-EDTA formed and it was rinsed with distilled water and left to dry 80 °C for 3 h. For chitosan preparation, 1% acetic acid was prepped by diluting stock 1 mL acetic acid with deionized water in 100 mL volumetric flask. Next, 400 mg medium molecular weight chitosan that was acquired from Aldrich was dissolved in 50 mL of 1% aqueous acetic acid and the solution vigorously stirring to ensure powder chitosan dissolved completely. To produce the nanostructured chitosan/maghemite (Cs/Fe$_2$O$_3$) composites, 30 mg Fe$_2$O$_3$ capped EDTA was dispersed in 10 mL of 0.1% in chitosan solution and sonicated in room temperature for 15 min. The Hg^{2+} standard solution with a concentration of 1000 ppm was diluted with deionized water to produce Hg^{2+} solutions with concentrations of 0.01, 0.05, 0.08, 0.1 and 0.5 ppm [55,56].

2.3. Preparation of Thin Film

To begin, glass slips (24 mm × 24 mm × 0.1 mm, Menzel-Glaser, Braunschweig, Germany), as a substrate, were coated with a thin layer of gold with thickness 50 nm using SC7640 sputter coater [57]. Next, approximately 0.55 mL of the chitosan, Fe$_2$O$_3$ and Cs/Fe$_2$O$_3$ composites solution was set separately on the surface of the gold coated glass slip. Then the glass slips were spun at 6000 rev min for 30 s using the Specialty Coating System, P-6708D (Inc. Medical Devices, Indianapolis, IN, USA) to produce the chitosan, Fe$_2$O$_3$ and Cs/Fe$_2$O$_3$ composites thin films.

2.4. Instrumental

Fourier transform infrared (FTIR) spectra for each surface modification of thin films were recorded in the transmittance mode using a Perkin-Elmer spectrophotometer (Waltham, MA, USA) under the wavelength range 400–4000 cm^{-1}. The absorbance spectra of the films were recorded from 200 to 500 nm using UV-Vis-NIR spectroscopy (UV-3600 Shimadzu, Kyoto, Japan). The optical band gap energy was calculated using the data obtained. Atomic force microscopy (AFM) analysis was carried out using Qscope 250, Qesant Instrument Corporation (Quesant, CA, USA) in intermittent mode to study the topography and height of Cs/Fe^{2}O^3 thin film. An optical-based sensing method based on surface plasmon resonance (SPR) was designed to identify the potential of the Cs/Fe$_2$O$_3$ thin film to detect Hg^{2+}. Figure 1 shows the schematic diagram of the SPR instrument setup [58–61]. The SPR experiment was carried out by inserting Hg^{2+} solutions with different concentration varied from 0.01 to 0.5 ppm. It was injected one after another into the cell to bind with Cs/Fe$_2$O$_3$ thin film coated onto gold surface thin film. The SPR curve and resonance angle for all concentrations was monitored and recorded.

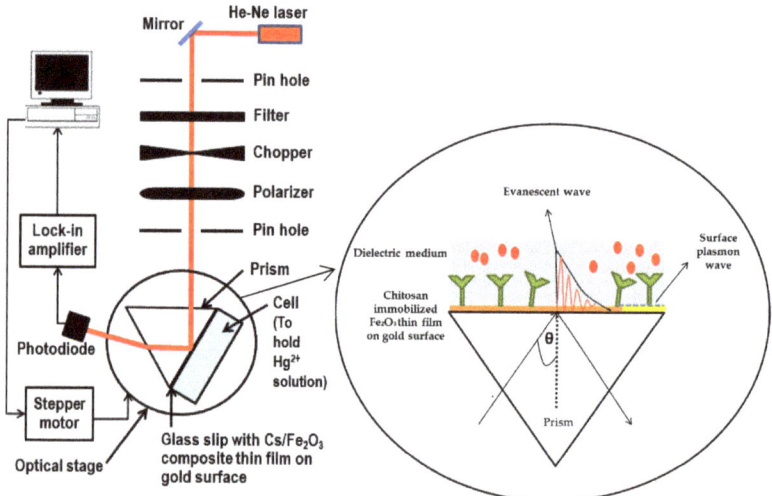

Figure 1. Optical setup of surface plasmon resonance spectroscopy.

3. Results and Discussion

3.1. FTIR Analysis

FTIR spectroscopy was used to identify the functional groups existed in Cs/Fe_2O_3 thin film. The spectrum of chitosan, Fe_2O_3 and Cs/Fe_2O_3 thin films in the range of 450–4000 cm^{-1} are represented in Figure 2. From the FTIR spectrum of chitosan thin film, the broad absorption band at 3386.43 cm^{-1} can be appointed to the stretching vibration of O–H. A weaker band found at 2901.26 cm^{-1} can be attributed to C–H stretching in chitosan. Another absorption band at 1655.48 cm^{-1} was associated with the presence of the C=O stretching bond. There is an absorption peak at 1084.47 cm^{-1} that corresponds to the C–O group, which indicates the presence of the –COOH group in chitosan thin film. Two more bands at 500.76 cm^{-1} and 458.22 cm^{-1} were assigned to the C–C bond and C–N bond respectively. This finding is well aligned to the previous study by Anas et al. [62].

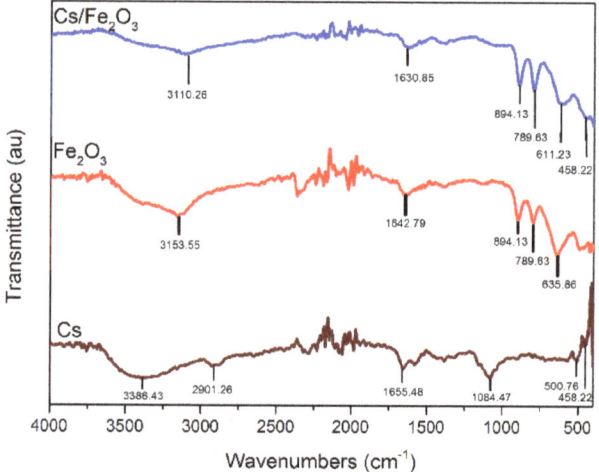

Figure 2. FTIR spectrum for chitosan, Fe_2O_3 and Cs/Fe_2O_3 thin films.

Next, a particular major peak in the Fe_2O_3 thin film was identified with the degree of cation vacancy, ordering between octahedral Fe cation and O atoms [63]. The absorption peak at 789.63 cm^{-1} is a characteristic of maghemite Fe–O stretching vibrations particles. This peak is solely attributed to the high degree of cationic vacancy ordering [64]. The broad band characteristic for bending vibration of water adsorbed on the maghemite's surface is at 2078.99 cm^{-1}. The intense bands at 1642.79 cm^{-1} and 3153.55 cm^{-1} were then assigned to CO_2 vibration and O–H vibrations, respectively, ratifying the presence of surface γ-Fe_2O_3 hydroxyl groups.

In the spectrum of Cs/Fe_2O_3, the chitosan does not provide clear absorption bands at a lower wavenumber. This is due to the low percentage of chitosan compared to maghemite in the synthesization process. However, the presence of chitosan can be observed based on the intensity peak. The peak intensity of Cs/Fe_2O_3 clearly increased after the sorption of chitosan and Fe_2O_3, i.e., at C–H stretching (458.22 cm^{-1}), C–C bond (611.23 cm^{-1}) and O=C=O stretching (1630.85 cm^{-1}). An increase in the peak intensity usually indicates an increase in the sum of the functional group (per unit volume) associated with the molecular bond [65]. On the other hand, a strong absorption band was observed at 789.63 cm^{-1}, confirmed the presence of Fe.O as the main phase of the Fe_2O_3 and a band at 3110.26 cm^{-1} that appointed to the O-H vibration of surface maghemite hydroxyl groups. Overall, the FTIR results showed the increasing peak intensity of Cs/Fe_2O_3, which confirmed the physical interaction of chitosan and γ-Fe_2O_3 in those composites.

3.2. Surface Morphology

The in situ atomic force microscopy (AFM) measurements enable the chitosan, Fe_2O_3 and Cs/Fe_2O_3 adsorption on thin films to be visualized in real time. The AFM images illustrate the topographical in the thin films as shown in Figures 3–5. The topographical can be observed by various parameters that exist to quantify the root mean square (rms) roughness of a surface. The RMS roughness value can be calculated from the cross-sectional profile or a surface area [66]. The RMS roughness obtained by chitosan, Fe_2O_3 and Cs/Fe_2O_3 thin film were 1.4 nm, 47 nm and 37.3 nm, respectively. The magnitude decreased in RMS roughness of Cs/Fe_2O_3 thin film compared to Fe_2O_3 thin film attributable to the association of two materials, which are chitosan and Fe_2O_3. The roughness implies that a smoothening mechanism by surface diffusion [67]. This result indicates that the presence of chitosan can enhance the surface of the thin film. The roughness introduced in the nanostructured maghemite in chitosan thin film intended appropriate form to enhance the thin film as sensing element [68]. This result is in line with the FTIR data, proving the presence of maghemite and chitosan in the Cs/Fe_2O_3 thin film based on the RMS roughness.

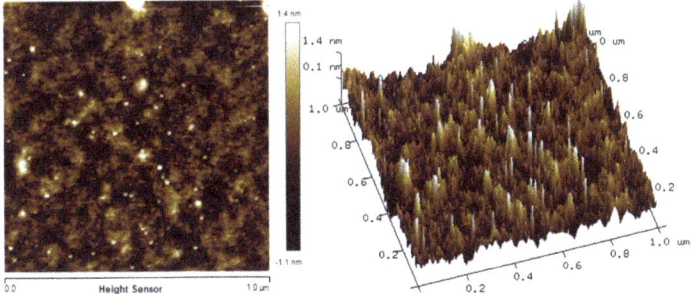

Figure 3. Atomic force microscopy (AFM) image of chitosan thin film.

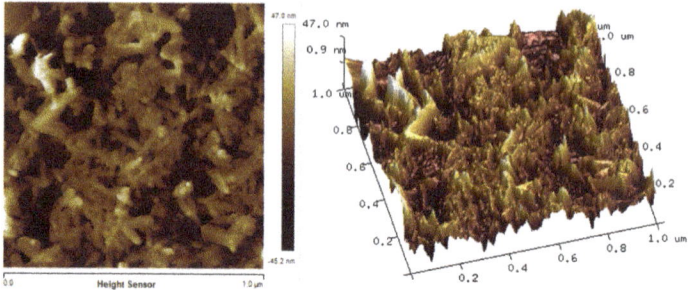

Figure 4. AFM image of Fe$_2$O$_3$ thin film.

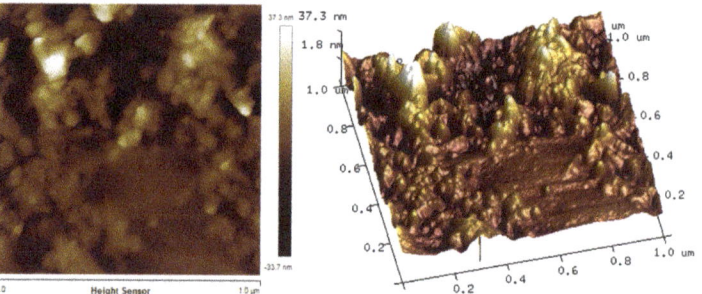

Figure 5. AFM image of Cs/Fe$_2$O$_3$ thin film.

3.3. Optical Studies

For the optical properties, the absorbance spectrum of the thin films was observed and measured at wavelength from 250 to 500 nm. The UV-Vis results of chitosan, Fe$_2$O$_3$ and Cs/Fe$_2$O$_3$ thin films are shown in Figure 6 it can be spotted that all of the thin film has diverse value of absorbance. From the graph, the absorbance spectra of Cs and Fe$_2$O$_3$ thin films were slightly higher as compared to the Cs/Fe$_2$O$_3$ thin film. The maximum absorption wavelength can be observed at 260–300 nm. The absorption peak about 300 nm corresponds to $\pi \to \pi^*$ transitions of C=O [69,70].

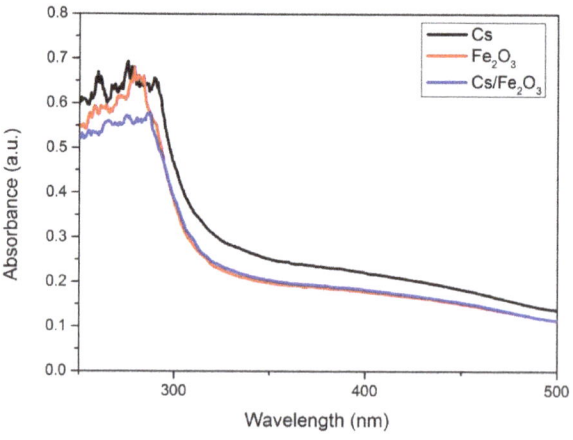

Figure 6. Absorbance spectrum chitosan, Fe$_2$O$_3$ and Cs/Fe$_2$O$_3$ thin films.

The UV-Vis absorbance spectrum was then quantitative analyzed based on the Beer–Lambert law theory. This law refers to a relation between the attenuation of light by a material and its properties, which the monochromatic light (single wavelength) is travelling. Since the amount of the emitted radiation intensity is only dependent on the thickness, t and concentration of the solution, the absorbance, A of the samples can be collected at a single wavelength, as follows [62]:

$$A = \log_{10} \frac{I_o}{I_t} \tag{1}$$

The transmittance, T of sample is given by the ratio of intensities of the presence I_t and the absence I_o of the sample:

$$T = \frac{I_t}{I_o} \tag{2}$$

Thus, the absorbance and transmittance can be related by:

$$A = -\log_{10} T \tag{3}$$

Apart from the absorbance, absorbance coefficient is a useful parameter to compare samples with varying thickness. The sample thickness was obtained by using atomic force microscopy. The absorbance coefficient, α (in unit of m^{-1}) is given by:

$$\alpha = 2.303 \frac{A}{t} \tag{4}$$

where t is the thickness of sample in unit of m. The absorbance coefficient and optical band gap can be related by:

$$\alpha = \frac{k(h\upsilon - E_g)^n}{h\upsilon} \tag{5}$$

Rearranging Equation (5) gives:

$$(\alpha h\upsilon)^{1/n} = k(h\upsilon - E_g) \tag{6}$$

where $h\upsilon$ is the photon energy, h is Plank's constant, E_g is the optical band gap, k is constant and n is the transition states, i.e., direct or indirect transitions. Direct transition is transition in which a photon excites an electron from the valence band to the conduction band directly if the momentum of electrons and holes is the same in both bands (conduction and valence). On the other hand, indirect transition is a photon cannot be emitted because the electron must pass through an intermediate state and transfer momentum to the crystal lattice. From these, it can be concluded that the absorption in the thin films corresponds to a direct energy gap. For direct transition, $n = 1/2$ and this value is substituted in Equation (6) and becomes:

$$(\alpha h\upsilon)^{1/n} = k(h\upsilon - E_g) \tag{7}$$

To evaluate the optical band gap, E_g of the chitosan, Fe_2O_3 and Cs/Fe_2O_3 thin films, the graphs of $(\alpha h\upsilon)^2$ against hv are plotted as shown in Figures 7–9, respectively. As a result, the intersection of straight line on the edge was obtained, indicating the direct transition of the optical band gap [71]. The calculated values of the optical band gap were 4.073 eV, 4.078 eV and 4.013 eV for chitosan, Fe_2O_3 and Cs/Fe_2O_3 thin films respectively (with the corresponding error of ±0.001 eV) [72,73]. This result indicated the maghemite had a band gap energy of 4.078 eV, which was higher than to the 2 eV bulk [64]. This might be due to the structure defects, that have changed the phase, strain and size of nanoparticles during heat treatment that led to the increase of band gap [74]. When Fe_2O_3 added on chitosan, the band gap became lower as compared to the individual band gap. It can be due to the increased of crystallite size attributed to the confinement effects that related to the rise amount of orbitals participating in the formation of valence bands and covalent bands through orbital overlap [75].

Thus, this showed that defects and confinement effects have a huge impact on the optical properties of a composite.

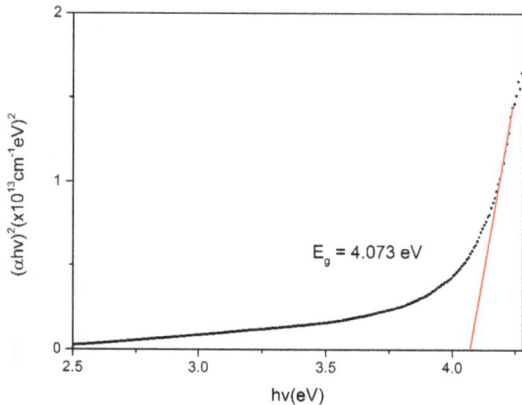

Figure 7. Optical band gap for chitosan thin film.

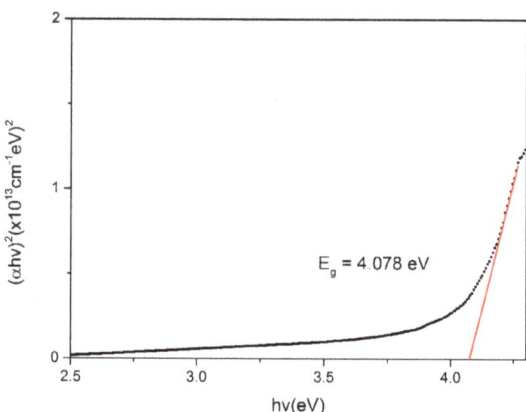

Figure 8. Optical band gap for Fe_2O_3 thin film.

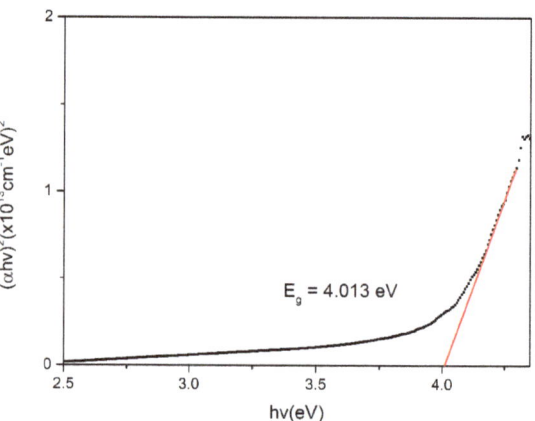

Figure 9. Optical band gap for Cs/Fe_2O_3 thin film.

3.4. Optical-Based Sensing of Hg^{2+}

The optical sensing based on surface plasmon resonance (SPR) phenomenon was conducted by using Cs/Fe_2O_3 thin film to identify the SPR angle for deionized water as a control experiment. The SPR angle of 55.225° was further applied to compare the SPR angle for different concentration of Hg^{2+} solution ranged from 0.01 to 0.5 ppm. The SPR reflectively curves for Cs/Fe_2O_3 thin film in contact with the different concentration of Hg^{2+} are shown in Figure 10. It can be seen that the SPR curves of Hg^{2+} solution shifted from 0 to 0.5 ppm as compared with the deionized water SPR curve. The SPR angle for 0.01, 0.05, 0.08, 0.1 and 0.5 ppm of Hg^{2+} were 54.615°, 54.398°, 54.212°, 54.027° and 53.836°, respectively, with the corresponding error of ±0.001° (the resolution of the stepping motor of the SPR). Overall, it was observed that the SPR shifted to the left with increasing concentration of Hg^{2+} solution. This finding can be attributed to the increase in binding between analyte–ligand, which resulted in the change of refractive index as well as the thickness of the Cs/Fe_2O_3 sensing layer [76–79]. Hence it is confirmed that Cs/Fe_2O_3 thin film has an affinity with Hg^{2+} and can be integrated with SPR optical-based sensing method for detection of Hg^{2+}.

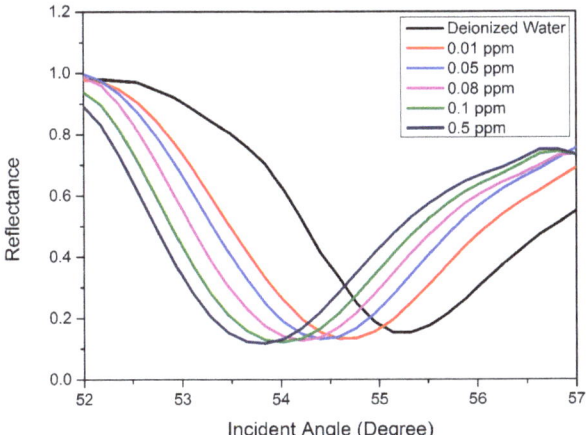

Figure 10. SPR curves for Cs/Fe_2O_3 thin film in contact with deionized water and Hg^{2+} solution with a concentration of 0.01–0.5 ppm.

4. Conclusions

In this study, a Cs/Fe_2O_3 thin film was successfully developed using the spin coating technique. The functional groups analysis from the FTIR results confirmed the correlation between chitosan and γ-Fe_2O_3, with the peak intensity of Cs/Fe_2O_3 clearly increasing after the sorption of chitosan and Fe_2O_3 at C–H stretching, C–C bond and O=C=O stretching. Next, the AFM result showed that the thin film was homogenous when the surface of chitosan on the thin film was covered by Fe_2O_3. Besides, the UV-Vis results confirmed that the Cs/Fe_2O_3 thin film had the lowest absorbance value compared its individual thin films with an optical band gap of 4.013 eV. The incorporation Cs/Fe_2O_3 thin film with the optical-based sensing method using the surface plasmon resonance technique provided positive response to the Hg^{2+} solution of different concentrations. This result demonstrated the enormous ability of Cs/Fe_2O_3 thin film for optical sensing of Hg^{2+} as low as 0.01 ppm.

Author Contributions: Conceptualization, methodology, writing—original draft preparation, N.I.M.F.; validation, supervision, writing—review and editing, funding acquisition, Y.W.F.; investigation, formal analysis, N.A.S.O.; software, S.S.; visualization, W.M.E.M.M.D. and H.S.H.; resources, M.N. All authors have read and agreed to the published version of the manuscript.

Funding: This research was funded and supported by the Ministry of Education Malaysia through the Fundamental Research Grant Scheme (FRGS) (FRGS/1/2019/STG02/UPM/02/1) and Putra Grant Universiti Putra Malaysia.

Acknowledgments: The authors acknowledged the laboratory facilities provided by the Institute of Advanced Technology, Department of Physics, and Department of Chemistry, Universiti Putra Malaysia.

Conflicts of Interest: The authors declare no conflict of interest.

References

1. Iwan, A.; Sek, D. Processible polyazomethines and polyketanils: From aerospace to light-emitting diodes and other advanced applications. *Prog. Polym. Sci.* **2008**, *33*, 289–345. [CrossRef]
2. Fen, Y.W.; Yunus, W.M.M.; Talib, Z.A.; Yusof, N.A. X-ray photoelectron spectroscopy and atomic force microscopy studies on crosslinked chitosan thin film. *Int. J. Phys. Sci.* **2011**, *6*, 2744–2749.
3. Martino, A.D.; Sittinger, M.; Risbud, M.V. Chitosan: A versatile biopolymer for orthopaedic tissue-engineering. *Biomaterials* **2005**, *26*, 5983–5990. [CrossRef]
4. Prashanth, K.V.H.; Tharanathan, R.N. Chitin/chitosan: Modifications and their unlimited application potential an-overview. *Trends Food Sci. Technol.* **2007**, *18*, 117–131. [CrossRef]
5. Divya, K.; Jisha, M.S. Chitosan nanoparticles preparation and applications. *Environ. Chem Lett.* **2017**, *16*, 101–112. [CrossRef]
6. Zargar, V.; Asghari, M.; Dashti, A. A review on chitin and chitosan polymers: Structure, chemistry, solubility, derivatives, and applications. *Chem. Bio. Eng.* **2015**, *2*, 204–226. [CrossRef]
7. Zhang, J.; Xia, W.; Liu, P.; Cheng, Q.; Tahirou, T.; Gu, W.; Li, B. Chitosan modification and pharmaceutical/biomedical applications. *Mar. Drugs* **2010**, *8*, 1962–1987. [CrossRef]
8. Fen, Y.W.; Yunus, W.M.M.; Yusof, N.A.; Ishak, N.S.; Omar, N.A.S.; Zainudin, A.A. Preparation, characterization and optical properties of ionophore doped chitosan biopolymer thin film and its potential application for sensing metal ion. *Optik* **2015**, *126*, 4688–4692. [CrossRef]
9. Ahmed, T.A.; Aljaeid, B.M. Preparation, characterization and potential application of chitosan, chitosan derivatives and chitosan metal nanoparticles in pharmaceutical drug delivery. *Drug Des. Devel. Ther.* **2016**, *10*, 483–507. [CrossRef]
10. Zhao, D.; Yu, S.; Sun, B.; Gao, S.; Guo, S.; Zhao, K. Biomedical applications of chitosan and its derivative nanoparticles. *Polymers* **2018**, *10*, 462. [CrossRef]
11. Saleviter, S.; Fen, Y.W.; Omar, N.A.; Zainudin, A.A.; Daniyal, W.M.E.M.M. Optical and structural characterization of immobilized 4-(2-pyridylazo) resorcinol in chitosan-graphene oxide composite thin film and its potential for Co^{2+} sensing using surface plasmon resonance technique. *Results Phys.* **2018**, *11*, 118–122. [CrossRef]
12. Shukla, S.K.; Mishra, A.K.; Arotiba, O.A.; Mamba, B.B. Chitosan-based nanomaterials: A state-of-the-art review. *Int. J. Biol. Macromol.* **2013**, *59*, 46–58. [CrossRef] [PubMed]
13. Zainudin, A.A.; Fen, Y.W.; Yusof, N.A.; Al-Rekabi, S.H.; Mahdi, M.A.; Omar, N.A.S. Incorporation of surface plasmon resonance with novel valinomycin doped chitosan-graphene oxide thin film for sensing potassium ion. *Spectrochim. Acta A Mol. Biomol. Spectrosc.* **2018**, *191*, 111–115. [CrossRef] [PubMed]
14. Hassanein, A.; Salahuddin, N.; Matsuda, A.; Kawamura, G.; Elfiky, M. Fabrication of biosensor based on chitosan-ZnO/Polypyrrole nanocomposite modified carbon paste electrode for electroanalytical application. *Mater. Sci. Eng. C* **2017**, *80*, 494–501. [CrossRef] [PubMed]
15. Saleviter, S.; Fen, Y.W.; Daniyal, W.M.E.M.M.; Abdullah, J.; Sadrolhosseini, A.R.; Omar, N.A.S. Design and analysis of surface plasmon resonance optical sensor for determining cobalt ion based on chitosan-graphene oxide decorated quantum dots-modified gold active layer. *Opt. Express* **2019**, *27*, 32294–32307. [CrossRef]
16. Mariani, F.Q.; Borth, K.W.; Müller, M.; Dalpasquale, M.; Anaissi, F.J. Sustainable innovative method to synthesize different shades of iron oxide pigments. *Dyes Pigments* **2017**, *137*, 403–409. [CrossRef]
17. Machala, L.; Zboril, R.; Gedanken, A. Amorphous Iron(III) Oxide- A review. *Phys. Chem. B* **2007**, *111*, 4003–4018. [CrossRef]
18. Machala, L.; Tucek, J.; Zboril, R. Polymorphous transformations of nanometric iron(III) oxide: A review. *Chem. Mater.* **2011**, *23*, 3255–3272. [CrossRef]

19. Zboril, R.; Mashlan, M.; Petridis, D. Iron(III) oxides from thermal processes- synthesis, structural, and magnetic properties, mossbauer spectroscopy characterization and applications. *Chem. Mater.* **2002**, *14*, 969–982. [CrossRef]
20. Al-Rekabi, S.H.; Kamil, Y.M.; Bakar, M.H.A.; Fen, Y.W.; Lim, H.N.; Kanagesan, S.; Mahdi, M.A. Hydrous ferric oxide-magnetite-reduced graphene oxide nanocomposite for optical detection of arsenic using surface plasmon resonance. *Opt. Laser Technol.* **2019**, *111*, 417–423. [CrossRef]
21. Kharisov, B.I.; Dias, H.V.; Kharissova, O.V. Mini-Review: Ferrite nanoparticles in the catalysis. *Arab. J. Chem.* **2019**, *12*, 1234–1246. [CrossRef]
22. Usman, U.A.; Yusoff, I.; Raoov, M.; Hodgkinson, J. The economic potential of the African iron-ore tailings: Synthesis of magnetite for the removal of trace metals in groundwater—A review. *Environ. Earth Sci.* **2019**, *615*, 1–22. [CrossRef]
23. Lapo, B.; Demey, H.; Zapata, J.; Romero, C.; Sastre, A.M. Sorption of Hg(II) and Pb(II) ions on chitosan-Iron(III) from aqueous solutions: Single and binary systems. *Polymers* **2018**, *10*, 367. [CrossRef] [PubMed]
24. Genchi, G.; Sinicropi, M.S.; Carocci, A.; Lauria, G. Mercury exposure and heart diseases. *Int. J. Environ. Res. Public Health* **2017**, *14*, 74. [CrossRef]
25. Li, P.; Feng, X.B.; Qiu, G.L.; Shang, L.H.; Li, Z.G. Mercury pollution in Asia: A review of the contaminated sites. *J. Hazard. Mater.* **2009**, *168*, 591–601. [CrossRef]
26. Reilly, S.B.; Lettmeier, B.; Matteucci, R.; Beinhoff, C.; Siebert, U.; Drasch, V. Mercury as a serious health hazard for children in gold mining areas. *Environ. Res.* **2008**, *107*, 89–97.
27. Miretzky, P.; Cirelli, A.F. Hg(II) removal from water by chitosan and chitosan derivatives: A review. *J. Hazard. Mater.* **2009**, *167*, 10–23. [CrossRef]
28. Roshidi, M.D.A.; Fen, Y.W.; Omar, N.A.S.; Saleviter, S.; Daniyal, W.M.E.M.M. Optical studies of graphene oxide/poly(amidoamine) dendrimer composite thin film and its potential for sensing Hg^{2+} using surface plasmon resonance spectroscopy. *Sens. Mater.* **2018**, *31*, 1147–1156. [CrossRef]
29. Ramdzan, N.S.M.; Fen, Y.W.; Omar, N.A.S.; Saleviter, S.; Zainudin, A.A. Optical and surface plasmon resonance sensing properties for chitosan/carboxyl-functionalized graphene quantum dots thin film. *Optik* **2019**, *178*, 802–812. [CrossRef]
30. Fen, Y.W.; Yunus, W.M.M.; Moksin, M.M.; Talib, Z.A.; Yusof, N.A. Surface plasmon resonance optical sensor for mercury ion detection by crosslinked chitosan thin film. *J. Optoelectron. Adv. Mater.* **2011**, *13*, 279–285.
31. Chen, Z.; Han, K.; Zhang, Y. Reflective fiber surface plasmon resonance sensor for high-sensitive mercury ion detection. *Appl. Sci.* **2019**, *9*, 1480. [CrossRef]
32. Uglov, A.N.; Bessmertnykh-Lemeune, A.; Guilard, R.; Averin, A.D.; Beletskaya, I.P. Optical methods for the detection of heavy metal ions. *Russ. Chem. Rev.* **2014**, *83*, 196–224. [CrossRef]
33. Tang, W.; Li, J.; Guo, Q.; Nie, G. An ultrasensitive electrochemiluminescence assay for Hg^{2+} through graphene quantum dots and poly(5-formylindole) nanocomposite. *Sens. Actuators B Chem.* **2019**, *282*, 824–830. [CrossRef]
34. Wang, B.B.; Jin, J.C.; Xu, Z.Q.; Jiang, Z.W.; Li, X.; Jiang, F.L.; Liu, Y. Single-step synthesis of highly photoluminescent carbon dots for rapid detection of Hg^{2+} with excellent sensitivity. *J. Colloid Interface Sci.* **2019**, *551*, 101–110. [CrossRef]
35. Xu, X.; Zhang, Y.; Wang, B.; Luo, L.; Xu, Z.; Tian, X. A novel surface plasmon resonance sensor based on a functionalized graphene oxide/molecularimprinted polymer composite for chiral recognition of L-tryptophan. *RSC Adv.* **2018**, *8*, 32538–32544. [CrossRef]
36. Kurihara, K.; Nakamura, K.; Suzuki, K. Asymmetric SPR sensor response curve-fitting equation for the accurate determination of SPR resonance angle. *Sens. Actuators B Chem.* **2002**, *86*, 49–57. [CrossRef]
37. Xue, T.; Qi, K.; Hu, C. Novel SPR sensing platform based on superstructure MoS_2 nanosheets for ultrasensitive detection of mercury ion. *Sens. Actuators B Chem.* **2019**, *284*, 589–594. [CrossRef]
38. Fen, Y.W.; Yunus, W.M.M.; Yusof, N.A. Detection of mercury and copper ions using surface plasmon resonance optical sensor. *Sens. Mater.* **2011**, *23*, 325–334.
39. Ramdzan, N.S.M.; Fen, Y.W.; Anas, N.A.A.; Omar, N.A.S.; Saleviter, S. Development of biopolymer and conducting polymer-based optical sensors for heavy metal ion detection. *Molecules* **2020**, *25*, 2548. [CrossRef]
40. Homola, J. Surface plasmon resonance sensors for detection of chemical and biological species. *Chem. Rev.* **2008**, *108*, 462–493. [CrossRef]

41. Lokman, N.F.; Azeman, N.H.; Suja, F.; Arsad, N.; Bakar, A.A. Sensitivity enhancement of Pb(II) ion detection in rivers using SPR-based Ag metallic layer coated with chitosan–graphene oxide nanocomposite. *Sensors* **2019**, *19*, 5159. [CrossRef] [PubMed]
42. Omar, N.A.S.; Fen, Y.W. Recent development of SPR spectroscopy as potential method for diagnosis of dengue virus E-protein. *Sens. Rev.* **2018**, *38*, 106–116. [CrossRef]
43. Fujii, E.; Koike, T.; Nakamura, K.; Sasaki, S.; Kurihara, K.; Citterio, D.; Suzuki, K. Application of an absorption-based surface plasmon resonance principle to the development of SPR ammonium ion and enzyme sensors. *Anal. Chem.* **2002**, *74*, 6106–6110. [CrossRef]
44. Eddin, F.B.K.; Fen, Y.W. Recent Advances in electrochemical and optical sensing of dopamine. *Sensors* **2020**, *20*, 1039.
45. Fen, Y.W.; Yunus, W.M.M. Surface plasmon resonance spectroscopy as an alternative for sensing heavy metal ions: A review. *Sens. Rev.* **2013**, *33*, 305–314.
46. Victoria, S. Application of surface plasmon resonance (SPR) for the detection of single viruses and single biological nano-objects. *J. Bacteriol. Parasitol.* **2012**, *3*, 1–3. [CrossRef]
47. Jia, Y.; Peng, Y.; Bai, J.; Zhang, X.; Cui, Y.; Ning, B.; Cui, J.; Gao, Z. Magnetic nanoparticle enhanced surface plasmon resonance sensor for estradiol analysis. *Sens. Actuators B Chem.* **2017**, *254*, 629–635. [CrossRef]
48. Daniyal, W.M.E.M.M.; Saleviter, S.; Fen, Y.W. Development of surface plasmon resonance spectroscopy for metal ion detection. *Sens. Mater.* **2018**, *30*, 2023–2038. [CrossRef]
49. Omar, N.A.S.; Fen, Y.W.; Saleviter, S.; Daniyal, W.M.E.M.M.; Anas, N.A.A.; Ramdzan, N.S.M.; Roshidi, M.D.A. Development of a graphene-based surface plasmon resonance optical sensor chip for potential biomedical application. *Materials* **2019**, *12*, 1928. [CrossRef]
50. Zhou, C.; Zou, H.; Li, M.; Sun, C.; Ren, D.; Li, Y. Fiber optic surface plasmon resonance sensor for detection of E. coli O157:H7 based on antimicrobial peptides and AgNPs-rGo. *Biosens. Bioelectron.* **2018**, *117*, 347–353. [CrossRef]
51. Anas, N.A.A.; Fen, Y.W.; Omar, N.A.S.; Daniyal, W.M.E.M.M.; Ramdzan, N.S.M.; Saleviter, S. Development of graphene quantum dots-based optical sensor for toxic metal ion detection. *Sensors* **2019**, *19*, 3850. [CrossRef] [PubMed]
52. Abdi, M.M.; Abdullah, L.C.; Sadrolhosseini, A.R.; Yunus, W.M.M.; Moksin, M.F.; Tahir, P.M. Surface Plasmon Resonance Sensing Detection of Mercury and Lead Ions Based on Conducting Polymer Composite. *PLoS ONE* **2011**, *6*, 24578. [CrossRef] [PubMed]
53. Kamaruddin, N.H.; Bakar, A.A.A.; Mobarak, N.N.; Zan, M.S.D.; Arsad, N. Binding affinity of a highly sensitive Au/Ag/Au/chitosan-graphene oxide sensor based on direct detection of Pb^{2+} and Hg^{2+} ions. *Sensors* **2017**, *17*, 2277. [CrossRef]
54. Sadrolhosseini, A.R.; Naseri, M.; Rasyid, S.A. Polypyrrole-chitosan/nickel-ferrite nanoparticle composite layer for detecting heavy metal ions using surface plasmon resonance technique. *Opt. Laser Technol.* **2017**, *93*, 216–223. [CrossRef]
55. Fen, Y.W.; Yunus, W.M.M.; Yusof, N.A. Surface plasmon resonance optical sensor for detection of essential heavy metal ions with potential for toxicity: Copper, zinc and manganese ions. *Sens. Lett.* **2011**, *9*, 1704–1711. [CrossRef]
56. Daniyal, W.M.E.M.M.; Fen, Y.W.; Abdullah, J.; Sadrolhosseini, A.R.; Saleviter, S.; Omar, N.A.S. Exploration of surface plasmon resonance for sensing copper ion based on nanocrystalline cellulose-modified thin film. *Opt. Express* **2018**, *26*, 34880–34893. [CrossRef]
57. Saleviter, S.; Fen, Y.W.; Omar, N.A.S.; Zainudin, A.A.; Yusof, N.A. Development of optical sensor for determination of Co(II) based on surface plasmon resonance phenomenon. *Sens. Lett.* **2017**, *15*, 862–867. [CrossRef]
58. Fen, Y.W.; Yunus, W.M.M.; Yusof, N.A. Surface plasmon resonance optical sensor for detection of Pb^{2+} based on immobilized p-tert-butylcalix[4]arene-tetrakis in chitosan thin film as an active layer. *Sens. Actuators B Chem.* **2012**, *171–172*, 287–293. [CrossRef]
59. Fen, Y.W.; Yunus, W.M.M. Characterization of the optical properties of heavy metal ions using surface plasmon resonance technique. *Opt. Photonics.* **2011**, *1*, 116–123. [CrossRef]
60. Fen, Y.W.; Yunus, W.M.M.; Talib, Z.A.; Yusof, N.A. Development of surface plasmon resonance sensor for determining zinc ion using novel active nanolayers as probe. *Spectrochim. Acta A Mol. Biomol. Spectrosc.* **2015**, *134*, 48–52. [CrossRef]

61. Roshidi, D.A.; Fen, Y.W.; Daniyal, W.M.E.M.M.; Omar, N.A.S.; Zulholinda, M. Structural and optical properties of chitosan–poly(amidoamine) dendrimer composite thin film for potential sensing Pb^{2+} using an optical spectroscopy. *Optik* **2019**, *185*, 351–358. [CrossRef]
62. Anas, N.A.A.; Fen, Y.W.; Omar, N.A.S.; Ramdzan, N.S.M.; Daniyal, W.M.E.M.M.; Saleviter, S.; Zainudin, A.A. Optical properties of chitosan/hydroxyl-functionalized graphene quantum dots thin film for potential optical detection of ferric(III) ion. *Opt. Laser Technol.* **2019**, *120*, 105724. [CrossRef]
63. Rajput, S.; Singh, L.P.; Pittman, C.U., Jr.; Mohan, D. Lead (Pb^{2+}) and copper (Cu^{2+}) remediation from water using superparamagnetic maghemite (γ-Fe_2O_3) nanoparticles synthesized by Flame Spray Pyrolysis (FSP). *J. Colloid Interface Sci.* **2016**, *492*, 176–190. [CrossRef] [PubMed]
64. Nasser, A.M.B. Synthesis and characterization of maghemite iron oxide (γ-Fe_2O_3) nanofibers: Novel semiconductor with magnetic feature. *J. Mater. Sci.* **2016**, *47*, 6237–6245.
65. Coates, J. *Encyclopedia of Analytical Chemistry*; John Wiley & Sons Ltd.: Chichester, NH, USA, 2006; pp. 10815–10837.
66. Zainudin, A.A.; Fen, Y.W.; Yusof, N.A.; Omar, N.A.S. Structural, optical and sensing properties of ionophore doped graphene based bionanocomposite thin film. *Optik* **2017**, *144*, 308–315. [CrossRef]
67. Vujtek, M.; Zboril, R.; Kubinek, R.; Mashlan, M. Ultrafine particles of iron(III) oxides by view of AFM–Novel route for study of polymorphism in nano-world. *Nanomaterials* **2003**, *3*, 1–8.
68. Daniyal, W.M.E.M.M.; Fen, Y.W.; Abdullah, J.; Saleviter, S.; Omar, N.A.S. Preparation and characterization of hexadecyltrimethyl-ammonium bromide modified nanocrystalline cellulose/graphene oxide composite thin film and its potential in sensing copper ion using surface plasmon resonance technique. *Optik* **2018**, *173*, 71–77. [CrossRef]
69. Rattana, T.; Chaiyakun, S.; Witit-anun, N.; Nuntawong, N.; Chindaudom, P. Preparation and characterization of graphene oxide nanosheets. *Procedia Eng.* **2012**, *32*, 759–764. [CrossRef]
70. Kumar, S.; Koh, J. Physiochemical and optical properties of chitosan based graphene oxide bionanocomposite. *Int. J. Biol. Macromol.* **2014**, *70*, 559–564. [CrossRef]
71. Abdulla, H.; Abbo, A. Optical and electrical properties of thin films of polyaniline and polypyrrole. *Int. J. Electrochem. Sci.* **2012**, *7*, 10666–10678.
72. Dhlamini, M.S.; Noto, L.L.; Mothudi, B.M.; Chithambo, M.; Mathevula, L.E. Structural and optical properties of sol-gel derived α-Fe_2O_3 nanoparticles. *J. Lumin.* **2017**, *192*, 879–887.
73. Al-Kuhaili, M.F.; Saleem, M.; Durrani, S.M.A. Optical properties of iron oxide (α-Fe_2O_3) thin films deposited by the reactive evaporation of iron. *J. Alloys Compd.* **2012**, *521*, 178–182. [CrossRef]
74. He, Y.P.; Miao, Y.M.; Li, C.R.; Wang, S.Q.; Cao, L.; Xie, S.S.; Yang, G.Z.; Zou, B.S.; Burda, C. Size and structure effect on optical transitions of iron oxide nanocrystals. *Rev. B Condens. Matter.* **2005**, *71*, 1–9. [CrossRef]
75. Deotale, A.J.; Nandedkar, R.V. Correlation between particle size, strain and band gap of iron oxide nanoparticles. *Mater. Today* **2016**, *3*, 2069–2076. [CrossRef]
76. Fen, Y.W.; Yunus, W.M.M.; Talib, Z.A. Analysis of Pb(II) ion sensing by crosslinked chitosan thin film using surface plasmon resonance spectroscopy. *Optik* **2013**, *124*, 126–133. [CrossRef]
77. Lokman, N.F.; Bakar, A.A.A.; Suja, F.; Abdullah, H.; Rahman, W.B.W.A.; Yaacob, M.H. Highly sensitive SPR response of Au/chitosan/graphene oxide nanostructured thin films toward Pb(II) ions. *Sens. Actuators B Chem.* **2014**, *195*, 459–466. [CrossRef]
78. Fen, Y.W.; Yunus, W.M.M. Utilization of chitosan-based sensor thin films for the detection of lead ion by surface plasmon resonance optical sensor. *IEEE Sens. J.* **2013**, *13*, 1413–1418. [CrossRef]
79. Fen, Y.W.; Yunus, W.M.M.; Yusof, N.A. Optical properties of crosslinked chitosan thin film as copper ion detection using surface plasmon resonance technique. *Opt. Appl.* **2011**, *41*, 999–1013.

© 2020 by the authors. Licensee MDPI, Basel, Switzerland. This article is an open access article distributed under the terms and conditions of the Creative Commons Attribution (CC BY) license (http://creativecommons.org/licenses/by/4.0/).

Article

Dynamics and Rheological Behavior of Chitosan-Grafted-Polyacrylamide in Aqueous Solution upon Heating

Mengjie Wang, Yonggang Shangguan *[ID] and Qiang Zheng

MOE Key Laboratory of Macromolecular Synthesis and Functionalization, Department of Polymer Science and Engineering, Zhejiang University, Hangzhou 310027, China; mengjiewang@zju.edu.cn (M.W.); zhengqiang@zju.edu.cn (Q.Z.)
* Correspondence: shangguan@zju.edu.cn

Received: 11 March 2020; Accepted: 13 April 2020; Published: 15 April 2020

Abstract: In this work, the transformation of chitosan-grafted-polyacrylamide (GPAM) aggregates in aqueous solution upon heating was explored by cryo-electron microscope (cryo-TEM) and dynamic light scattering (DLS), and larger aggregates were formed in GPAM aqueous solution upon heating, which were responsible for the thermo-thickening behavior of GPAM aqueous solution during the heating process. The heating initiates a transformation from H-bonding aggregates to a large-sized cluster formed by self-assembled hydrophobic chitosan backbones. The acetic acid (HAc) concentration has a significant effect on the thermo-thickening behavior of GPAM aqueous solution; there is a critical value of the concentration (>0.005 M) for the thermo-thickening of 10 mg/mL GPAM solution. The concentration of HAc will affect the protonation degree of GPAM, and affect the strength of the electrostatic repulsion between GPAM molecular segments, which will have a significant effect on the state of the aggregates in solution. Other factors that have an influence on the thermo-thickening behavior of GPAM aqueous solution upon heating were investigated and discussed in detail, including the heating rate and shear rate.

Keywords: chitosan-grafted-polyacrylamide; thermo-thickening; rheological; dynamic light scattering; cryo-electron microscope

---

1. Introduction

Thermo-responsive polymers have been extensively studied over the past few decades owing to their industrial and biomedical applications [1–7]. Among them, thermo-thickening has received much attention from academic and industrial fields because of its huge application potential in many fields [3,7–12], especially in oil recovery [9,11]. It is always a great challenge in oil recovery to keep high viscosity of polymer displacement agent at a high temperature because the viscosity of most current oil displacement agents will decrease upon heating or in a high-temperature environment [9,11]. Differing from most of the commonly used polymers, which can hardly solve this problem [7,13], the thermo-thickening polymers whose viscosity can be enhanced during heating process have great potential to be used in enhanced oil recovery as well as water treatment and paper manufacturing [14]. In addition, because the thermo-thickening polymers with high concentration usually present a sol–gel transition temperature, they also have great potential in controlled permeation [7], tissue engineering [13], drug delivery [2,6], and so on.

Chitosan (CS) has received tremendous interest owing to its intrinsic properties, biocompatibility, nontoxicity, amphipathic, accessibility, and abundance in nature [15,16]. It is the precursor for other applications such as drug delivery [17–19], emulsifier [20], water treatment [21–24], and so on.

Furthermore, it was found that CS modified by chemical or physical methods [25,26] could present thermo-thickening behavior and can be roughly divided into two kinds: CS complexes [27–29] and the derivatives of CS [30–32]. For CS complexes, thermo-thickening behaviors were less reported thus far except chitosan/poly(vinyl alcohol) (CS/PVA) [27] and chitosan/glycerophosphate (CS/β-GP) [29]. For CS derivatives, modification by poly-N-isopropylacrylamide (PNIPAM) is a common way to realize thermo-thickening [13,31], and recently, the carboxymethyl chitin also shows a novel thermo-thickening, which is pH- and temperature-dependent [33]. As there are many kinds of molecular interactions involved in these aqueous solutions mentioned above, it is difficult to elucidate the molecular mechanism of the thermo-thickening behavior of these complex systems. For CS/PVA, Schuetz et al. [27] thought that CS linked with PVA through hydrogen bonds at a low temperature. With temperature increasing, hydrogen bonds were broken and hydrophobic association among hydrophobic segments of CS chains gradually increased, as a result the gel-like structure formed. In CS/β-GP, there is no clear graph presented just excluding the influence of the hydrogen bond [29]. Among CS derivatives, the mechanism of thermo-thickening for modification by PNIPAM has been acknowledged [13,31,34] universally. Owing to the lower critical solution temperature (LCST) of PNIAPM, a phase separation will happen once the temperature rises to ~33 °C. Recently, the carboxymethyl chitin also showed a novel thermo-thickening associated with pH-dependence [33]. However, the mechanism is also not clear. By far, except CS-g-PNIPAM, there is no straight evidence to illustrate their assumption or figure out the mechanism.

Chitosan-grafted-polyacrylamide (GPAM) is one of the CS derivatives that has been reported to be a high-efficiency flocculating agent [35–38] and a potential oil-displacing agent [39]. Recently, we observed the thermo-thickening behavior of GPAM aqueous solution and proposed a preliminary outline of the molecular mechanism based on nuclear magnetic resonance (NMR) and transmission electron microscope (TEM) results. It was found the transformation from a hydrogen bonding (H-bonding) aggregate to a hydrophobic aggregate upon heating was responsible for the thermo-thickening [12]. However, some details during the thermo-thickening of GPAM solution are still unclear, such as the effect of acids on thermo-thickening, among others. In this work, we focus on the dynamics and rheological behavior of GPAM aqueous solution upon heating. GPAM samples with various grafting ratios, as shown in Table S1, are used to explore the thermo-induced structure in aqueous solutions using dynamic light scattering (DLS) and cryo-electron microscope (cryo-TEM). The influences of acid concentration, ramp rate, and shear rate on thermo-thickening of GPAM aqueous solutions are addressed.

2. Experimental Section

2.1. Materials

Chitosan (CS) powder and acetic acid-d_4 (99.9%) were purchased from Sigma and Aldrich, Shanghai, China. Acetone (99.5%) and sodium hydroxide (96.0%) were purchased from Sinopharm Chemical Reagent Co., Ltd., Shanghai, China. Ammonium ceric nitrate (CAN, 99.0%), acetic acid (98.0%), acrylamide (99.0%), and deuterium oxide (D_2O, 99.0%) were purchased from Aladdin, Shanghai, China. All chemicals and reagents were used without further purification.

2.2. Sample Preparation

Highly deacetylated chitosan was obtained by intermittent alkali treatment. Twenty (20) grams (g) of CS with a deacetylation of 78.9% was added into a mixture of distilled water (600 mL) and sodium hydroxide (300 g), which was stirred for 20 min at 110 °C. Then, the mixture was heated to 110 °C with mechanical agitation. After 1 h of mechanical agitation, the mixture was cooled down to room temperature and the precipitates were filtrated and washed by ultrapure water to neutrality. The chitosan, after being washed in water, was treated again in the alkaline solution for further deacetylation. The collected precipitate was dried at 50 °C until reaching a constant weight. The deacetylation degree of CS was increased by the intermittent alkali treatment to 98.0%, as determined by proton nuclear magnetic resonance (^{1}H-NMR) using a Bruker 500 spectrometer (500 MHz) (Bruker, Karlsruhe, Germany) at room temperature (see Figure S1 in Supporting Information), and the calculation method of the degree of deacetylation is shown in Equation S1 in Supporting Information. The synthesis of chitosan-*g*-polyacrylamide (GPAM) has been presented in the previous work [12]. The graft ratio of GPAM was characterized by ^{1}H-NMR (see Figure S2 in Supporting Information), and the calculation method of the graft ratio is shown in Equation S2. In this work, we synthesized GPAM samples with three kinds of grafting rate, as shown in Table S1.

2.3. Rheological Measurements

Rheological experiments were performed on a stress-controlled rotational rheometer, Discovery Hybrid Rheometer-2 (DHR, TA Instruments, Newcastle, DE, USA). A 40 mm cone-plate geometry with a 2° cone angle and a 50 mm gap size was chosen for the steady shear tests. A 40 mm parallel plate geometry with a 500 mm gap size was chosen for all of the dynamic rheological tests. A defined amount of sample solution was directly poured very slowly onto the Peltier region in order to avoid the shear thinning effect and small air bubbles caused by using pipettes. Liquid paraffin was coasted around the margin of the solution sample to prevent the evaporation of solvent. Oscillatory temperature sweep tests were carried out using a strain amplitude of 0.2% (within the linear viscoelastic region, LVR), an oscillatory frequency of 6.283 rad·s^{-1}, and a heating rate of 5 °C/min, unless otherwise stated.

2.4. Dynamic Light Scattering (DLS) and Ultraviolet and Visible Spectrum (UV)

The hydrodynamic radius R_h and size distribution were measured by dynamic light scattering (DLS) at the scattering angle of 90° using a 90 plus particle size analyzer (Brookhaven Instruments Corp., Holtsville, NY, USA) The wavelength of laser light was 635 nm. The CONTIN program was used for the analysis of dynamic light scattering data. The sample was filtered through the Millipore filters of 1 μm and 0.45 μm, successively, which was repeated at least three times.

The transmittance of GPAM solution was analyzed by a Lambda 35 UV/vis absorption spectrometer (PerkinElmer, Waltham, MA, USA) at 20 °C.

3. Results and Discussion

3.1. Transformation of GPAM Aggregates upon Heating

The structural transformation of GPAM in the solution sample upon heating was investigated by TEM and DLS in our previous work [12]. However, the microstructures of GPAM aqueous solution observed by TEM, which were obtained from dried solution samples, could not be the same as those in water. In addition, the concentration of GPAM used for DLS in the previous work is too low to observe the thermo-thickening process, because DLS is not appropriate to investigate high concentration polymer solution samples. Considering the possible impact of the above facts, in this work, we made appropriate improvements to investigate the macromolecular mechanism of thermo-thickening. Compared with ordinary TEM, the sample preparation method for cryo-TEM is to rapidly freeze the solution sample with liquid ethane, and then observe the frozen sample under low temperature and vacuum conditions [40,41]; as a result, the true structural form of GPAM in solution is preserved to the greatest extent. So, we used cryo-TEM instead of TEM to examine the structure of GPAM in aqueous solution. In addition, the transformation of GPAM aggregates in solution upon the heating process was investigated by increasing the concentration of GPAM solution used for DLS measurement while guaranteeing the existence of the thermo-thickening phenomenon.

As the viscosity of the GPAM solution may start to increase when the temperature is higher than 20 °C, as reported previously [12], here, we investigate the macromolecular architecture in solution at 10 °C. To understand the evolution of the macromolecular architecture of GPAM upon heating, a thermal treatment at 40 °C for 10 min was applied to the solution sample. In addition, the thermo-thickening curve of the GPAM solution sample held at 40 °C was also investigated for comparison, as shown in the inset of Figure 1a. It can be observed that the viscosity of the solution sample basically tends to constant value after 10 min at 40 °C. As no thermo-thickening appears at 10 °C, the size distribution of GPAM is shown in Figure 1a. Only a single peak at about 200 nm appears for the original GPAM solution at 10 °C, which should be attributed to the aggregations of serval macromolecules rather than a single chain. After the sample was heated and held at 40 °C for 10 min, a bimodal distribution arises at 10 °C; a peak of about 400 nm and a peak of about several thousand nm. These results suggest that the aggregates with larger size form upon heating compared with the original sample.

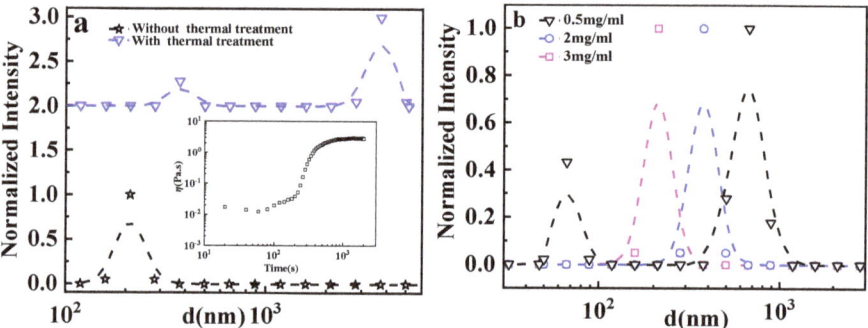

Figure 1. (a) Size distribution at 10 °C of 3 mg/mL chitosan–grafted–polyacrylamide 1 (GPAM1) solutions (0.01 M HAc) without and with a thermal treatment of holding at 40 °C for 3 min; inset displays the viscosity evolution of 3 mg/mL GPAM1 solutions with 0.01 M HAc at 40 °C. (b) Dependence of size distribution for GPAM1 solutions with 0.01 M HAc on concentration at 10 °C.

It should be pointed out that the size distribution of GPAM aggregates in the initial sample measured here is about several hundred nanometers, and more aggregations with larger size appear during the thermo-thickening process. These results are because of the fact that we use a higher solution concentration of 3 mg/mL, which is closer to the truth of the GPAM molecular mechanism of the thermo-thickening process, rather than using a lower concentration solution to demonstrate the conformation and aggregation changes of GPAM molecules in the previous report [12]. Figure 1b gives the aggregation information of GPAM at different concentrations. When the concentration is low enough, about 0.5 mg/mL, it shows a bimodal distribution. The peak with a smaller size distribution, about ~80 nm, corresponds to single chain conformation, and the peak with a larger size, about ~700 nm, corresponds to macromolecular aggregates. In the cellulose solution and other solutions, the double peaks distribution was also reported [42]. With the increase of concentration, the single chain conformation disappears and the size of aggregations becomes smaller.

The size evolution of GPAM in aqueous solution at 40 °C is given in detail in Figure 2. The size distribution of GPAM always presents a single peak, while the size and peak width increase gradually until 6 min, indicating the formation of larger aggregations. After 11 min, the peak gradually evolves into a bimodal distribution; a larger single peak at about 3000 nm and a small peak at about 350 nm. The result is in accordance with Figure 1a. As time goes by, another smaller single peak at about 70 nm appears at 16 min and 21min, which corresponds to the single chain size. This is a very interesting result, because it means that some molecules not only do not contribute to thickening, but exist in the solution as single molecules. This indicated that the newborn lager aggregations upon heating were unstable with the increasing thermal treatment time; when the size of the association continues to increase, a small number of GPAM molecules are separated from the aggregates owing to the damage of hydrogen bonding, and exist as single molecules When the time reaches 21 min, there are even three peaks in the GPAM solution, indicating the macromolecular do exist in complex and heterogeneous forms during thermo-thickening.

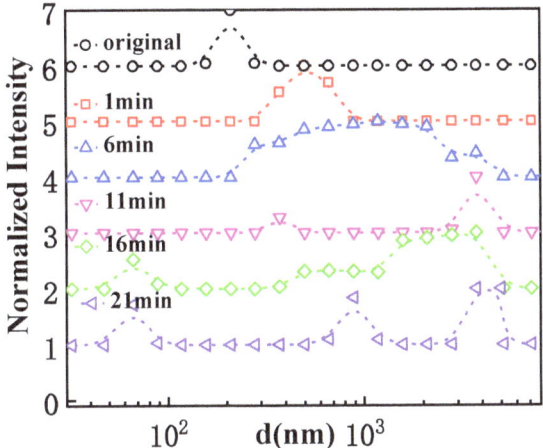

Figure 2. Size distribution of 3 mg/mL GPAM1 solutions with 0.01 M HAc at different times at 40 °C.

During ordinary TEM sample preparation, there is a process of volatilization of the solvent and the solute in the solution will accumulate, so it is necessary to use a lower concentration sample for observation. The GPAM samples for cryo-TEM observation are obtained by rapidly freezing with liquid ethane, so it could realistically show the morphological feature of GPAM aggregates in solution. When the sample concentration is very low, the target content in the observation field is so low that it is difficult to observe. When the GPAM concentration reaches 10 mg/mL, the viscosity is relatively large, which is not conducive for sample preparation; therefore, we selected 6 mg/mL GPAM aqueous solution with 0.02 M HAc for cryo-TEM observation.

Figure 3a,b are the cryo-TEM images of the samples of GPAM solutions aged for 30 min at 10 °C and 40 °C respectively. The cryo-TEM images exhibit a dark domain and a sparse dark region. We simply consider the dark domain as aggregates and the sparse dark region as loose structures. Compared with Figure 3a, these aggregates gathered together and formed a much larger cluster structure. This result is consistent with the results measured by DLS above and confirms the formation of larger-size aggregates in GPAM solution during the heating process.

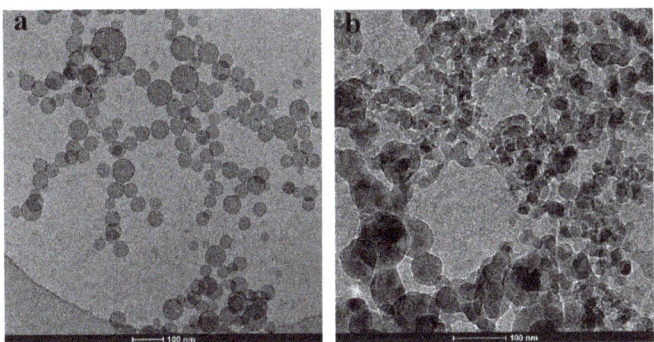

Figure 3. Cryo–electron microscope (cryo–TEM) observations of 6 mg/mL GPAM2 in 0.02 M HAc solution at (**a**) 10 °C and (**b**) 40 °C.

3.2. Effects of HAc on GPAM Solution

As GPAM aqueous solution in the presence of HAc, the effect of HAc on the existence of GPAM must be considered. When GPAM is dispersed in acetic acid solution at different concentrations, its ionization in water will obey the following law:

$$GPAM-NH_2 + HAc \xrightarrow{K} GPAM-NH_3^+ + Ac^- \quad (1)$$

Usually, the pKa value of CS is about 6.5 [43], and the pka value of HAc is about 4.76 [44]:

$$K = \frac{[-NH_3^+][Ac^-]}{[-NH_2][HAc]} = \frac{[Ac^-][H^+]}{[HAc]} \times \frac{[-NH_3^+]}{[-NH_2][H^+]} = \frac{Ka_{HAc}}{Ka_{CS}} \quad (2)$$

According to Equation (1) and (2), the relationship between the degree of protonation of the amino group and the concentration of HAc can be obtained, as shown in Figure S3 in Supporting Information. As the concentration of HAc increases, the degree of protonation of the amino group increases.

Figure 4 gives the transparency for the GPAM solution with the increasing HAc concentration. It is stable when the concentration of HAc is higher than 0.035 M. The macromolecules and aggregates in GPAM solution decrease gradually with the increase of HAc concentration, as shown in the inset of Figure 4. This is because the protonation of –NH$_2$ is enhanced with the increasing HAc; the electrostatic repulsion between GPAM molecular segments increases accordingly; and, consequently, the hydrogen bond interaction among aggregations is gradually weakened, resulting in a decrease of the aggregation's size.

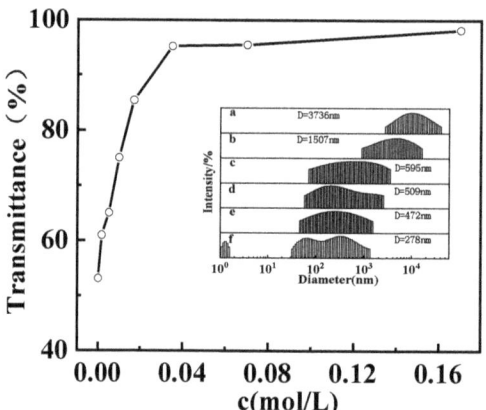

Figure 4. Optical transmittance at 600 nm for 10 mg/mL GPAM2 aqueous solutions as a function of HAc concentration at 20 °C. Inset displays the particle size of 10 mg/mL GPAM2 aqueous solutions with the HAc concentration of (a) 0.00175 M, (b) 0.0105 M, (c) 0.0175 M, (d) 0.0351 M, (e) 0.0702 M, and (f) 0.175 M.

Figure 5a gives the steady flow results of GPAM at different HAc concentrations. All samples show obvious shear thinning. With the increase of HAc, the viscosity of GPAM solutions decreases, especially at a low shear rate. All GPAM solution samples present a constant viscosity in the low shear rate region, so zero-shear-rate viscosity (η_0) can be obtained from steady flow curves, as listed in Table 1. η_0 decrease with the increasing concentration of HAc indicates the decrease of the aggregation's size, which is in accordance with the above results. However, irreversible thermo-thickening can be found in the GPAM solution with 0.01 M HAc after a thermal treatment (holding at 60 °C for 15 min) rather than the GPAM solutions without HAc. As shown in Figure 5b, the GPAM solution presents a slight increase of viscosity at a low shear rate after thermal treatment, while viscosity for the GPAM solution with 0.01 M HAc increases sharply at a low shear rate. Those results indicate that HAc plays an important role in the thermo-thickening behavior of the GPAM solution.

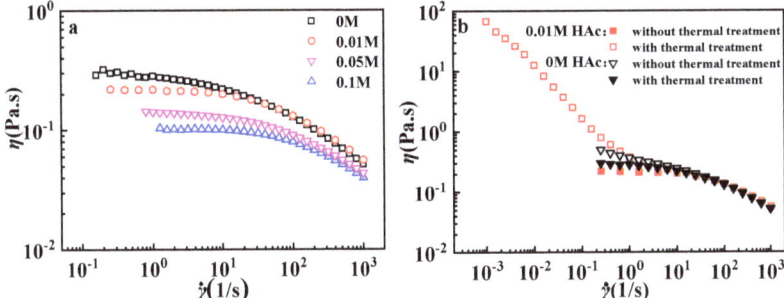

Figure 5. Steady flow curves of 10 mg/mL GPAM2 solutions (a) with different HAc concentration and (b) with and without a thermal treatment (holding at 60 °C for 15 min). All tests were conducted at 10 °C.

Table 1. η_0 of 10 mg/mL chitosan-grafted-polyacrylamide 2 (GPAM2) solutions with different HAc concentration at 10 °C.

HAc (mol/L)	η_0 (Pa·s)
0	0.28
0.01	0.22
0.05	0.15
0.1	0.10

To further investigated the effect of acid on the structure evolution of the GPAM solution during the above thermo-cycle, frequency sweeps for GPAM solutions subjected to thermal cycle were conducted at 10 °C. In Figure 6a, moduli for the two GPAM solutions without HAc have the similar tendency: G' lower than G'' in the low frequency regime and G' larger than G'' in the high frequency regime, while their G'' are close in the whole investigation range. In general, $G' \sim \omega^2$ and $G'' \sim \omega$ mean the existence of a homogenous structure; the decreased exponent indicates the existence of a physical network or aggregations [27]. In Figure 6a, the exponent is smaller than the theoretical value, indicating the existence of aggregations in the GPAM solution without HAc in a sense. Figure 6b gives the frequency sweep results for solution samples containing HAc. When the sample is subjected to thermo-cycle, G' is larger than G'' in the low frequency regime and the modules present a weak frequency dependency. This suggested a sol-to-gel transformation had taken place. On the contrary, the GPAM solutions without thermo-cycle also seem sol-like. Therefore, GPAM solution containing HAc presents an irreversible structure evolution from aggregations to a more structured fluid in the thermo-cycle for GPAM solutions.

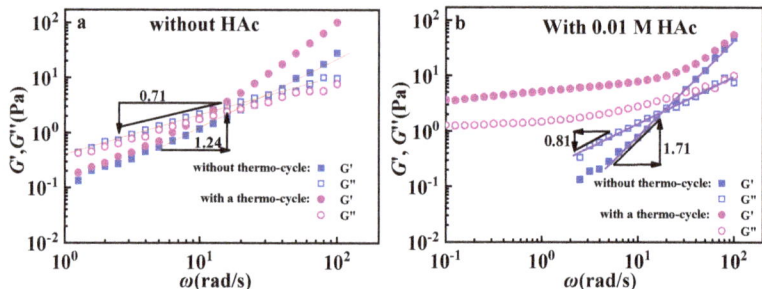

Figure 6. Frequency sweeps of 10 mg/mL GPAM2 solutions (**a**) without HAc and (**b**) with 0.01 M HAc subjected to thermo–cycle (temperature increases from 10 °C to 60 °C, then decreases to 10 °C, ramp rate: 5 °C/min) at 10 °C.

The thermo-thickening process of GPAM solutions was investigated as a function of HAc concentration; Figure 7 gives the temperature sweep results of GPAM solution samples with different HAc concentrations and the mass concentration of the GPAM solution is fixed at 10 mg/mL. The different HAc concentrations determine the content of the protonation of $-NH_2$, that is, in the range of the HAc concentration of 0.00175 M to 0.175 M, the larger the concentration of HAc, the larger amount of the protonation of $-NH_2$ of the moiety. As shown in the inset of Figure 7, the onset temperature (the critical temperature at the onset of an increase in viscosity) increases with the concentration of HAc. This may be attributed to stronger hydrophilicity and electrostatic repulsion induced by more protonation of $-NH_2$, whichs inhibit the hydrophobic aggregation. The onset temperature obviously decreases when the concentration of HAc decreases to 0.01 M. Moreover, both GPAM solutions containing 0.0017 M and 0.0052 M HAc present little thermo-thickening, indicating that there is a critical value of the concentration (>0.005 M) for the thermo-thickening of 10 mg/mL GPAM solution. This is concentration of HAc may be too low, leading to the protonation of GPAM being significantly less and a correspondingly small amount of positive charges, so the electrostatic repulsive force between GPAM molecules is very weak and the water solubility is poor, which leads to the shrinkage of molecular segments, and the molecules are tightly bonded through hydrogen bonding. These densely structured hydrogen-bonded associations cannot be destroyed easily during the heating process, so GPAM molecules cannot be reorganized in large-sized association structures through hydrophobic association, and the thermo-thickening behavior cannot be exhibited. When the HAc concentration reaches a certain value, the water solubility of GPAM increases and the electrostatic repulsion will destroy some of the intermolecular hydrogen bonds, so the structure of the GPAM association will be loose in the aqueous solution. The molecular chain can be extracted from the hydrogen-bonded association and form a large-sized association structure through hydrophobic association, thereby exhibiting thermo-thickening behavior. It is worth mentioning that when the HAc concentration is 0.01 M, the onset temperature of thermo-thickening is the lowest, which means the thermo-thickening structures can be formed easily; therefore, we chose the GPAM aqueous solution with 0.01 M HAc for the following rheological experiments.

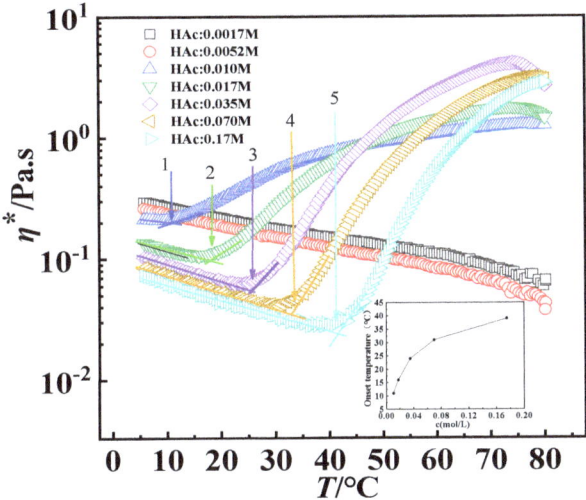

Figure 7. Influence of HAc concentration on thermo–thickening for the GPAM aqueous solution with 10 mg/mL concentration. Inset displays the relationship between onset temperature and concentration of HAc in the 10 mg/mL GPAM aqueous solution. The strain is 0.2% and the oscillatory frequency is 6.283 rad·s^{-1}. The heating rate of the samples is 5 °C/min.

3.3. Influence of Heating Rate and Shear Rate

Figure 8a gives the influences of heating rate on the thermo-thickening behavior. The sample presents a lower T_{trans} (the temperature for G'–G'' crossover) at a slow heating rate, indicating T_{trans} has a dependence of heating rate. When the heating rate was 0.5 °C, G' and G'' of the GPAM aqueous solution after thermo-cycle treatment were significantly greater than those of 5 °C and 10 °C. These results indicate that a slower heating rate is conducive to the formation of thermo-thickening structures. The thermo-thickening behavior of the GPAM aqueous solution is dependent on the heating time and temperature; the thermo-thickening structure of the GPAM aqueous solution will be more complete at a longer heating time or a higher temperature [12]. When the heating rate is lower, it means that the heating time and the residence time in the high temperature region are longer, so the T_{trans} will be lower and G', G'' will be greater. Furthermore, it can be found from Figure 8a,b that modulus variation with temperature and ω were consistent with each other for 5 °C/min and 10 °C/min. The possible reason may lie in that structure evolution cannot keep up with the ramp rate to cause no apparent discrepancy in a higher ramp rate.

In order to study the shear-resistance of GPAM after thermo-thickening, a serious of shear recover experiments were carried out by fixing different maximal shear rates. As shown in Figure 9, the increased viscosity of the GPAM solution after thermo-treatment can remain. Furthermore, with the increase of shear rate, the viscosity shear thinning occurs, while its viscosity can recover well as the shear rate decreases in 0.1~0.001 s^{-1} (Figure 9a). With the shear rate range increasing, the viscosity cannot recover in time as the shear rate decreases in 10~0.001 s^{-1}, meaning that the damaged associated structure of GPAM in solution needs more time to achieve complete recovery (Figure 9b). When the maximum shear rate reached 100 s^{-1} or 1000^{-1}, the structure was broken thoroughly and came back to the original structure without thermal treatment (Figure 8c,d).

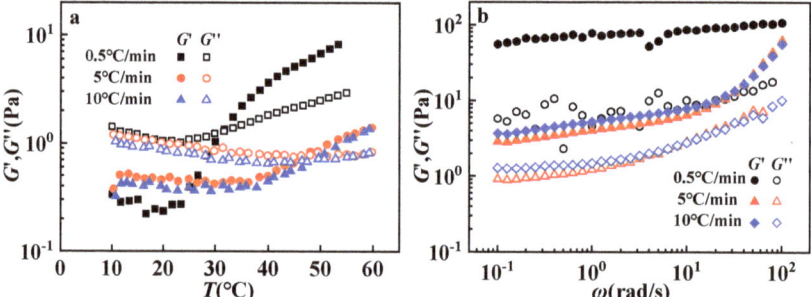

Figure 8. (**a**) Temperature sweeps of 10 mg/mL GPAM2 solutions with 0.01 M HAc during the heating process at different ramp rates; (**b**) frequency sweeps at 10 °C for 10 mg/mL GPAM2 solutions with 0.01 M HAc after a thermal cycle with different ramp rates (from 10 to 60 to 10 °C).

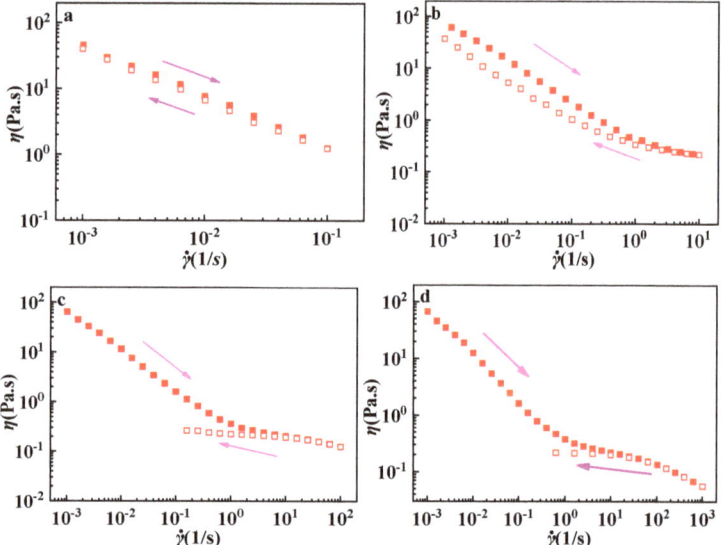

Figure 9. Shear and recovery rates of 10 mg/mL GPAM2 solutions with 0.01 M HAc in different limits of the upper shear rate of (**a**) 0.1 s^{-1}, (**b**) 10 s^{-1}, (**c**) 100 s^{-1}, and (**d**) 1000 s^{-1}, with a thermal treatment (holding at 60 °C for 15 min). All tests were conducted at 10 °C.

4. Conclusions

During the heating process, large-size aggregates were formed in the GPAM aqueous solution through hydrophobic association from the hydrophobic groups on GPAM, which were responsible for the thermo-thickening of the GPAM aqueous solution. As the concentration of HAc in the GPAM aqueous solution increased, the protonation degree of GPAM increased and the electrostatic repulsive force between GPAM molecules would gradually increase, so the size of the GPAM aggregates in the solution gradually decreased. When the concentration of HAc was less than 0.05 M, the protonation degree of GPAM was very low and the solubility was very poor, and then a dense hydrogen bonding association was formed in the solution, which cannot be destroyed during the heating process. As a result, the thermo-thickening behavior disappeared. Furthermore, a higher HAc concentration leads to stronger hydrophilicity and electrostatic repulsion induced by more protonation of –NH$_2$, which inhibit the hydrophobic aggregation. In addition, a slower heating rate is conducive to the formation of thermo-thickening structures and a strong shear rate will destroy the thermo-thickening structure of GPAM.

Supplementary Materials: The following are available online at http://www.mdpi.com/2073-4360/12/4/916/s1, Figure S1: ^{1}H-NMR spectrums of 0.5 wt.% CS solution in an acetic acid-d$_4$/water-d$_2$ mixture at room temperature (the DDA value of samples of (a) and (b) are 79% and 98% respectively). Figure S2: ^{1}H-NMR spectrums of CS and GPAM in an acetic acid-d4/water-d2 mixture at room temperature. Figure S3: Relationship between the degree of protonation of the amino group and the concentration. Table S1: Molecular parameters of GPAM.

Author Contributions: Conceptualization, M.W.; methodology, M.W.; software, M.W.; validation, M.W.; formal analysis, M.W.; investigation, M.W.; resources, Y.S.; data curation, M.W.; writing—original draft preparation, M.W.; writing—review and editing, Q.Z.; visualization, M.W.; supervision, Y.S.; project administration, Y.S.; funding acquisition, Y.S. All authors have read and agreed to the published version of the manuscript.

Acknowledgments: This work was supported by the National Nature Science Foundation of China (Grant No. 51473145, 51773174, 51973467) and Zhejiang Provincial Natural Science Foundation of China (No. LR16E030002).

Conflicts of Interest: The authors declare no conflict of interest.

References

1. Cammas, S.; Suzuki, K.; Sone, C.; Sakurai, Y.; Kataoka, K.; Okano, T. Thermo-responsive polymer nanoparticles with a core-shell micelle structure as site-specific drug carriers. *J. Control. Release* **1997**, *48*, 157–164. [CrossRef]
2. Censi, R.; Vermonden, T.; van Steenbergen, M.J.; Deschout, H.; Braeckmans, K.; De Smedt, S.C.; van Nostrum, C.F.; di Martino, P.; Hennink, W.E. Photopolymerized thermosensitive hydrogels for tailorable diffusion-controlled protein delivery. *J. Control. Release* **2009**, *140*, 230–236. [CrossRef] [PubMed]
3. Chen, Y.; Shull, K.R. Thermothickening behavior of self-stabilized colloids formed from associating polymers. *Macromolecules* **2019**, *52*, 4926–4933. [CrossRef]
4. Hourdet, D.; L'Alloret, F.; Audebert, R. Reversible thermothickening of aqueous polymer solutions. *Polymer* **1994**, *35*, 2624–2630. [CrossRef]
5. Liu, H.; Yang, Q.; Zhang, L.; Zhuo, R.; Jiang, X. Synthesis of carboxymethyl chitin in aqueous solution and its thermo- and pH-sensitive behaviors. *Carbohydr. Polym.* **2016**, *137*, 600–607. [CrossRef] [PubMed]
6. Liu, J.; Huang, W.; Pang, Y.; Zhu, X.; Zhou, Y.; Yan, D. Self-assembled micelles from an amphiphilic hyperbranched copolymer with polyphosphate arms for drug delivery. *Langmuir* **2010**, *26*, 10585–10592. [CrossRef]
7. McCormick, C.L.; Nonaka, T.; Johnson, C.B. Water-soluble copolymers: 27. Synthesis and aqueous solution behaviour of associative acrylamideN-alkylacrylamide copolymers. *Polymer* **1988**, *29*, 731–739. [CrossRef]
8. Hourdet, D.; Gadgil, J.; Podhajecka, K.; Badiger, M.V.; Brûlet, A.; Wadgaonkar, P.P. Thermoreversible behavior of associating polymer solutions: Thermothinning versus thermothickening. *Macromolecules* **2005**, *38*, 8512–8521. [CrossRef]
9. Sabhapondit, A.; Borthakur, A.; Haque, I. Characterization of acrylamide polymers for enhanced oil recovery. *J. Appl. Polym. Sci.* **2003**, *87*, 1869–1878. [CrossRef]
10. Schmaljohann, D. Thermo- and pH-responsive polymers in drug delivery. *Adv. Drug Deliv. Rev.* **2006**, *58*, 1655–1670. [CrossRef]
11. Wever, D.A.Z.; Picchioni, F.; Broekhuis, A.A. Polymers for enhanced oil recovery: A paradigm for structure–property relationship in aqueous solution. *Prog. Polym. Sci.* **2011**, *36*, 1558–1628. [CrossRef]
12. Shangguan, Y.G.; Liu, M.G.; Jin, L.; Wang, M.J.; Wang, Z.K.; Wu, Q.; Zheng, Q. Thermo-thickening behavior and its mechanism in a chitosan-graft-polyacrylamide aqueous solution. *Soft Matter* **2018**, *14*, 6667–6677. [CrossRef] [PubMed]
13. Seetapan, N.; Mai-ngam, K.; Plucktaveesak, N.; Sirivat, A. Linear viscoelasticity of thermoassociative chitosan-g-poly(N-isopropylacrylamide) copolymer. *Rheol. Acta* **2006**, *45*, 1011–1018. [CrossRef]
14. Wang, J.; Chen, Z.; Mauk, M.; Hong, K.S.; Li, M.; Yang, S.; Bau, H.H. Self-actuated, thermo-responsive hydrogel valves for lab on a chip. *Biomed. Microdevices* **2005**, *7*, 313–322. [CrossRef]
15. Huang, X.; Pang, Y.; Liu, Y.; Zhou, Y.; Wang, Z.; Hu, Q. Green synthesis of silver nanoparticles with high antimicrobial activity and low cytotoxicity using catechol-conjugated chitosan. *RSC Adv.* **2016**, *6*, 64357–64363. [CrossRef]

16. Bhattarai, N.; Gunn, J.; Zhang, M. Chitosan-based hydrogels for controlled, localized drug delivery. *Adv. Drug Deliv. Rev.* **2010**, *62*, 83–99. [CrossRef]
17. Yu, S.; Zhang, X.; Tan, G.; Tian, L.; Liu, D.; Liu, Y.; Yang, X.; Pan, W. A novel pH-induced thermosensitive hydrogel composed of carboxymethyl chitosan and poloxamer cross-linked by glutaraldehyde for ophthalmic drug delivery. *Carbohydr. Polym.* **2017**, *155*, 208–217. [CrossRef]
18. Li, X.; Tsibouklis, J.; Weng, T.; Zhang, B.; Yin, G.; Feng, G.; Cui, Y.; Savina, I.N.; Mikhalovska, L.; Sandeman, S. Nano carriers for drug transport across the blood–brain barrier. *J. Drug Target.* **2017**, *25*, 17–28. [CrossRef]
19. Li, L.; Jiang, G.; Yu, W.; Liu, D.; Chen, H.; Liu, Y.; Tong, Z.; Kong, X.; Yao, J. Preparation of chitosan-based multifunctional nanocarriers overcoming multiple barriers for oral delivery of insulin. *Mater. Sci. Eng. C* **2017**, *70*, 278–286. [CrossRef]
20. Mwangi, W.W.; Ho, K.W.; Tey, B.T.; Chan, E.-S. Effects of environmental factors on the physical stability of pickering-emulsions stabilized by chitosan particles. *Food Hydrocolloid* **2016**, *60*, 543–550. [CrossRef]
21. Subramani, S.E.; Thinakaran, T. Isotherm, kinetic and thermodynamic studies on the adsorption behaviour of textile dyes onto chitosan. *Process Saf. Environ. Prot.* **2017**, *106*, 1–10.
22. Shariful, M.I.; Sharif, S.B.; Lee, J.J.L.; Habiba, U.; Ang, B.C.; Amalina, M.A. Adsorption of divalent heavy metal ion by mesoporous-high surface area chitosan/poly (ethylene oxide) nanofibrous membrane. *Carbohydr. Polym.* **2017**, *157*, 57–64. [CrossRef] [PubMed]
23. Habiba, U.; Afifi, A.M.; Salleh, A.; Ang, B.C. Chitosan/(polyvinyl alcohol)/zeolite electrospun composite nanofibrous membrane for adsorption of Cr^{6+}, Fe^{3+} and Ni^{2+}. *J. Hazard. Mater.* **2017**, *322*, 182–194. [CrossRef] [PubMed]
24. Albadarin, A.B.; Collins, M.N.; Naushad, M.; Shirazian, S.; Walker, G.; Mangwandi, C. Activated lignin-chitosan extruded blends for efficient adsorption of methylene blue. *Chem. Eng. J.* **2017**, *307*, 264–272. [CrossRef]
25. Zia, K.M.; Tabasum, S.; Nasif, M.; Sultan, N.; Aslam, N.; Noreen, A.; Zuber, M. A review on synthesis, properties and applications of natural polymer based carrageenan blends and composites. *Int. J. Biol. Macromol.* **2017**, *96*, 282–301. [CrossRef]
26. Rinaudo, M. Chitin and chitosan: Properties and applications. *Prog. Polym. Sci.* **2006**, *31*, 603–632. [CrossRef]
27. Schuetz, Y.B.; Gurny, R.; Jordan, O. A novel thermoresponsive hydrogel based on chitosan. *Eur. J. Pharm. Biopharm.* **2008**, *68*, 19–25. [CrossRef]
28. Pakravan, M.; Heuzey, M.-C.; Ajji, A. Determination of phase behavior of poly(ethylene oxide) and chitosan solution blends using rheometry. *Macromolecules* **2012**, *45*, 7621–7633. [CrossRef]
29. Chenite, A.; Buschmann, M.; Wang, D.; Chaput, C.; Kandani, N. Rheological characterisation of thermogelling chitosan/glycerol-phosphate solutions. *Carbohydr. Polym.* **2001**, *46*, 39–47. [CrossRef]
30. Cho, J.; Heuzey, M.-C.; Bégin, A.; Carreau, P.J. Effect of urea on solution behavior and heat-induced gelationof chitosan-β-glycerophosphate. *Carbohydr. Polym.* **2006**, *63*, 507–518. [CrossRef]
31. Recillas, M.; Silva, L.L.; Peniche, C.; Goycoolea, F.M.; Rinaudo, M.; Argüelles-Monal, W.M. Thermoresponsive behavior of chitosan-g-n-isopropylacrylamide copolymer solutions. *Biomacromolecules* **2009**, *10*, 1633–1641. [CrossRef] [PubMed]
32. Bhattarai, N.; Ramay, H.R.; Gunn, J.; Matsen, F.A.; Zhang, M. PEG-grafted chitosan as an injectable thermosensitive hydrogel for sustained protein release. *J. Control. Release* **2005**, *103*, 609–624. [CrossRef] [PubMed]
33. Niang, P.M.; Huang, Z.; Dulong, V.; Souguir, Z.; Le Cerf, D.; Picton, L. Thermo-controlled rheology of electro-assembled polyanionic polysaccharide (alginate) and polycationic thermo-sensitive polymers. *Carbohydr. Polym.* **2016**, *139*, 67–74. [CrossRef] [PubMed]
34. Bao, H.; Li, L.; Leong, W.C.; Gan, L.H. Thermo-responsive association of chitosan-graft-poly(n-isopropylacrylamide) in aqueous solutions. *J. Phys. Chem. B* **2010**, *114*, 10666–10673. [CrossRef]
35. Lu, Y.; Shang, Y.; Huang, X.; Chen, A.; Yang, Z.; Jiang, Y.; Cai, J.; Gu, W.; Qian, X.; Yang, H.; et al. Preparation of strong cationic chitosan-graft-polyacrylamide flocculants and their flocculating properties. *Ind. Eng. Chem. Res.* **2011**, *50*, 7141–7149. [CrossRef]
36. Sokker, H.H.; El-Sawy, N.M.; Hassan, M.A.; El-Anadouli, B.E. Adsorption of crude oil from aqueous solution by hydrogel of chitosan based polyacrylamide prepared by radiation induced graft polymerization. *J. Hazard. Mater.* **2011**, *190*, 359–365. [CrossRef]

37. Wang, J.-P.; Chen, Y.-Z.; Zhang, S.-J.; Yu, H.-Q. A chitosan-based flocculant prepared with gamma-irradiation-induced grafting. *Bioresour. Technol.* **2008**, *99*, 3397–3402. [CrossRef]
38. Yang, Z.; Yuan, B.; Huang, X.; Zhou, J.; Cai, J.; Yang, H.; Li, A.; Cheng, R. Evaluation of the flocculation performance of carboxymethyl chitosan-graft-polyacrylamide, a novel amphoteric chemically bonded composite flocculant. *Water Res.* **2012**, *46*, 107–114. [CrossRef]
39. Yuan, B.; Shang, Y.; Lu, Y.; Qin, Z.; Jiang, Y.; Chen, A.; Qian, X.; Wang, G.; Yang, H.; Cheng, R. The flocculating properties of chitosan-graft-polyacrylamide flocculants (I)—Effect of the grafting ratio. *J. Appl. Polym. Sci.* **2010**, *117*, 1876–1882. [CrossRef]
40. Dubochet, J.; Lepault, J.; Freeman, R.; Berriman, J.A.; Homo, J.-C. Electron microscopy of frozen water and aqueous solutions. *J. Microsc.* **1982**, *128*, 219–237. [CrossRef]
41. Milne, J.L.S.; Borgnia, M.J.; Bartesaghi, A.; Tran, E.E.H.; Earl, L.A.; Schauder, D.M.; Lengyel, J.; Pierson, J.; Patwardhan, A.; Subramaniam, S. Cryo-electron microscopy—A primer for the non-microscopist. *FEBS J.* **2013**, *280*, 28–45. [CrossRef] [PubMed]
42. Korchagina, E.V.; Philippova, O.E. Multichain aggregates in dilute solutions of associating polyelectrolyte keeping a constant size at the increase in the chain length of individual macromolecules. *Biomacromolecules* **2010**, *11*, 3457–3466. [CrossRef]
43. Pillai, C.K.S.; Paul, W.; Sharma, C.P. Chitin and chitosan polymers: Chemistry, solubility and fiber formation. *Prog. Polym. Sci.* **2009**, *34*, 641–678. [CrossRef]
44. Rinaudo, M.; Pavlov, G.; Desbrières, J. Influence of acetic acid concentration on the solubilization of chitosan. *Polymer* **1999**, *40*, 7029–7032. [CrossRef]

© 2020 by the authors. Licensee MDPI, Basel, Switzerland. This article is an open access article distributed under the terms and conditions of the Creative Commons Attribution (CC BY) license (http://creativecommons.org/licenses/by/4.0/).

Article

Antioxidant and Moisturizing Properties of Carboxymethyl Chitosan with Different Molecular Weights

Nareekan Chaiwong [1], Pimporn Leelapornpisid [2], Kittisak Jantanasakulwong [1,3], Pornchai Rachtanapun [1,3,4], Phisit Seesuriyachan [1,3], Vinyoo Sakdatorn [1], Noppol Leksawasdi [1,3] and Yuthana Phimolsiripol [1,3,*]

[1] Faculty of Agro-Industry, Chiang Mai University, Chiang Mai 50100, Thailand; meen.nareekan@gmail.com (N.C.); kittisak.jan@cmu.ac.th (K.J.); pornchai.r@cmu.ac.th (P.R.); phisit.seesuriyachan@gmail.com (P.S.); design_by_yu@hotmail.com (V.S.); noppol@hotmail.com (N.L.)
[2] Faculty of Pharmacy, Chiang Mai University, Chiang Mai 50200, Thailand; pimporn.lee@cmu.ac.th
[3] Cluster of Agro Bio-Circular-Green Industry, Chiang Mai University, Chiang Mai 50100, Thailand
[4] Center of Excellence in Materials Science and Technology, Faculty of Science, Chiang Mai University, Chiang Mai 50200, Thailand
* Correspondence: yuthana.p@cmu.ac.th; Tel.: +665-394-8236

Received: 3 June 2020; Accepted: 23 June 2020; Published: 28 June 2020

Abstract: This research aimed to synthesize carboxymethyl chitosan (CMCH) from different molecular weights of chitosan including low MW (L, 50–190 kDa), medium MW (M, 210–300 kDa) and high MW (H, 310–375 kDa) on the antioxidant and moisturizing properties. The L-CMCH, M-CMCH and H-CMCH improved the water solubility by about 96%, 90% and 89%, respectively when compared to native chitosan. Higher MW resulted in more viscous of CMCH. For antioxidant properties, IC_{50} values of DPPH and ABTS radical scavenging activity for L-CMCH were 1.70 and 1.37 mg/mL, respectively. The L-CMCH had higher antioxidant properties by DPPH and ABTS radical scavenging assay and FRAP. The moisturizing properties on pig skin using a Corneometer® showed that 0.5% H-CMCH significantly presented ($p \leq 0.05$) greater moisturizing effect than that of untreated-skin, distilled water, propylene glycol and pure chitosan from three molecular weights.

Keywords: chitosan; carboxymethyl chitosan; molecular weight; antioxidant properties; skin moisturizing

1. Introduction

Chitosan was generally considered in the way that it has low toxicity, biodegradable, accelerates wound-healing, antibacterial properties and gel-forming properties [1]. Chitosan is cheap and inexhaustible material with numerous applications in cosmetics, pharmaceuticals, nourishment science and biotechnology [2,3]. The uses of chitosan are restricted because of its insolubility at neutral or basic region. Hence the solubility of chitosan must be improved. Carboxymethylation is a chemical modification which can improve water solubility. The water solubility properties and applications of carboxymethyl chitosan (CMCH) strongly depended on its structural characteristics, the average degree of substitution (DS), the position of the carboxymethylation (grafting to amino or hydroxyl groups) and the average number of hydroxyl groups substituted by carboxymethyl groups [4]. The CMCH is prepared by the replacement of –OH groups of chitosan with –CH_2COOH groups with the alternative functional groups such as *O*-, *N*- and *N,O*- carboxymethyl chitosan [2]. Substitution of *N*- and *O*-carboxymethyl chitosan derivatives take place when chitosan reacts with monohalocarboxylic acids using different reaction conditions to control the selectivity of reaction such as temperature and ratios of chitosan, pH as well as monochloroacetic acid. Promotion of *O*-substitution occurs when the reaction is carried out at low temperature such as 0–10 °C [5,6], but *N*-substitution is dominated at high temperature [7]. The optimal reaction temperature of N-CMCH synthesized from chitosan was 90 °C [8]. The solubility of cellulose derivatives did not only depend on the DS, but also on the distribution of the substituents for glucose units along the cellulose chain [9]. The –COOH and –NH_2 groups replacement indicate capacity of the chemical modifications to improve their physical properties [3]. CMCH is dissolvable in a wide pH range with several advantages and low harmfulness [10]. CMCH not only has a good solubility in water, but also has unique chemical, physical and biologic properties such as high viscosity, large hydrodynamic volume, biocompatibility, good ability to form films, fibers and hydrogels [11–13]. Hence, it was widely utilized in numerous biomedical fields, for example, a moisture-retention agent, wound dressing agent, artificial bone and skin, blood anticoagulant and as a component in different drug delivery [14]. Chitosan and CMCH were investigated for coating and film forming abilities to extend product shelf life. The effects of different chitosan types and molecular sizes on properties of CMCH films to plastic replacement were also studied [15]. Zhang et al. [16] found that chitosan modification could improve the antioxidant activity by addition of quaternium on amino groups. Ying et al. [17] prepared various Schiff base typed chitosan saccharide derivatives to enhance the ability of DPPH scavenging radical and also water solubility in comparison to native chitosan. Moreover, antioxidant activities of *N*-carboxymethylchitosan oligosaccharides with different DS (0.28, 0.41 and 0.54) were also evaluated by the scavenging of DPPH radical, superoxide anion and the determination of reducing power. The increase in DS of *N*-CMCH resulted in decreased DPPH radical scavenging activity with increased reducing power [18].

The antioxidant activities of chitosan and CMCH are evidence that the active hydroxyl and amino groups within the polymer chains may participate in free radical scavenging which were varied with MW [17]. Zhao et al. [19] reported that CMCH was a better antioxidant than native chitosan, especially in terms of its reducing power, scavenging ability towards DPPH and superoxide radicals as well as chelating ability of ferrous ions. In case of native chitosan, the moisture-absorption and moisture-retention capacities of chitosan depended on the MW and DS. The ability to absorb moisture increased when the MW was decreased [20]. Humectant property of chitosan improved with increasing MW. For the CMCH, Jimtaisong et al. [21] reported that MW and DS could also affect the exhibitions of the moisture-retention capacity of CMCH. Water-holding capacity of CMCH is related to the presence of positive electrical charges and high molecular weight that facilitate adherence onto the skin when implemented as a skin moisturizer. Muzzarelli et al. [22] revealed that 0.25% CMCH solution was comparable with 20% propylene glycol in terms of moisture-retention capacity with equivalent viscosity to hyaluronic acid (HA), a compound with excellent moisture-retention property. Furthermore, moisture absorption and moisture retention capacities of CMCH could also be significantly improved by utilizing higher MW with the presence of intermolecular hydrogen bonds within molecular chains. Gel formation resulting from addition of CMCH as hydrating agent in cosmetics is also ideal for the skin as it asserts positive feeling of the customers. In cosmetic products, humectants are used to increase the amount of water in the top layers of the skin. The activity of humectant polymers depends on cationic charges, molecular weight and hydrophobicity of polymer. The positively charged ions facilitate neutralization of negatively charged ions on the skin [23].

However, the antioxidant and moisturizing properties of CMCH prepared from different molecular weights of chitosan have not yet been investigated. Therefore, this research aimed to synthesize CMCH from different molecular weights of chitosan including low MW (L, 50–190 kDa), medium MW (M, 210–300 kDa) and high MW (H, 310–375 kDa) and characterized their respective antioxidant and moisturizing properties.

2. Materials and Methods

2.1. Materials

Three different molecular weights of chitosan including low MW (L, 50–190 kDa), medium MW (M, 210–300 kDa) and high MW (H, 310–375 kDa) with degree of deacetylation above 90% were obtained from Ta Ming Enterprises Co., Ltd.; Samutsakon, Thailand. Ethanol, methanol, isopropanol, sodium hydroxide and glacial acetic acid were purchased from RCI Labscan (Bangkok, Thailand). Monochloroacetic acid was obtained from Sigma-Aldrich (Darmstadt, Germany). All other reagents were of analytical grade.

2.2. Synthesis of CMCH

CMCH was synthesized by following method of Tantala et al. [14]. Chitosan flake was grounded and sieved to obtain particle size under 60-mesh (Endecotts, UK). The chitosan (25 g) was suspended in 50% (w/v) sodium hydroxide solution (400 mL) and 100 mL of isopropanol was added and mixed well at 50 °C for 1 h. Monochloroacetic acid (50 g) was dissolved in isopropanol (50 mL), gradually dropped into the reaction for 30 min and the system was allowed to continuously react at 50 °C for 4 h. The reaction was stopped by adding 70% (v/v) methanol. The pH of the sample was later adjusted to 7.0 by 1% (v/v) glacial acetic acid. From that point, the solid was separated and washed in 70–90% ethanol for desalting and dried in a hot air oven (Binder, Germany) at 80 °C for 12 h. The mass yield of CMCH was calculated using Equation (1).

$$\text{Yield (\%)} = \frac{\text{chitosan (g)} - \text{CMCH (g)}}{\text{chitosan (g)}} \times 100 \quad (1)$$

2.3. Moisture Content, pH and Viscosity Measurement

Moisture content was determined according to the Association of Official Analytical Chemists (AOAC) standard method no. 930.15 [24]. The pH values were measured by a pH meter (FiveEasy F20, Metter Toledo, Switzerland). The viscosity of 1% (w/v) solution of chitosan and CMCH were estimated by Brookfield viscometer (DV-II, Brookfield Engineering Labs Inc., Stoughton, MA, USA) using a spindle No. 28 at 100 rpm.

2.4. Water Solubility of CMCH

The water solubility of CMCH samples at 25 °C was tested by using the method of Rachtanapun et al. [15]. After addition of 0.3 g samples (initial dried weight) into 10 mL water (3% w/v), the solutions were filtered with Whatman filter paper No. 4 (Sigma-Aldrich, Germany) which was previously dried at 105 °C for 24 h before use. The mass of dried CMCH residues was obtained by weight difference to obtain final dry weight. The tests were performed in triplicate to detect random error and the solubility was determined using Equation (2).

$$\text{Water solubility (\%)} = \frac{\text{initial dried weight of CMCH (g)} - \text{final dried weight of CMCH (g)}}{\text{initial dried weight of CMCH (g)}} \times 100 \quad (2)$$

2.5. FTIR Analysis

The FTIR spectra of chitosan and CMCH were obtained using a Fourier transform infrared spectrometer (Frontier, PerkinElmer, Waltham, MA, USA). All spectra were recorded in the range of 500–4000 cm^{-1} as described by Surin et al. [25].

2.6. Antioxidant Properties

The stock solution of L, M and H (stock 5-mg/mL in 0.2% (*v/v*) acetic acid), L-CMCH, M-CMCH and H-CMCH (stock 5-mg/mL in distilled water) at different concentrations of 1, 2, 3, 4 and 5-mg/mL were prepared and used for DPPH, ABTS and FRAP assays.

2.6.1. DPPH Radical Scavenging Assay

The ability of antioxidants to scavenge the 2,2-diphenyl-1-picrylhydrazyl (DPPH) free radical was completed by modified method of Hu et al. [26]. After that, 100 μL of the stock samples (as described above) were blended with 100 μL of 0.2-mM DPPH reagent (Sigma-Aldrich, Singapore) and incubated at 25 °C for 30 min in the dark. Absorbance was measured at 517 nm in a 96-wells microplate reader (SpectraMax® i3x, Molecular Devices, San Jose, CA, USA). The radical scavenging activity of the sample was calculated based on the gallic acid (Sigma-Aldrich, Schnelldorf, Germany). Results were expressed as milligram gallic equivalent per gram of sample (mgGAE/g sample). The percentage of DPPH radical scavenging activity can be calculated as shown in Equation (3) before plotting of IC_{50} against respective concentration.

$$\text{DPPH radical scavenging activity (\%)} = [(A_{517}\text{ control} - A_{517}\text{ sample})/A_{517}\text{ control}] \times 100 \quad (3)$$

2.6.2. ABTS Radical Scavenging Assay

The 2,2-Azino-bis-(3-ethylbenzothiazoline-6-sulfonic acid) (ABTS) radical scavenging activity was conducted according to method described by Xie et al. [27]. ABTS (Sigma-Aldrich, Singapore) reagent was freshly prepared by mixing 8 mL of 7-mM ABTS stock solution with 12.5 mL of 2.45-mM potassium persulfate (Sigma-Aldrich, Singapore). ABTS powder and potassium persulfate powder were individually dissolved with water to the required concentration and then combined together in a bottle. After 16 h of incubation in the dark at 25 °C, the resultant dark blue color of ABTS reagent solution was diluted with ethanol until the absorbance reading reached 0.7 ± 0.2. The solution of L, M, H, L-CMCH, M-CMCH and H-CMCH were prepared as described previously in Section 2.6.1. Each sample solution (0.5 mL) was mixed with 1.0 mL of ABTS stock solution and incubated for 6 min in the dark. Absorbance was measured at 734 nm in the 96-well microplate reader. The ABTS radical scavenging activity was expressed as milligram gallic equivalent per gram of sample (mgGAE/g sample). The percentage of ABTS radical scavenging activity can be calculated as shown in Equation (4) with plotting of IC_{50} against respective concentration.

$$\text{ABTS radical scavenging activity (\%)} = [(A_{734}\text{ control} - A_{734}\text{ sample})/A_{734}\text{ control}] \times 100 \quad (4)$$

2.6.3. FRAP Assay

The ferric reducing antioxidant power (FRAP) assay was carried out according to the technique of Woranuch et al. [28]. The FRAP reagent was prepared by mixing 25 mL of 0.3-M acetate buffer (pH 3.6), 2.5 mL of 4,6-tripyridyl-s-triazine (TPTZ) (Sigma-Aldrich, Schnelldorf, Germany) solution in 40-mM HCl (RCI Labscan, Bangkok, Thailand) and 2.5 mL of 20-mM ferrous sulfate (Loba Chemie, India). Thus, 50 μL of samples were mixed with 950 μL of FRAP reagent and incubated in dark for 30 min. Absorbance was measured at 593 nm in 96-well microplate. The ferric reducing antioxidant power of sample was determined based on the ferrous sulfate (Merck, Darmstadt, Germany). Results were expressed as ferrous sulfate equivalent antioxidant capacity, with μmol Fe^{2+}/g sample.

2.7. Moisturizing Properties on Pork Skin

The skin moisturizing of the 0.5% (w/v) L, M, H, L-CMCH, M-CMCH and H-CMCH solutions were examined on pork skin and compared with untreated-skin, water and propylene glycol. The pork skins were prepared from back side of the pig ear obtained from three different market sources including the Mae Hia fresh market, the Ton Payom fresh market and the Hangdong fresh market (Chiang Mai, Thailand). The samples were washed and cleaned with removal of the fat layer prior to cutting into 3 × 3 cm. Each sample (100 µL) was applied on the skin surface. The skin without any substance was used as a control. The skin moisturizing was measured before applying on samples and after application at 0, 15 and 30 min intervals using Corneometer® (Courage + Khazaka Electronic, Germany). Before applying the sample and recording the parameter, the pig skins were kept at 25 °C for 30 min. This method was adapted from Kassakul et al. [29]. The degree of skin moisturizing (%) was tested in triplicate to detect random error and calculated using Equation (5).

$$\text{Degree of skin moisturizing (\%)} = \frac{\text{after applying} - \text{before applying}}{\text{before applying}} \times 100 \qquad (5)$$

2.8. Statistical Analysis

All data were analyzed by one-way ANOVA. Mean separation was performed by Duncan's multiple range tests with significance level ($p \leq 0.05$). Statistical analyses were performed with the SPSS 17.0 (SPSS, Inc.; IBM Corp.; Chicago, IL, USA).

3. Results and Discussion

3.1. Effect of CMCH Synthesis

CMCH was prepared at three different molecular weights of chitosan (L, M and H). The yield, moisture content, water solubility, viscosity and pH of chitosan products (L-CMCH, M-CMCH and H-CMCH) were reported in Table 1. L-CMCH had the highest yield, water solubility and viscosity, while moisture content and pH of L-CMCH, M-CMCH and H-CMCH were not significantly different ($p > 0.05$) among the range of 6.36–6.87% and 7.27–7.33%, respectively. The solubility is a significant property of CMCH that measures their resistance to water. Table 1 shows water solubility of the L-CMCH M-CMCH and H-CMCH which indicates the significant effect ($p \leq 0.05$) of larger MW on decreased water solubility. The decreasing trend was 96.87% for L-CMCH, 90.06% for M-CMCH and 89.49% for H-CMCH compared to the L, M and H. The solubility and conformation of CMCH happens from the deacetylation, pH and MW of native chitosan. The solubilization process of CMCH related to functionalized polymers, different types of chemical and physical interactions such as hydrogen bonds, hydrophobic interactions and van der Waals forces. high water solubility suggests that CMCH is moisture absorption and more helpful ability to bind with water than chitosan. higher solubility is due to forming hydrogen bonding with carboxylic groups of CMCH with water molecules. This causes the hydrated water molecules that around the chain of CMCH are more than that surrounding the chitosan chains, resulting in higher water solubility [30]. This results also are consistent with the report of Siahaan et al. [31] who found that the temperature and NaOH concentration affected to CMCH synthesis. The interactions between NaOH and monochloroacetic acid resulted in reduced CMCH forming and lower solubility. The mitigation in solubility may stem from the loss of free amino-functional groups that enhance hydrophobic nature of the compounds [32]. The greater solubility also corresponded to the decrease in viscosity L-CMCH and M-CMCH are slightly different, but H-CMCH requires significantly higher viscosity. This could be explained that CMCHs with chains longer or higher MW were contributing to the gel.

FTIR spectra of chitosan and CMCH are presented in Figure 1. The essential characteristic peaks of chitosan are at 3288 cm^{-1} (O–H stretch), 2875 cm^{-1} (C–H stretch), 1591–1645 cm^{-1} (N–H bend), 1059 cm^{-1} (bridge-O stretch) and 1023 cm^{-1} (C–O stretch) [2]. For CMCH, the spectrum was different from the spectrum of chitosan (Figure 1 and Table 2). The IR spectrum of CMCH showed the intrinsic peak at 1747 cm^{-1}, the most visible difference was the appearance of a new peak which belonged to C = O stretching vibration (amide I) Putra et al. [33] identified C = O peak on CMCH whose wave number could be 1600–1850 cm^{-1}, 1660–1680 cm^{-1} and also 1606 cm^{-1}. The CMCH showed the disappearance of the –NH$_2$ associated band at 1647 cm^{-1} which could be associated with characteristic vibration deformation of the primary amine N–H and the combination N–H peak with new peak at 1583. The appearance of some new intensive peaks at 2922–2853 and 1583 cm^{-1} could be attributed to the methyl groups and the long carbon segment of the quaternary ammonium salt [34]. Compared to the peaks of chitosan, the new bands at 1583 cm^{-1} and 1411 cm^{-1} corresponded to the carboxy group (overlapped with N–H bending) and –CH$_2$COOH group, respectively. The intense spectrum of CMCH indicating carboxymethylation on both the amino and hydroxyl groups of chitosan [2]. Characteristic peaks of the first C–O and the second C–O groups between 1052 and 1024 cm^{-1} (C–O stretch) did not change. It was confirmed that chitosan was converted to CMCH by new transmission peaks of -COO groups at 1583 and 1747 cm^{-1}. These new -COO groups enhanced hydrophilic properties of the CMCH which enhanced solubility of the compound.

(a)

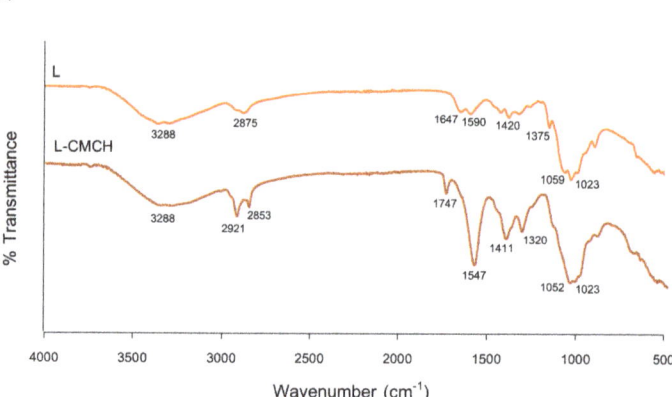

(b)

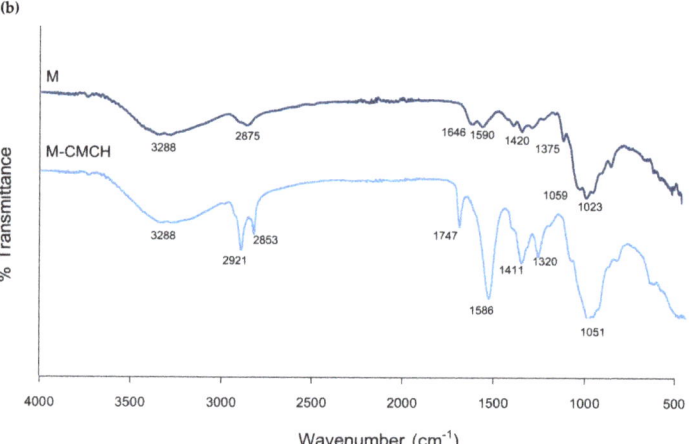

Figure 1. *Cont.*

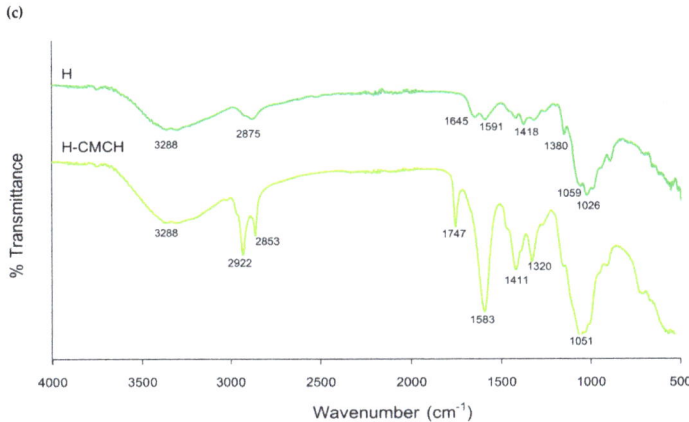

Figure 1. FT-IR spectra of (**a**) L and L-CMCH; (**b**) M and M-CMCH; (**c**) H and H-CMCH.

Table 1. Yield, moisture content, water solubility, viscosity and pH of carboxymethyl chitosan (CMCH) with different MW.

CMCH	Yield (%)	Moisture Content (%) [ns]	Water Solubility (%) *	Viscosity (cP)	pH [ns]
L-CMCH	41.33 [b] ± 0.34	6.87 ± 0.12	96.87 [a] ± 0.29	325.74 [c] ± 0.32	7.27 ± 0.29
M-CMCH	43.70 [a] ± 1.55	6.36 ± 0.51	90.06 [ab] ± 3.30	336.83 [b] ± 0.16	7.32 ± 0.22
H-CMCH	45.36 [a] ± 0.65	6.56 ± 0.60	89.49 [b] ± 3.72	360.05 [a] ± 0.84	7.33 ± 0.13

* Water solubility (%) of L-CMCH, M-CMCH and H-CMCH indicates the comparison to native chitosan. Different letters (a–c) in each column indicate significant differences ($p \leq 0.05$). ns means no significant difference.

Table 2. Functional groups and wave number (cm^{-1}) of chitosan and CMCH.

Functional Groups	Wave Number (cm^{-1})	
	Chitosan	CMCH
NH$_2$ association in primary amines, OH association in pyranose ring	3288	3289
CH$_2$ in CH$_2$OH group	2873	2917
C = O in NHCOCH$_3$ group (amide I)	1647	1747
N–H bending (amide II)	1591	1583
CH$_3$ in CH$_2$OH group	1421	1411
C–N stretching (amide II)	1319	1320
C–O, C–O–C stretching	1023	1052

3.2. Antioxidant Properties

The results from DPPH assay of L, M, H, L-CMCH, M-CMCH and H-CMCH are shown in Figure 2. L-CMCH showed the highest ($p \leq 0.05$) scavenging activity. The DPPH scavenging activities of L-CMCH, M-CMCH and H-CMCH were higher than those of L, M and H. IC$_{50}$ values of DPPH and ABTS radical scavenging activities of L-CMCH were 1.70 and 1.37 mg/mL, respectively. However, no significant differences in DPPH scavenging potential were found among the L, M and H. Younes et al. [35] also found that IC$_{50}$ value was determined between 1.62- 2.20 mg/mL for shrimp chitosan (*Metapenaeus monoceros*) at different concentrations (0–5 mg/mL). The DPPH radicals scavenging ability of chitosan and its derivatives (CMCH) increased as the concentration increased [36]. This is probably due to the relatively poor hydrogen-donating ability of chitosan that prevent chain breaking [37]. Some studies suggested that DPPH radical scavenging of chitosan increased with decreasing MW [38]. Again, it is confirmed that the CMCH have strong antioxidant activity, which is dependent on the particle size [34]. In addition, chitosan chains possess active hydroxyl and amino groups that can react with free radicals [39]. Scavenging activity of chitosan is related to the extent of reaction between free radicals and protonated amino groups [18]. Our results indicated that the DPPH scavenging ability of CMCH synthesized with low, medium or high MW was higher than that of pure chitosan from three MW. Elbarbary & Mostafa [40] also confirmed that the antioxidant activities of CMCH could be enhanced by decreasing MW of CMCH. high MW can contribute to a more compact structure and relatively stronger effect of intramolecular hydrogen bond. The antioxidant activity for ABTS radical was similar to those of DPPH assay even though ABTS radicals are more reactive than DPPH radicals [19]. Hence, these showed that antioxidant activity is expanded with decreasing MW with L-CMCH, M-CMCH and H-CMCH, respectively, compared to the L, M and H Figure 3.

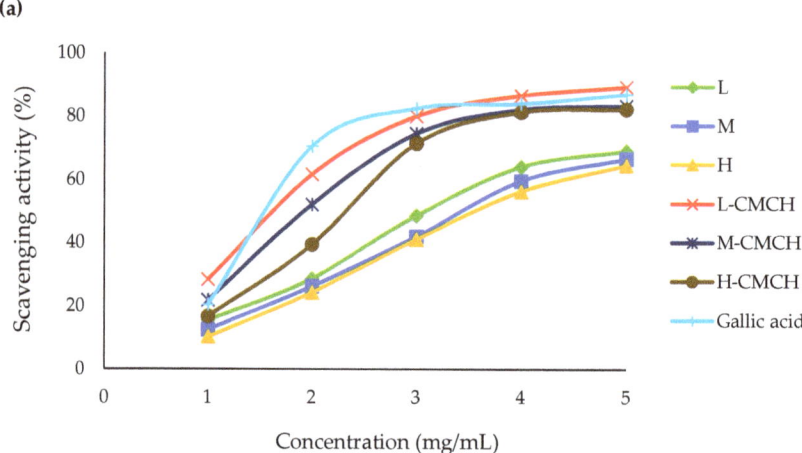

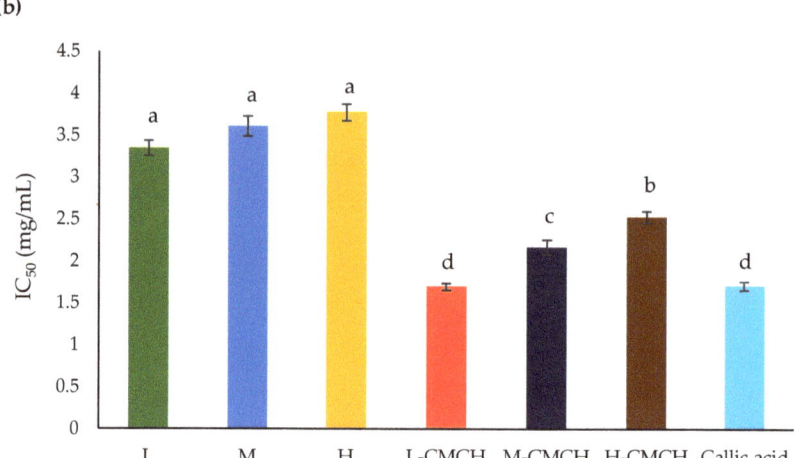

Figure 2. (a) DPPH radical scavenging activity (%) and (b) IC_{50} of L, M, H, L-CMCH, M-CMCH and H-CMCH. Different letters (a–d) indicate significant difference between treatments ($p \leq 0.05$).

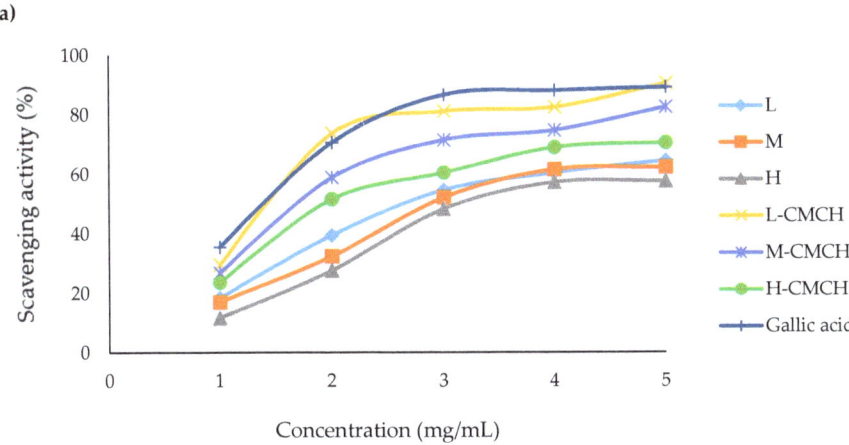

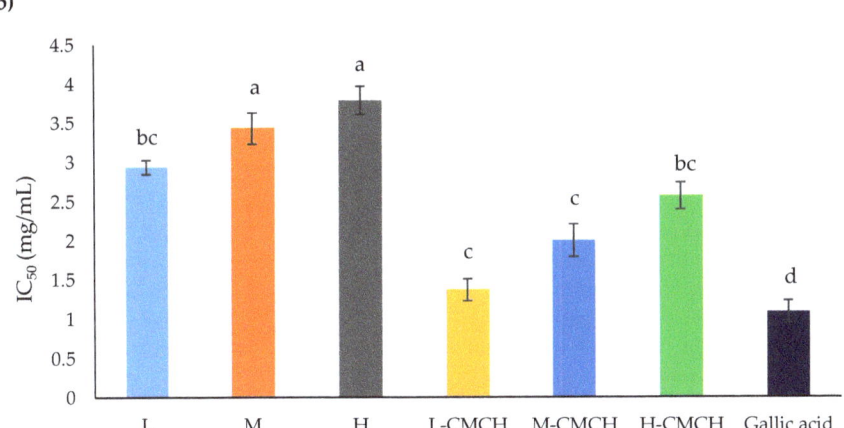

Figure 3. (**a**) ABTS radical scavenging activity (%) and (**b**) IC_{50} of L, M, H, L-CMCH, M-CMCH and H-CMCH. Different letters (a–d) indicate significant difference between treatments ($p \leq 0.05$).

FRAP assay in Figure 4 revealed the variation of antioxidant capacity with corresponding concentration levels [41]. In similar manner of DPPH assay, control could slightly reduce ferric to ferrous ions. In this assay, L-CMCH had the highest ($p \leq 0.05$) ability to reduce ferric to ferrous ion [42], followed by M-CMCH and H-CMCH, while L, M and H showed the lowest ability ($p \leq 0.05$). The replacement of -NH groups by -COO groups in the CMCH structure was previously reported to be beneficial not only to level of solubility, but also antioxidant activities [43]. Although some studies suggested the effects of molecular weights to FRAP antioxidant activities [44], such effect was not evident in current study.

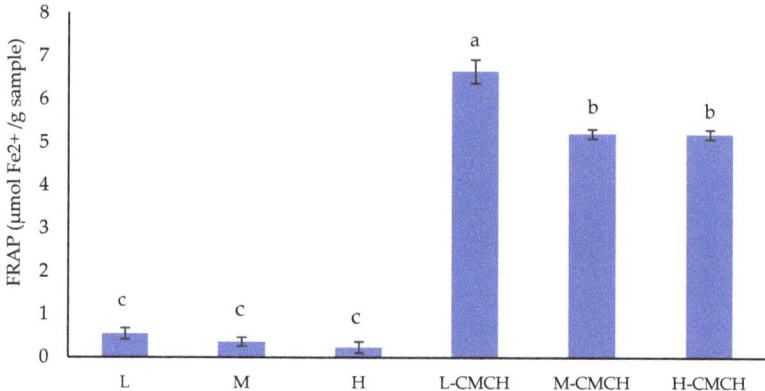

Figure 4. Ferric reducing antioxidant power (FRAP) of L, M, H, L-CMCH, M-CMCH and H-CMCH. Different letters (a–c) indicate significant difference between treatments ($p \leq 0.05$).

3.3. Skin Moisturizing Properties

The degree of skin moisturizing indicates the water-holding capacity of the skin which can be tested by the Corneometer method. The Corneometer® measures the changes of electrical capacitance related to the moisture contents of the skin before and after applying the solutions [29]. The degree of skin moisturizing of the L, M, H, L-CMCH, M-CMCH and H-CMCH solutions were examined on pork skin and compared with untreated skin, water and propylene glycol at 15 and 30 min as presented in Figure 5. The effect degree of moisturizing on time at 15 and 30 min showed that the degree of skin moisturizing of solutions decreased with increasing time after applying solutions, except 0.5% H-CMCH. The degree of skin moisturizing of H-CMCH had no significant difference after applying between 15 and 30 min. Applying H-CMCH solution for 15 and 30 min were the highest degree of skin moisturizing, showing high moisturizing effect (more than 200%). While the degree of skin moisturizing of untreated skin, water propylene glycol, L, M, H, L-CMCH and M-CMCH solutions applying on pork skin for 30 min were significantly decreased from 15 min. This confirms that the H-CMCH solution provided a good moisture absorption. In fact, the skin moisturizing effects appeared to decrease with increasing time due to lack of mechanisms to maintain skin moisturizing and dryness of pork skin cells [45]. The higher molecular weight CMCH also had the superior moisture retention capacity. Kassakul et al. [29] found that 0.2% *Hibiscus rosa sinensis* mucilage as natural ingredient provided good results of skin moisturizing after applying for 30 min by about 130%. The results showed that moisturizing products could increase the water content of the skin while maintaining softness and smoothness [20]. After applying solutions containing different MW of water-soluble CMCH (L-CMCH, M-CMCH, H-CMCH), the moisture content of the skin increased. The mechanism of moisturizing effect is based on the formation of water film of skin surface after dissolution of CMCH and subsequent stage of water evaporation could further prevent water evaporation from the skin [46]. Positive electrical charges and relatively high MW facilitates prolong skin adherence [21]. Our results also showed that H-CMCH decreased the loss of water while elevating skin humidity. The higher apparent viscosity of H-CMCH can improve the stability and enhance skin hydration. In fact, 0.5% H-CMCH was superior to untreated skin, water and propylene glycol in terms of degree of skin moisturizing effect. The higher MW of CMCH also indicates potential for film forming and coating to multilayer of the skin. Subsequently, it could be used in cosmetic preparation with suggested further studies of the testing skin irritation in human subjects.

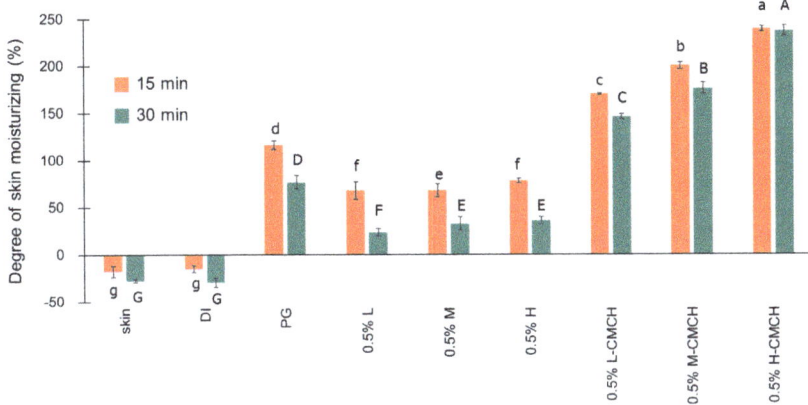

Figure 5. Degree of skin moisturizing (%) as affected by time (15 and 30 min) and different treatments (skin, DI, PG, L, M, H, L-CMCH, M-CMCH and H-CMCH) on pork skin Different lowercase letters (a–g) indicate significant differences between solutions at 15 min and different uppercase letters (A–G) indicate significant differences between solutions at 30 min.

4. Conclusions

Carboxymethyl chitosan (CMCH) was effectively synthesized and characterized by FTIR. The modifications in biologic properties including water solubility, antioxidant properties as well as efficacy of moisturizing property of CMCH were evident. It is clearly seen that higher MW of chitosan and CMCH resulted in lower antioxidant properties but provided greater moisturizing property. The H-CMCH improved the water solubility by about 89%, when compared to chitosan. The higher levels of DPPH, ABTS and FRAP were also detected. The moisturizing effect was at the highest level when 0.5% H-CMCH was applied to pig skin. H-CMCH is an effective water-soluble polymer with high viscosity which could be successfully utilized in pharmaceuticals and cosmetics as emulsion stabilizers and thickening agents. Future work is required to investigate this biopolymer for skin irritation in human subjects.

Author Contributions: N.C. designed and performed the experiments and participated in the interpretation of the results and the writing of the paper. P.L., K.J., P.R., P.S., V.S. and N.L. interpreted the results and edited the manuscript. Y.P. supervised and discussed the research and edited the manuscript. All the authors contributed to the realization of the manuscript. All authors have read and agreed to the published version of the manuscript.

Acknowledgments: The authors acknowledge the financial support provided by the National Research Council of Thailand (NRCT) and Thailand Institute of Scientific and Technological Research (TISTR) for microbial strains support for this project. We wish to thank Center of Excellence in Materials Science and Technology, Chiang Mai University for financial support under the administration of Materials Science Research Center, Faculty of Science, Chiang Mai University. This research work and APC was also partially supported by Chiang Mai University under the Cluster of Agro Bio-Circular-Green Industry under Grant number CMU-8392(10)/W.152-12032020.

Conflicts of Interest: The authors declare no conflict of interest.

References

1. Matica, M.A.; Aachmann, F.L.; Tøndervik, A.; Sletta, H.; Ostafe, V. Chitosan as a Wound Dressing Starting Material: Antimicrobial Properties and Mode of Action. *Int. J. Mol. Sci.* **2019**, *20*, 5889. [CrossRef] [PubMed]
2. Mourya, V.K.; Inamdar, N.N.; Tiwari, A. Carboxymethyl chitosan and its applications. *Adv. Mater. Lett.* **2010**, *1*, 11–33. [CrossRef]

3. Farag, R.K.; Mohamed, R.R. Synthesis and Characterization of Carboxymethyl chitosan nanogels for swelling studies and antimicrobial activity. *Molecules* **2013**, *18*, 190–203. [CrossRef]
4. Zhou, X.; Yang, J.; Qu, G. Study on synthesis and properties of modified starch binder for foundry. *J. Mater. Process. Technol.* **2007**, *183*, 407–411. [CrossRef]
5. Zheng, M.; Han, B.; Yang, Y.; Liu, W. Synthesis, characterization and biological safety of O-carboxymethyl chitosan used to treat Sarcoma 180 tumor. *Carbohydr. Polym.* **2011**, *86*, 231–238. [CrossRef]
6. Chen, X.; Park, H. Chemical characteristics of O-carboxymethyl chitosan related to its preparation conditions. *Carbohydr. Polym.* **2003**, *53*, 355–359. [CrossRef]
7. Bukzem, A.; Signini, R.; Martins, D.; Lião, L.; Ascheri, D. Optimization of carboxymethyl chitosan synthesis using response surface methodology and desirability function. *Int. J. Biol. Macromol.* **2016**, *85*. [CrossRef]
8. Song, Q.; Zhang, Z.; Gao, J.; Ding, C. Synthesis and Property Studies of N-Carboxymethyl Chitosan. *J. Appl. Polym. Sci.* **2011**, *119*. [CrossRef]
9. Choi, Y.; Maken, S.; Lee, S.; Chung, E.; Park, J.; Min, B. Characteristics of water-soluble fiber manufactured from carboxymethylcellulose synthesis. *Korean J. Chem. Eng.* **2007**, *24*, 288–293. [CrossRef]
10. Jaidee, A.; Luangkamin, S.; Rachtanapun, P. ^{1}H-NMR analysis of degree of substitution in N,O-carboxymethyl chitosan from various sources and types. *Adv. Mater. Res.* **2012**, *506*, 158–161. [CrossRef]
11. Jayakumar, R.; Prabaharan, M.; Nair, S.V.; Tokura, S.; Tamura, H.; Selvamurugan, N. Novel carboxymethyl derivatives of chitin and chitosan materials and their biomedical applications. *Prog. Mater. Sci.* **2010**, *55*, 675–709. [CrossRef]
12. Du, H.; Shi, S.; Liu, W.; Teng, H.; Piao, M. Processing and modification of hydrogel and its application in emerging contaminant adsorption and in catalyst immobilization: A review. *Environ. Sci. Pollut. Res.* **2020**, *27*, 12967–12994. [CrossRef]
13. Liu, H.; Wang, C.; Li, C.; Qin, Y.; Wang, Z.; Yang, F.; Li, Z.; Wang, J. A functional chitosan-based hydrogel as a wound dressing and drug delivery system in the treatment of wound healing. *RSC Adv.* **2018**, *8*, 7533–7549. [CrossRef]
14. Tantala, J.; Thongngam, M.; Rachtanapun, P.; Rachtanapun, C. Antibacterial activity of chitosan and carboxymethyl chitosan from different types and sources of chitosan. *Ital. J. Food Sci.* **2012**, *24*, 97–101.
15. Rachtanapun, P.; Jakkaew, M.; Suriyatem, R. Characterization of chitosan and carboxymethyl chitosan films from various sources and molecular sizes. *Adv. Mater. Res.* **2012**, *506*, 417–420. [CrossRef]
16. Zhang, X.; Geng, X.D.; Jiang, H.J.; Li, J.R.; Huang, J.Y. Synthesis and characteristics of chitin and chitosan with the (2- hydroxy-3-trimethylammonium) propyl functionality, and evaluation of their antioxidant activity in vitro. *Carbohydr. Polym.* **2012**, *89*, 486–491. [CrossRef]
17. Ying, G.Q.; Xiong, W.Y.; Wang, H.; Sun, Y.; Liu, H.Z. Preparation, water solubility and antioxidant activity of branched-chain chitosan derivatives. *Carbohydr. Polym.* **2010**, *83*, 1787–1796. [CrossRef]
18. Zimoch-Korzycka, A.; Bobak, Ł.; Jarmoluk, A. Antimicrobial and antioxidant activity of chitosan/hydroxypropyl methylcellulose film-forming hydrosols hydrolyzed by cellulase. *Int. J. Mol. Sci.* **2016**, *17*, 1436. [CrossRef] [PubMed]
19. Zhao, D.; Huang, J.; Hu, S.; Mao, J.; Mei, L. Biochemical activities of N,O-carboxymethyl chitosan from squid cartilage. *Carbohydr. Polym.* **2011**, *85*, 832–837. [CrossRef]
20. Aranaz, I.; Acosta, N.; Civera, C.; Elorza, B.; Mingo, J.; Castro, C.; Gandía, M.; Heras Caballero, A. Cosmetics and cosmeceutical applications of chitin, chitosan and their derivatives. *Polymers* **2018**, *10*, 213. [CrossRef] [PubMed]
21. Jimtaisong, A.; Saewan, N. Utilization of carboxymethyl chitosan in cosmetics. *Int. J. Cosmet. Sci.* **2014**, *36*, 12–21. [CrossRef] [PubMed]
22. Muzzarelli, R.A.; Muzzarelli, C. Chitosan chemistry: Relevance to the biomedical sciences. *Adv. Polym. Sci.* **2005**, *186*, 151–209. [CrossRef]
23. Gautier, S.; Xhauflaire-Uhoda, E.; Gonry, P.; Piérard, G.E. Chitin-glucan, a natural cell scaffold for skin moisturization and rejuvenation. *Int. J. Cosmet. Sci.* **2008**, *30*, 459–469. [CrossRef]
24. Association of Official Analytical Chemists (AOAC). *Official Method of Analysis of AOAC International*, 17th ed.; The Association of Official Analytical Chemists: Gaithersburg, MD, USA, 2000.

25. Surin, S.; Seesuriyachan, P.; Thakeow, P.; You, S.G.; Phimolsiripol, Y. Antioxidant and antimicrobial properties of polysaccharides from rice brans. *Chiang Mai J. Sci.* **2018**, *45*, 1372–1382.
26. Hu, Q.; Wang, T.; Zhou, M.; Xue, J.; Luo, Y. In vitro antioxidant-activity evaluation of gallic-acid-grafted chitosan conjugate synthesized by free-radical-induced grafting method. *J. Agric. Food Chem.* **2016**, *64*, 5893–5900. [CrossRef]
27. Xie, M.; Hu, B.; Wang, Y.; Zeng, X. Grafting of gallic acid onto chitosan enhances antioxidant activities and alters rheological properties of the copolymer. *J. Agric. Food Chem.* **2014**, *62*, 9128–9136. [CrossRef]
28. Woranuch, S.; Yoksan, R. Preparation characterization and antioxidant property of water-soluble ferulic acid grafted chitosan. *Carbohydr. Polym.* **2013**, *9*, 495–502. [CrossRef] [PubMed]
29. Kassakul, W.; Praznik, W.; Viernstein, H.; Hongwiset, D.; Phrutivorapongkul, A.; Leelapornpisid, P. Characterization of the mucilages extracted from hibiscus *rosa-sinensis linn* and *hibiscus mutabilis linn* and their skin moisturizing effect. *Int. J. Pharm. Pharm. Sci.* **2014**, *6*, 453–457.
30. Silva, S.; Kaisler, M.; Broek, L.; Frissen, A.; Boeriu, C. Water-soluble chitosan derivatives and pH-responsive hydrogels by selective C-6 oxidation mediated by TEMPO-laccase redox system. *Carbohydr. Polym.* **2018**, *186*, 299–309. [CrossRef]
31. Siahaan, P.; Mentari, N.; Wiedyanto, U.; Hudiyanti, D.; Hildayani, S.; Laksitorini, M. The optimum conditions of carboxymethyl chitosan synthesis on drug delivery application and its release of kinetics study. *Indones. J. Chem.* **2017**, *17*, 291–300. [CrossRef]
32. Samar, M.M.; El-Kalyoubi, M.H.; Khalaf, M.M.; El-Razik, M.M.A. Physicochemical, functional, antioxidant and antibacterial properties of chitosan extracted from shrimp wastes by microwave technique. *Ann. Agric. Sci.* **2013**, *58*, 33–41. [CrossRef]
33. Putra, P.; Husni, A.; Puspita, I.D. Characterization and Application of *N,O*-Carboxy Methyl Chitosan Produced at Different Tempererature of Etherification. *Int. J. Pharm. Clin. Res.* **2016**, *8*, 1493–1498.
34. Xie, W.; Xu, P.; Liu, Q. Antioxidant activity of water-soluble chitosan derivatives. *Bioorganic Med. Chem. Lett.* **2001**, *11*, 1699–1701. [CrossRef]
35. Younes, I.; Hajji, S.; Frachet, V.; Rinaudo, M.; Jellouli, K.; Nasri, M. Chitin extraction from shrimp shell using enzymatic treatment. Antitumor, antioxidant and antimicrobial activities of chitosan. *Int. J. Biol. Macromol.* **2014**, *69*, 489–498. [CrossRef] [PubMed]
36. Asan-Ozusaglam, M.; Cakmak, Y.; Kaya, M.; Erdoğan, S.; Baran, T.; Menteş, A.; Saman, I. Antimicrobial and antioxidant properties of Ceriodaphnia quadrangula ephippia chitosan. *Rom. Biotechnol. Lett.* **2016**, *21*, 11881–11890.
37. Wu, C.; Wang, L.; Fang, Z.; Hu, Y.; Chen, S.; Sugawara, T.; Ye, X. The Effect of the Molecular Architecture on the Antioxidant Properties of Chitosan Gallate. *Mar. Drugs* **2016**, *14*, 95. [CrossRef] [PubMed]
38. Chen, W.; Li, Y.; Yang, S.; Yue, L.; Jiang, Q.; Xia, W. Synthesis and antioxidant properties of chitosan and carboxymethyl chitosan-stabilized selenium nanoparticles. *Carbohydr. Polym.* **2015**, *132*, 574–581. [CrossRef] [PubMed]
39. Chang, S.; Wu, C.; Tsai, G.J. Effects of chitosan molecular weight on its antioxidant and antimutagenic properties. *Carbohydr. Polym.* **2018**, *181*, 1026–1032. [CrossRef] [PubMed]
40. Elbarbary, A.M.; Mostafa, T.B. Effect of γ-rays on carboxymethyl chitosan for use as antioxidant and preservative coating for peach fruit. *Carbohydr. Polym.* **2014**, *104*, 109–117. [CrossRef]
41. Avelelas, F.; Horta, A.; Pinto, L.; Cotrim Marques, S.; Marques Nunes, P.; Pedrosa, R.; Leandro, S.M. Antifungal and antioxidant properties of chitosan polymers obtained from nontraditional *Polybius henslowii* Sources. *Mar. Drugs* **2016**, *17*, 239. [CrossRef]
42. Tamer, T.M.; Hassan, M.A.; Omer, A.M.; Baset, W.M.; Hassan, M.E.; El-Shafeey, M.; Eldin, M.M. Synthesis, characterization and antibacterial evaluation of two aromatic chitosan Schiff base derivatives. *Process Biochem.* **2016**, *51*, 1721–1730. [CrossRef]
43. Kwaszewska, A.; Sobiś-Glinkowska, M.; Szewczyk, E.M. Cohabitation- relationships of *corynebacteria* and *staphylococci* on human skin. *Folia Microbiol.* **2014**, *59*, 495–502. [CrossRef]
44. Chaiyana, W.; Leelapornpisid, P.; Phongpradist, R.; Kiattisin, K. Enhancement of antioxidant and skin moisturizing effects of olive oil by incorporation into microemulsions. *Nanomater. Nanotechnol.* **2016**, *6*, 1–8. [CrossRef]

45. Tamer, T.; Valachová, K.; Mohyeldin, M.; Soltes, L. Free radical scavenger activity of chitosan and its animated derivative. *J. Appl. Pharm. Sci.* **2016**, *6*, 195–201. [CrossRef]
46. Szymańska, E.; Winnicka, K. Stability of chitosan-a challenge for pharmaceutical and biomedical applications. *Mar. Drugs* **2015**, *13*, 1819–1846. [CrossRef] [PubMed]

 © 2020 by the authors. Licensee MDPI, Basel, Switzerland. This article is an open access article distributed under the terms and conditions of the Creative Commons Attribution (CC BY) license (http://creativecommons.org/licenses/by/4.0/).

Article

A Theoretical Multifractal Model for Assessing Urea Release from Chitosan Based Formulations

Manuela Maria Iftime [1], Stefan Andrei Irimiciuc [2], Maricel Agop [3], Marian Angheloiu [4], Lacramioara Ochiuz [5,*] and Decebal Vasincu [5]

1. Romanian Academy of Sciences, Petru Poni Institute of Macromolecular Chemistry, 41A Grigore Ghica Voda Alley, 700487 Iasi, Romania; ciobanum@icmpp.ro
2. National Institute for Laser, Plasma and Radiation Physics—NILPRP, 409 Atomistilor Street, 077125 Bucharest, Romania; stefan.irimiciuc@inflpr.ro
3. Department of Physics, "Gh. Asachi" Technical University of Iasi, 700050 Iasi, Romania; m.agop@yahoo.com
4. Center for Services and Research in Advanced Biotechnologies, Calugareni, Sanimed International Impex SRL, Road Bucuresti-Magurele, no. 70 F, sector 5, 077125 Bucharest, Romania; marian.angheloiu@sanimed.ro
5. University of Medicine and Farmacy Grigore T. Popa Iasi, 700115 Iași, Romania; deci_vas@yahoo.com
* Correspondence: ochiuzd@yahoo.com

Received: 25 April 2020; Accepted: 25 May 2020; Published: 1 June 2020

Abstract: This paper reports the calibration of a theoretical multifractal model based on empirical data on the urea release from a series of soil conditioner systems. To do this, a series of formulations was prepared by in situ hydrogelation of chitosan with salicylaldehyde in the presence of different urea amounts. The formulations were morphologically characterized by scanning electron microscopy and polarized light microscopy. The in vitro urea release was investigated in an environmentally simulated medium. The release data were fitted on five different mathematical models, Korsmeyer–Peppas, Zero order, First order, Higuchi and Hixson–Crowell, which allowed the establishment of a mechanism of urea release. Furthermore, a multifractal model, used for the fertilizer release for the first time, was calibrated using these empirical data. The resulting fit was in good agreement with the experimental data, validating the multifractal theoretical model.

Keywords: chitosan; multifunctional materials; multifractal theoretical model

---

1. Introduction

In recent years, fertilizer release has become an important topic in the field of agriculture. With advances in material design and engineering, new multifunctional materials have been introduced for the development of soil conditioners, particularly in fertilizer delivery systems [1]. Hydrogels are an important class of materials suitable for this purpose; they have substantial applicability in various domains such as medicine, agriculture, food industry, water treatments and so on [2]. Hydrogels obtained from both natural and synthetic macromolecules were extensively used as a matrix for controlled drug release with the aim to maximize the bio-efficacy, simplify clinical applicability and improve quality of life [2,3]. In recent years, the concept of hydrogel matrix has been translated to agriculture, being used as a matrix for different fertilizers aiming to increase their efficiency by controlled release [4]. Among the hydrogels, those obtained from renewable resources such as chitosan, present suitable properties which make them very important for delivery systems. They are biocompatible and biodegradable, and they present antifungal and antiviral activity [5]. Moreover, the hydrogels can swell and keep the moisture in soil for a longer time, and have the ability to encapsulate fertilizers by strong chemical or physical forces, further favoring their release in a controlled prolonged manner [2].

The beneficial properties of the chitosan hydrogels can be further improved by a proper choice of the crosslinker [5,6]. In this context, our group succeeded in preparing chitosan hydrogels by crosslinking with nontoxic monoaldehydes, some of them of natural origin [7–13]. The advantage of such a method proved to be the easy tuning of the hydrogel properties by an appropriate choice of the aldehyde. Accordingly, the use of salicylaldehyde led to biodegradable and biocompatible hydrogels with excellent mechanical, swelling and self-healing properties [9]. Taking into consideration these particular properties, the system was further investigated as a multifunctional matrix capable of releasing the fertilizer in a controlled manner [14]. As model fertilizer, urea was used, considering its high nitrogen content and low cost and also the possibility to improve its efficiency by minimizing loss by volatilization, denitrification or leaching processes [15]. By varying the crosslinking density and the urea amount, a large series of formulations was prepared, and the prolonged release ability was investigated. It was proved that these systems are promising soil conditioners, which deserves deeper investigation into the morphology–release behavior relationship for a better understanding of the mechanisms which govern the urea release, to allow further improvements of the design.

It is known that the fertilizer release from different matrix polymers is affected by multiple complex factors, such as the matrix structure, which further influences the swelling capacity and degradation, the release medium (pH, temperature, ionic strength) and the possible interaction between the fertilizer and carrying matrix [16]. Consequently, for a better understanding of the urea release mechanism from the salicyl-imine-chitosan matrix, we propose to assess the fertilizer release kinetics using both empiric and multifractal type laws.

2. Materials and Methods

2.1. Materials

Chitosan of low molecular weight (314 kDa, DA = 87%), salicylaldehyde of 98% purity, urea of 98% purity, ethanol, and glacial acetic acid were purchased from Aldrich and used as received. Bidistilled water was obtained in our laboratory.

2.2. Preparation of the Urea Release Systems

The formulations used in this paper as urea release systems were prepared according to a procedure mainly based on the in situ encapsulation of urea during the hydrogelation process of chitosan with salicylaldehyde (SA) [14]. By varying the molar ratio between amino groups of chitosan and aldehyde groups of salicylaldehyde, and the amount of urea, a series of 8 formulations was prepared (Table 1). The urea amount was calculated to be half, equal or double compared to the matrix amount, to give a final content in the formulation of 0%, 33%, 50% and 66% w/w (Table 1). The formulations were obtained as hydrogels, which were further lyophilized to give the dry formulations in the form of xerogels, which were used for investigations. They were coded 1.5-Ux and 2-Ux, where the 1.5 and 2 numbers reflect the molar ratio between the functional groups (NH_2/CHO = 1.5/1 and 2/1, respectively) and x in Ux indicates the mass ratio of the urea to the blank matrix, giving a different percent of urea in different formulations. The 1.5-U0 and 2-U0 samples represent the blank matrix, without urea, which were used as references. Table 1 presents the amounts of the reagents used for the preparation of the urea release systems and their codes.

Table 1. The compositions of the urea release systems and their codes.

Code	NH$_2$/CHO Molar Ratio	Chitosan (mg)	SA (mg)	Matrix (mg)	Urea (mg)	Urea (%)	Formulation (mg)
1.5-U0	1.5/1	100	41	141	0	0	141
1.5-U0.5	1.5/1	100	41	141	70.5	33	211.5
1.5-U1	1.5/1	100	41	141	141	50	282
1.5-U2	1.5/1	100	41	141	282	66	423
2-U0	2/1	100	31	131	0	0	131
2-U0.5	2/1	100	31	131	65	33	196
2-U1	2/1	100	31	131	131	50	262
2-U2	2/1	100	31	131	262	66	393

2.3. Methods

The xerogels formulations were obtained by the lyophilization of corresponding hydrogel formualtions using Labconco FreeZone Freeze Dry System equipment, for 24 h at −54 °C and 1.512 mbar.

The morphology of the formulations was investigated on the corresponding xerogels, using a field emission Scanning Electron Microscope (SEM) EDAX–Quanta 200 at accelerated electron energy of 20 KeV.

The supramolecular ordering of the xerogels formulations was observed with a polarized light microscopy (POM) with a Leica DM 2500 microscope.

2.4. The In Vitro Urea Release Protocol

The in vitro urea release was investigated for 35 days, at room temperature, using distilled water as the release medium. For a proper comparison, the amounts of formulations used in this investigation were previously weighted to contain the same amount of urea (50 mg). The formulations were immersed into vials containing 10 mL of distilled water. At fixed intervals, at each hour on the first day, and on each day over the next 35 days, 1 mL of supernatant was withdrawn from the vials and replaced with 1 mL of distilled water. The supernatant samples (1 mL each) were collected, lyophilized and the quantity of released urea was measured by ^{1}H-NMR spectroscopy, by fitting on a calibration curve [14]. The proton spectra were recorded on a Bruker Avance NEO 400 MHz spectrometer equipped with a 5 mm broadband inverse detection z-gradient probe. Chemical shifts were described in δ units (ppm) and were referenced to sodium 3-(trimethylsilyl)-[2,2,3,3-d4]-1-propionate (TSP) as external standard at 0.0 ppm. The experiments were performed in duplicate. The calibration curve was realized by graphical representation of the integral value of urea protons vs. concentration, as obtained for 8 urea solutions in dimethyl sulfoxide-d$_6$ (DMSO-d$_6$) of known concentration. For the NMR study of urea released from samples, the certain quantities (1 mL) of supernatant were lyophilized and then dissolved in 0.6 mL DMSO-d$_6$. The obtained solutions were transferred in NMR tubes containing capillaries with known concentrations of TSP in D$_2$O.

2.5. Evaluation of the Release Kinetics

In order to investigate the mechanism of the fertilizer release, the release data of the studied formulations were fitted on the 5 different mathematical models: Korsmeyer–Peppas, Zero order, First order, Higuchi and Hixson–Crowell [17,18]:

- Zero order model: $Q_t = k_0 \cdot t$, where Q_t is the amount of urea dissolved in the time t and K_0 is the zero order release constant.

- First order model: $\log Q_t = k \cdot t/2.303$, where Q_t is the amount of urea released in the time t and K is the first order release constant.
- Korsmeyer–Peppas model: $M_t/M_\infty = K \cdot t^n$, where M_t/M_∞ is the fraction of urea released at the time t, K is the release rate constant and n is the release exponent.
- Higuchi model: $Q_t = k_H \cdot t^{1/2}$, where Q_t is the amount of urea released in the time t and K_H is the Higuchi dissolution constant.
- Hixson–Crowell model: $W_0^{1/3} - W_t^{1/3} = k \cdot t$, where W_0 is the initial amount of urea in the formulation, W_t is the remaining amount of urea in formulation at time t and K is a constant.

2.6. Theoretical Model

Our fundamental hypothesis is that the structural units' dynamics in the polymer–drug complex systems are described by continuous but non-differentiable curves (multifractal curves). In such a context the drug release dynamics will be described through the multifractal theory of motion in the form of hydrodynamic regimes at various resolution scales (multifractal hydrodynamic model [19–24]).

Therefore, let us consider one-dimensional multifractal hydrodynamic equations S (18) and S (19) from Supplementary Material:

$$\partial_t V_D + V_D \partial_x V_D = -\partial_x \left[-2\lambda^2 (dt)^{(\frac{4}{f(\alpha)})-2} \frac{\partial_x \partial_x \sqrt{\rho}}{\sqrt{\rho}} \right] \tag{1}$$

$$\partial_t \rho + \partial_x (\rho V_D) = 0 \tag{2}$$

In Equations (1) and (2) V_D is the differentiable velocity, ρ is the state density, λ is a coefficient associated to the multifractal-non-multifractal transition, dt is the scale resolution, t is the nonfractal temporal coordinate and the affine parameter of the movement curve, x is the spatial fractal coordinate and $f(\alpha)$ is the singularity spectrum of fractal dimension [21–24].

These equations for initial and boundary conditions [19,20]:

$$V_D(x, t=0) = V_0, \rho(x, t=0) = \frac{1}{\sqrt{\pi} \alpha} \exp\left[-\left(\frac{x}{\alpha}\right)^2\right] \tag{3}$$

$$V_D(x = V_0 t, t) = V_0, \rho(x = -\infty, t) = \rho(x = +\infty, t) = 0 \tag{4}$$

with V_0 the initial velocity and α the parameter of the gaussian distribution of positions, using the mathematical procedure from [25–28], provide the following solution:

$$V_D(x, t, \sigma, \alpha) = \frac{V_0 \alpha^2 + \left(\frac{\sigma}{\alpha}\right)^2 xt}{\alpha^2 + \left(\frac{\sigma}{\alpha}\right)^2 t^2} \tag{5}$$

$$\rho(x, t, \sigma, \alpha) = \frac{(\pi)^{-1/2}}{\left[\alpha^2 + \left(\frac{\sigma}{\alpha}\right)^2 t^2\right]^{1/2}} \exp\left[-\frac{(x - V_0 t)}{\alpha^2 + \left(\frac{\sigma}{\alpha}\right)^2 t^2}\right] \tag{6}$$

with

$$\sigma = \lambda (dt)^{[\frac{2}{f(\alpha)}]-1} \tag{7}$$

Introducing the non-dimensional variables

$$\frac{x}{V_0 \tau_0} = \xi, \quad \frac{t}{\tau_0} = \eta \tag{8}$$

and non-dimensional parameters

$$\frac{\sigma\tau_0}{a^2} = \mu, \quad \frac{\alpha}{V_0\tau_0} = \phi \tag{9}$$

with τ_0 the specific time, Equations (5) and (6) become

$$V_D(\mu,\xi,\eta) = \frac{V_D(x,t)}{V_0} = \frac{1+\mu^2\xi\eta}{1+\mu^2\eta^2} \tag{10}$$

$$\rho(\mu,\xi,\eta) = \pi^{1/2}\,a\rho(x,t) = \left(1+\mu^2\eta^2\right)^{-1/2}\exp\left[-\frac{(\xi-\eta)^2}{\phi^2(1+\mu^2\eta^2)}\right] \tag{11}$$

In such a context, since the state density ($\rho(\mu,\xi,\eta)$) defines the number of structural units in the polymer–fertilizer complex system and considering that $\overline{m}$ is the non-dimensional rest mass of the polymer–fertilizer structural units, then the non-dimensional mass variation (with respect to non-dimensional time η of the fertilizer release mechanism $\frac{d\overline{M}}{d\eta}$) is represented by means of the following relation:

$$\frac{d\overline{M}}{d\eta} = -m_0 \frac{d\rho(\mu,\xi,\eta)}{d\eta} \tag{12}$$

This relation will be used to validate our theoretical model based on the empirical data which will be presented in Section 3.

3. Results and Discussion

In view of modeling the urea release characteristics, eight formulations based on chitosan, salicylaldehyde and urea (Table 1) were prepared applying the procedure of the in situ hydrogelation described in the Experimental section. It should be remarked that the in situ procedure allowed for efficient encapsulation of a large amount of fertilizer [29–32]. The formulations were firstly investigated by scanning electron microscopy (SEM) measurements to observe the influence of both the crosslinking density and urea content on their morphology. As can be seen in Figure 1a, the formulations were porous. Compared to the blank matrix (1.5-U0, 2-U0), the formulations displayed larger pores (approx. 50 µm compared to approx. 25 µm) and visibly thicker pore walls. Moreover, in the pore walls, acicular crystals can be distinguished, characteristic of the urea crystals [33]. Compared to the scale bar, their size can be appreciated at the micrometric level. On the other hand, considering the large amount of urea compared to the matrix amount (i.e., half, equal or double), the fraction of visible micrometric crystals is quite low. This enables the visualization of a large fraction of urea crystals encapsulated at the sub-micrometric level and even at the nano-metric level. As expected, the density of the micrometric crystals seems to increase along with urea content in formulation, a feature also observed for such systems with content of bioactive components [34]. This observation was further supported by polarized light microscopy (POM) which displayed more homogeneous birefringent textures with a lower content of urea and crystalline shapes of a higher dimension as the urea content increased (Figure 1b). The birefringent crystalline shapes were attributed to the urea sub-micrometric and micrometric crystals, encapsulated in the chitosan hydrogel matrix by physical forces, developed due to the strong polycationic nature of chitosan in hydrogel state [35]. The continuous birefringence with a particular banded texture was correlated with the layered supramolecular ordering of the hydrogels [36,37].

When discussing the theoretical models used for drug-release mechanisms in the literature the homogeneity assumption in its various forms (homogenous kinetic space, law of mass etc.) is at their core. The functionality of such a hypothesis allowed the development of a class of differentiable models in the description of drug release dynamics in such systems. However, biological systems are nowadays understood as inherently non-differential (fractal). Specifically, in the microenvironments where any drug molecules with membrane interface, metabolic enzymes or pharmacological receptors are unanimously recognized as unstirred, space-restricted, heterogeneous and geometrically fractal. It is thus necessary to define a new class of models, this time non-differentiable, in describing biological system dynamics and particularly drug release dynamics in such systems. Usually, such an approach is known as Fractal Pharmacokinetics (PK) and implies the use of fractional calculus, expanding on the notion of dimension. This complex analysis allowed the modeling of processes such as drug dissolution, absorption, distribution, and kinetics with bio-molecular reactions. Our mathematical approach, in the context of "compartmental analysis", presents itself as a new method for describing drug release dynamics in complex systems (evidently discarding fractional derivative and other standard "procedures" used in PK), considering the proposal that drug release dynamics can be described through continuous but non-differentiable curves (multifractal curves). Then, instead of "working" with a single variable described by a strict, non-differentiable function, it is possible to "operate" only with approximations of these mathematical functions, obtained by averaging them on different scale resolutions.

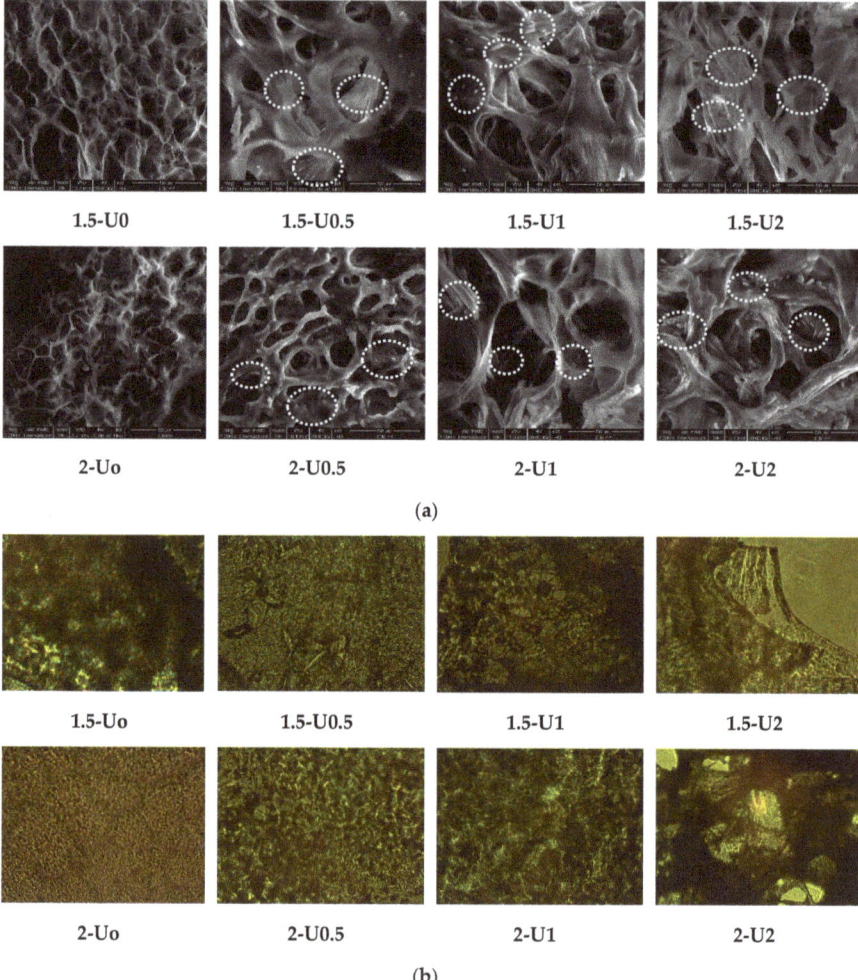

Figure 1. Representative (**a**) SEM (scale bar: 50 μm) and (**b**) polarized light microscopy (POM) (magnification: 400×) images of the 1.5-Ux and 2-Ux formulations. The crystals in the SEM images were marked with circles.

The graphical representation of the urea release from the understudy formulations is depicted in Figure 2. As can be seen, the urea release advanced in three stages and was significantly affected by the encapsulation pathway: (1) a burst release in the first 5 h (release up to 46% of urea), (2) a slower release in the next 11 days (up to 75% released urea) and (3) a continuous slow release in the next 23 days (almost all urea was released in the water medium).

From the 2-U2 and 1.5-U2 samples containing larger urea crystals, the release occurred faster, while from the other samples in which the urea was encapsulated as smaller crystals, the release produced slower. Moreover, the samples with a higher crosslinking degree (1.5-Ux) appeared to release slightly faster compared to the ones with a lower crosslinking degree (2-Ux).

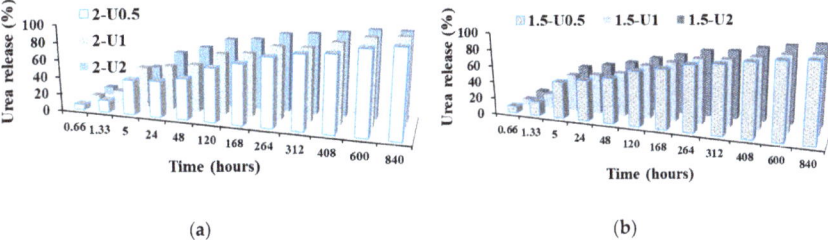

Figure 2. Graphical representation of the urea release from **1.5-Ux** (**b**) and **2-Ux** (**a**) formulations.

In order to understand the kinetics release of the urea on each of the three stages, from 1.5-Ux and 2-Ux samples, the release data were fitted on the mathematical equations of the Korsmeyer–Peppas, Zero order, First order, Higuchi and Hixson–Crowell (Figure 3a–c). As can be seen in Figure 3a,b, the obtained in vitro release data proved a good fitting in the *first stage* (Figure 3a) and *second stage* (Figure 3b) on all five mathematical models. This good fitting indicates that the urea release mechanism is controlled by both dissolution and diffusion through the hydrogel matrix. Considering the morphology of the urea release systems, this mechanism correlates well with the faster dissolution of the micrometric crystals in the first stage, less anchored into the matrix, followed by the submicrometric ones in the second stage. In the *third second stage*, except with Korsmeyer–Peppas, the fitting of all mathematical models failed for almost all the samples (Figure 3c) indicating that heterogeneous erosions of the matrix occurred, which favored the release of the urea encapsulated at the nanometric level or even the molecular level, and were very well anchored into the matrix.

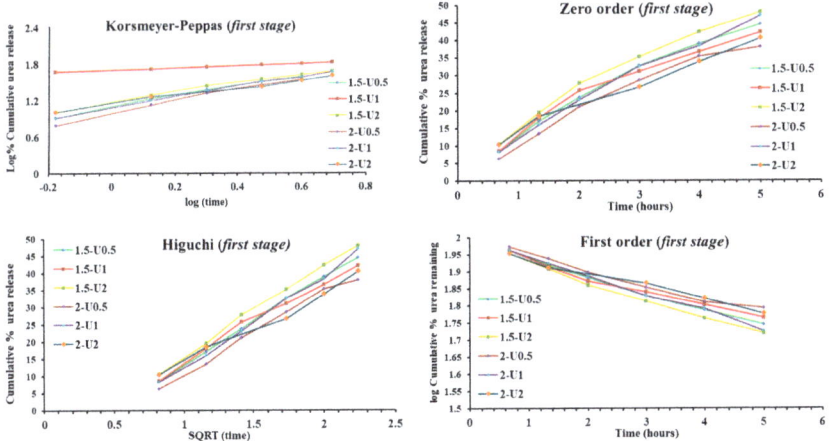

Figure 3. *Cont.*

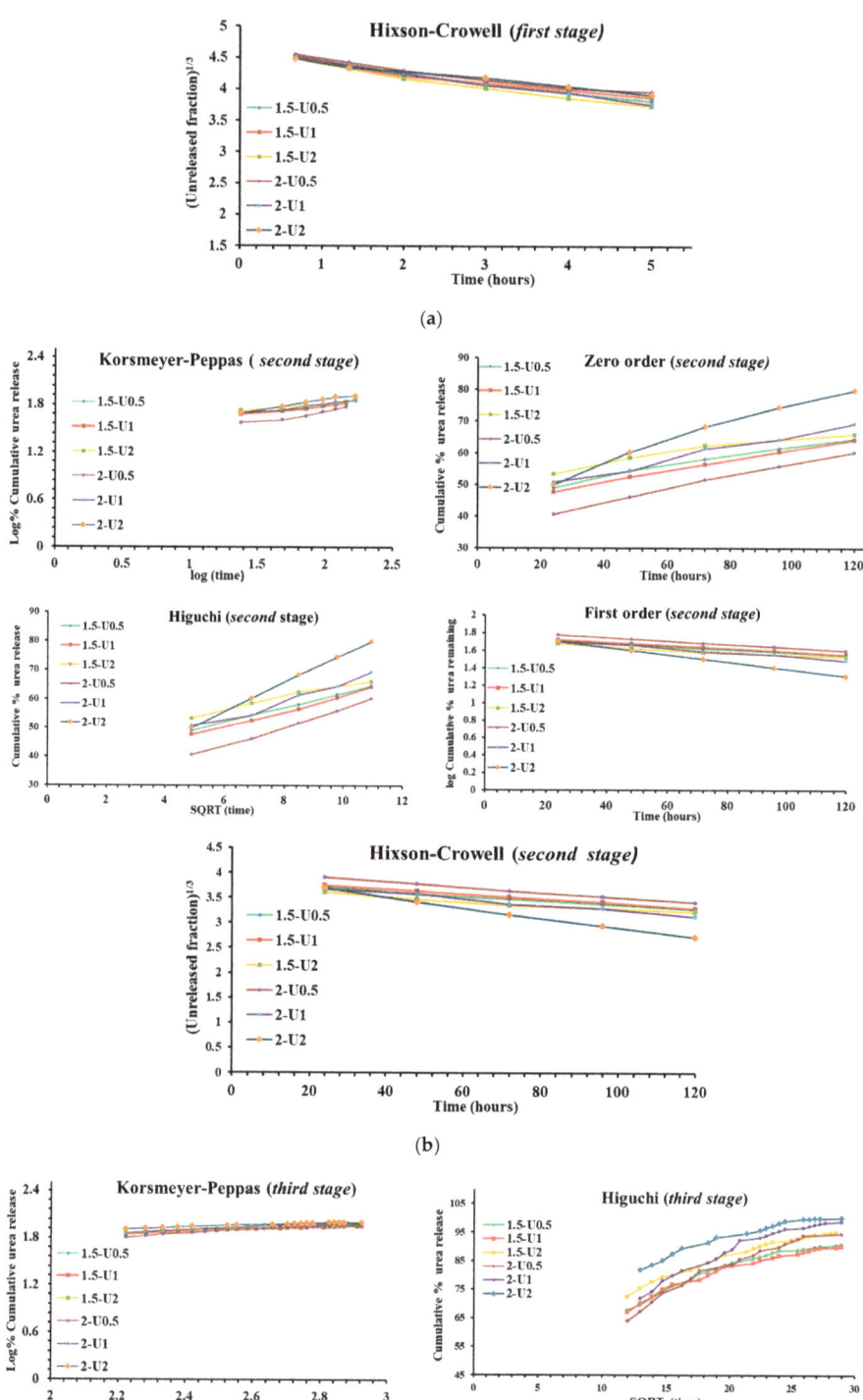

Figure 3. Cont.

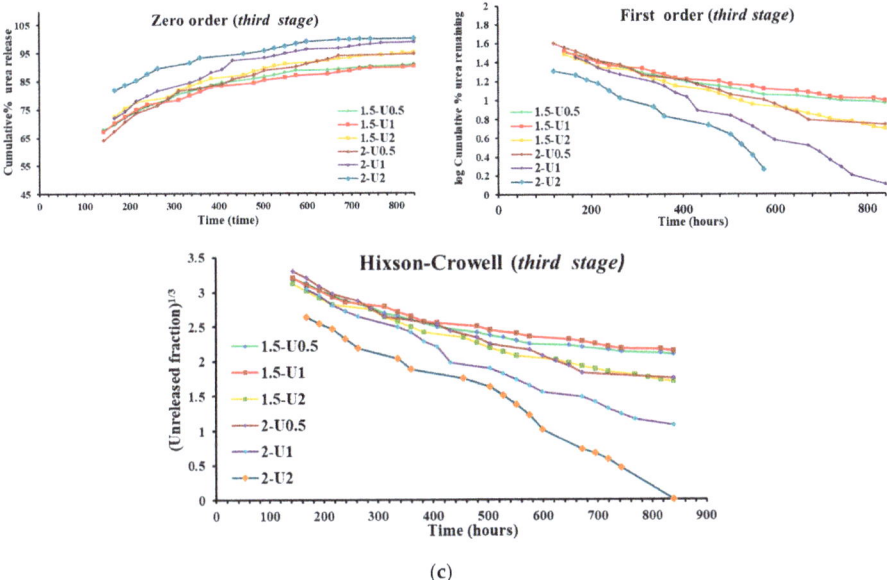

Figure 3. Linear forms of the all five mathematical models applied for the release of urea from 1.5-Ux and 2-Ux in (**a**) *first stage*, (**b**) *second stage* and (**c**) *third stage*.

To further understand the forces which drive the urea release, the multifractal model presented in Section 2.6 was calibrated on the empirical data presented in the previous section. In this case, the evaluation of the release kinetics has been conducted through Equation (12). In Figure 4, the 3D representation of the release mass variation in time and space and the fit of the experimental data using our model are presented. The model was calibrated [22–24] to fit the empirical data presented in the previous section. It can be observed that the multifractal model accurately predicts the behavior seen empirically with a steep increase for a short moment of time and a saturation plateau for considerably longer periods. The fitting when using the multifractal functions allowed the determination of the fractal degree [22,23] for each stage of the urea release scenario. The multifractal model worked at each time-scale as the inherent characteristic of the model was the possibility to transcend various scales in the framework of the same mathematical apparatus. It was observed that in the first release stage the fractality of the system was high, which meant that the release was a highly energetic one. The fractality decreased as the release advanced in time (second stage), a fact which reflects a decrease overall in the urea mass released. It should be noted that in the third stage, where the fractality degree is small, there is no significant dependence on the amount of initio urea percentage, meaning that with a slow release behavior the initial values do not affect the late time-scale behavior.

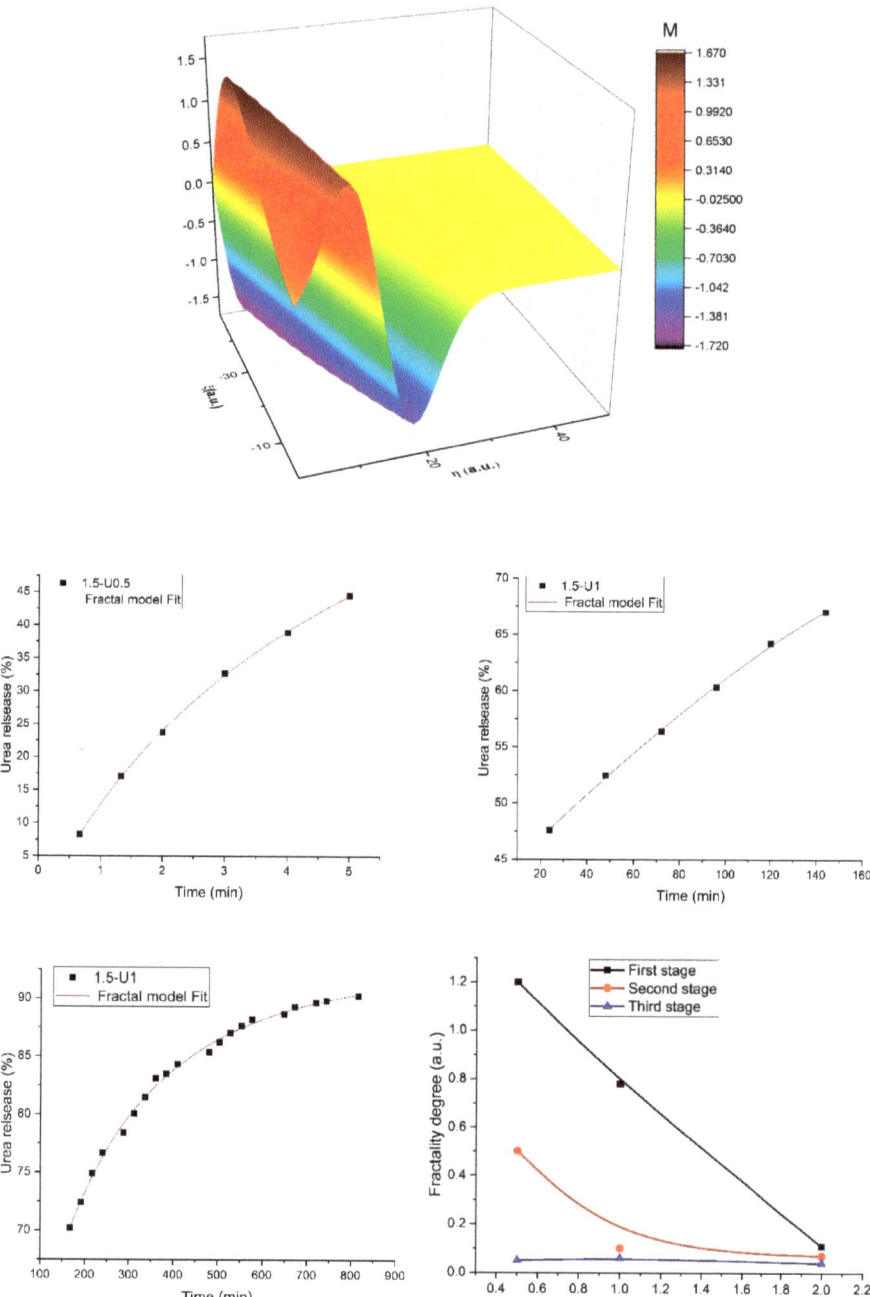

Figure 4. 3D representation of the urea release mass evolution in time and the theoretical fit of each of the expansion stages for the 1.5-U1 case and the fractality degree evolution with the amounts of encapsulated urea.

4. Conclusions

A number of formulations were prepared by in situ dispersion of different amounts of urea into hydrogels based on chitosan and salicylaldehyde eco-reagents in different molar ratios. SEM and POM indicated that urea was encapsulated in the form of crystals of different sizes: microcrystals, submicrometric crystals and even at a molecular level. The in vitro release data showed that urea release took place in three different stages during a 35-day period, corresponding to the different dissolution rate functions of the crystal size: (i) a faster release was favored by the rapid dissolution of the bigger crystals which were less anchored into the matrix in the first stage, followed by (ii) a slower release of the smaller crystals better anchored in the second stage, and further by (iii) the slower release of the smallest crystals during the third stage when erosion of the matrix occurred. A theoretical multifractal model has been fitted with the empirical data of the urea release from the formulations. The calibration of the theoretical multifractal model entirely confirmed this release profile, suggesting this simple model as an important tool for understanding the morphology–release relationship of the complex release systems. These good results encourage the further application of this model on other fertilizer release systems and even others such as drug release systems. The advantages of this multifractal approach need to be viewed as a more general implementation, not being directly related to one particular drug–polymer matrix. Through the scale resolution parameter, the model can navigate and describe different configurations for the drug release mechanisms.

Supplementary Materials: For details on the multifractal model used for assessing urea release from chitosan based formulations please check our supplementary material at http://www.mdpi.com/2073-4360/12/6/1264/s1.

Author Contributions: Conceptualization, M.M.I. and M.A. (Maricel Agop); methodology, M.M.I., L.O. and M.A. (Marian Angheloiu); investigation, M.M.I., S.A.I., M.A. (Maricel Agop); writing—original draft preparation, L.O., M.A. (Marian Angheloiu), S.A.I.; writing—review and editing, D.V. and M.M.I.; visualization, S.A.I.; supervision, M.A. (Maricel Agop), L.O., D.V. All authors have read and agreed to the published version of the manuscript.

Funding: The research leading to these results has received funding from the Romanian National Authority for Scientific Research, MEN-UEFISCDI grant, project number PN-III-P1-1.2-PCCDI-2017-0569, Contract nr. 10PCCDI/2018. Access to research infrastructure developed in the "Petru Poni" Institute of Macromolecular Chemistry through the European Social Fund for Regional Development, Competitiveness Operational Programme Axis 1, Project InoMatPol (ID P_36_570, Contract 142/10.10.2016, cod MySMIS: 107464) is gratefully acknowledged.

Conflicts of Interest: The authors declare no conflict of interest.

References

1. León, P.O.; Muñoz-Bonilla, A.; Soto, D.; Ramirez, J.; Marquez, Y.; Colina, M.; Fernández-García, M. Preparation of Oxidized and Grafted Chitosan Superabsorbents for Urea Delivery. *J. Polym. Environ.* **2018**, *26*, 728–739. [CrossRef]
2. Abobatta, W. Impact of hydrogel polymer in agricultural sector. *Adv. Agric. Environ. Sci. Open Access* **2018**, *1*, 59–64. [CrossRef]
3. Fu, Y.; Kao, W.J. Drug release kinetics and transport mechanisms of non-degradable and degradable polymeric delivery systems. *Expert Opin. Drug Deliv.* **2010**, *7*, 429–444. [CrossRef] [PubMed]
4. Campos, E.V.R.; Oliveira, J.Z.; Fraceto, L.F.; Singh, B. Polysaccharides as saffer release systems for agrochemicals. *Agron. Sustain. Dev.* **2015**, *35*, 47–66. [CrossRef]
5. Ahmadi, F.; Oveisi, Z.; Mohammadi Samani, S.; Amoozgar, Z. Chitosan based hydrogels: Characteristics and pharmaceutical applications. *Res. Pharm. Sci.* **2015**, *10*, 1–16. [PubMed]
6. Casettari, L.; Vllasaliu, D.; Lam, J.; Soliman, M.; Illum, L. Biomedical applications of amino acid-modified chitosans: A review. *Biomaterials* **2012**, *33*, 7565–7583. [CrossRef]
7. Marin, L.; Ailincai, D.; Mares, M.; Paslaru, E.; Cristea, M.; Nica, V.; Simionescu, B.C. Iminochitosan biopolymeric films. Obtaining, self-assembling, surface and antimicrobial properties. *Carbohydr. Polym.* **2015**, *117*, 762–770. [CrossRef]
8. Ailincai, D.; Marin, L.; Morariu, S.; Mares, M.; Bostanaru, A.C.; Pinteala, M.; Simionescu, B.C.; Barboiu, M. Dual crosslinked iminoboronate-chitosan hydrogels with strong antifungal activity against Candida planktonic yeasts and biofilms. *Carbohydr. Polym.* **2016**, *152*, 306–316. [CrossRef]

9. Iftime, M.M.; Morariu, S.; Marin, L. Salicyl- imine-chitosan hydrogels: Supramolecular architecturing as a crosslinking method toward multifunctional hydrogels. *Carbohydr. Polym.* **2017**, *165*, 39–50. [CrossRef]
10. Marin, L.; Ailincai, D.; Morariu, S.; Tartau-Mititelu, L. Development of biocom-patible glycodynameric hydrogels joining two natural motifs by dynamic constitutional chemis-try. *Carbohydr. Polym.* **2017**, *170*, 60–71. [CrossRef]
11. Olaru, A.M.; Marin, L.; Morariu, S.; Pricope, G.; Pinteala, M.; Tartau-Mititelu, L. Biocompatible based hydrogels for potential application in local tumour therapy. *Carbohydr. Polym.* **2018** *179*, 59–70. [CrossRef]
12. Bejan, A.; Ailincai, D.; Simionescu, B.C.; Marin, L. Chitosan hydrogelation with a phenothiazine based aldehyde–toward highly luminescent biomaterials. *Polym. Chem.* **2018**, *9*, 2359–2369. [CrossRef]
13. Iftime, M.M.; Marin, L. Chiral betulin- imino-chitosan hydrogels by dynamic covalent sonochemistry. *Ultrason. Sonochem.* **2018**, *45*, 238–247. [CrossRef] [PubMed]
14. Iftime, M.M.; Ailiesei, G.L.; Ungureanu, E.; Marin, L. Designing chitosan based eco-friendly multifunctional soil conditioner systems with urea controlled release and water re-tention. *Carbohydr. Polym.* **2019**, *223*, 115040. [CrossRef]
15. Azeem, B.; Kushaari, K.Z.; Man, Z.B.; Basit, A.; Thanh, T.H. Review on materials& methods to produce controlled release coated urea fertilizer. *J. Control. Release* **2014**, *181*, 11–21. [PubMed]
16. Sempeho, S.I.; Kim, H.T.; Mubofu, E.; Hilonga, A. Meticulous Overview on the Controlled Release Fertilizers. *Adv. Chem.* **2014**, *2014*, 16. [CrossRef]
17. Lin, C.C.; Metters, A.T. Hydrogels in controlled release formulations: Network design and mathematical modelling. *Adv. Drug Deliv. Rev.* **2006**, *58*, 1379–1408. [CrossRef]
18. Craciun, A.M.; Mititelu Tartau, L.; Pinteala, M.; Marin, L. Nitrosalicyl-imine-chitosan hydrogels based drug delivery systems for long term sustained release in local therapy. *J. Colloid Interface Sci.* **2019**, *536*, 196–207. [CrossRef]
19. Agop, M.; Nica, P.; Ioannou, P.D.; Malandraki, O.; Gavanas-Pahomi, I. El Naschie's epsilon(∞) space-time, hydrodynamic model of scale relativity theory and some applications. *Chaos Solitons Fractals* **2007**, *34*, 1704–1723. [CrossRef]
20. Agop, M.; Nica, P.; Girtu, M. On the vacuum status in Weyl-Dirac theory. *Gen. Relativ. Gravit.* **2008**, *40*, 35–55. [CrossRef]
21. Gottlieb, I.; Agop, M.; Jarcau, M. El Naschie's Cantorian space-time and general relativity by means of Barbilian's group. A Cantorian fractal axiomatic model of space-time. *Chaos Solitons Fractals* **2004**, *19*, 705–730. [CrossRef]
22. Irimiciuc, S.A.; Bulai, G.; Gurlui, S.; Agop, M. On the separation of particle flow during pulse laser deposition of heterogeneous materials-A multi-fractal approach. *Powder Technol.* **2018**, *339*, 273–280. [CrossRef]
23. Irimiciuc, S.A.; Gurlui, S.; Agop, M. Particle distribution in transient plasmas generated by ns-laser ablation on ternary metallic alloys. *Appl. Phys. B* **2019**, *125*, 190. [CrossRef]
24. Agop, M.; Mihaila, I.; Nedeff, F.; Irimiciuc, S.A. Charged Particle Oscillations in Transient Plasmas Generated by Nanosecond Laser Ablation on Mg Target. *Symmetry* **2020**, *12*, 292. [CrossRef]
25. Merches, I.; Agop, M. *Differentiability and Fractality in Dynamics of Physical Systems*; World Scientific: Hackensack, NJ, USA, 2016.
26. Mandelbrot, B.B. *The Fractal Geometry of Nature*; W.H. Freeman and Co.: San Fracisco, CA, USA, 1982.
27. Jackson, E.A. *Perspectives of Nonlinear Dynamics*; Cambridge University Press: New York, NY, USA, 1993; Volume 1.
28. Cobzeanu, B.M.; Irimiciuc, S.; Vaideanu, D.; Gregorovici, A.; Popa, O. Possible Dynamics of Polymer Chains by Means of a Ricatti's Procedure - an Exploitation for Drug Release at Large Time Intervals. *Mater. Past. Environ.* **2017**, *54*, 531–534.
29. Araújo, B.R.; Romaoa, L.P.C.; Doumer, M.E.; Mangrich, A.S. Evaluation of the interactions between chitosan and humics in media for the controlled release of nitrogen fertilizer. *J. Environ. Manag.* **2017**, *190*, 122–131. [CrossRef] [PubMed]
30. Hussain, M.R.; Devi, R.R.; Maji, T.K. Controlled release of urea from chitosan microspheres prepared by emulsification and cross-linking method. *Iran. Polym. J.* **2012**, *21*, 473–479. [CrossRef]
31. Rattanamanee, A.; Niamsup, H.; Srisombat, L.; Punyodom, W.; Watanesk, R.; Watanesk, S. Role of chitosan on some physical properties and the urea controlled release of the silk fibroin/gelatin hydrogel. *J. Polym. Environ.* **2015**, *23*, 334–340. [CrossRef]

32. Narayanan, A.; Dhamodharan, R. Super water-absorbing new material from chitosan, EDTA and urea. *Carbohydr. Polym.* **2015**, *134*, 337–343. [CrossRef]
33. Cuadra, I.A.; Cabañas, A.; Cheda, J.A.R.; Türk, M.; Pando, C. Cocrystallization of the anticancer drug 5-fluorouracil and coformers urea, thiourea or pyrazinamide using supercritical CO_2 as an antisolvent (SAS) and as a solvent (CSS). *J. Supercrit. Fluids* **2020**, *160*, 104813. [CrossRef]
34. Ailincai, D.; Mititelu Tartau, L.; Marin, L. Drug delivery systems based on biocompatible imino-chitosan hydrogels for local anticancer therapy. *Drug Deliv.* **2018**, *25*, 1080–1090. [CrossRef] [PubMed]
35. Marin, L.; Popescu, M.C.; Zabulica, A.; Uji-I, H.; Fron, E. Chitosan as a matrix for biopolymer dispersed liquid crystal systems. *Carbohydr. Polym.* **2013**, *95*, 16–24. [CrossRef] [PubMed]
36. Kasch, N.; Dierking, I.; Turner, M.; Romero-Hasler, P.; Sotobustamante, E.A. Liquid crystalline textures and polymer morphologies resulting from electropolymerisation in liquid crystal phases. *J. Mater. Chem. C* **2015**, *3*, 8018–8023. [CrossRef]
37. Marin, L.; Zabulica, A.; Sava, M. Symmetric liquid crystal dimers containing a luminescent mesogen: Synthesis, mesomorphic behavior, and optical properties. *Soft Mater.* **2013**, *11*, 32–39. [CrossRef]

© 2020 by the authors. Licensee MDPI, Basel, Switzerland. This article is an open access article distributed under the terms and conditions of the Creative Commons Attribution (CC BY) license (http://creativecommons.org/licenses/by/4.0/).

Article

Development and Performance of Bioactive Compounds-Loaded Cellulose/Collagen/ Polyurethane Materials

Iuliana Spiridon [1], Narcis Anghel [1,*], Maria Valentina Dinu [1], Stelian Vlad [1], Adrian Bele [1], Bianca Iulia Ciubotaru [2], Liliana Verestiuc [2] and Daniela Pamfil [1]

[1] "Petru Poni" Institute of Macromolecular Chemistry, Grigore Ghica–Vodă 41, 700487 Iași, Romania; spiridon@icmpp.ro (I.S.); vdinu@icmpp.ro (M.V.D.); vladus@icmpp.ro (S.V.); bele.adrian@icmpp.ro (A.B.); pamfil.daniela@icmpp.ro (D.P.)

[2] Faculty of Medical Bioengineering, Grigore T. Popa University of Medicine and Pharmacy, 9-13 Kogălniceanu Street, 700454 Iași, Romania; ciubotaru.bianca@icmpp.ro (B.I.C.); liliana.verestiuc@bioinginerie.ro (L.V.)

* Correspondence: anghel.narcis@icmpp.ro

Received: 30 April 2020; Accepted: 20 May 2020; Published: 23 May 2020

Abstract: Here we present a new biomaterial based on cellulose, collagen and polyurethane, obtained by dissolving in butyl imidazole chloride. This material served as a matrix for the incorporation of tannin and lipoic acid, as well as bioactive substances with antioxidant properties. The introduction of these bioactive principles into the base matrix led to an increase of the compressive strength in the range 105–139 kPa. An increase of 29.85% of the mucoadhesiveness of the film containing tannin, as compared to the reference, prolongs the bioavailability of the active substance; a fact also demonstrated by the controlled release studies. The presence of bioactive principles, as well as tannins and lipoic acid, gives biomaterials an antioxidant capacity on average 40%–50% higher compared to the base matrix. The results of the tests of the mechanical resistance, mucoadhesiveness, bioadhesiveness, water absorption and antioxidant capacity of active principles recommend these biomaterials for the manufacture of cosmetic masks or patches.

Keywords: cellulose; collagen; biomaterials; tannins; lipoic acid; *Quercus robur* L.

1. Introduction

Natural polymers present a large variety of biological applications due to their low cost, biodegradability and biocompatibility, and have become, in recent years, an important starting point for biomaterials with applications in medicine, as delivery systems for drugs and cell therapies, or as scaffolds for tissue engineering, implants and wound dressings [1]. It is well known that controlled biodegradability and structural integrity in physiological conditions are very important properties for improved biomaterials. Herein, an environmentally acceptable and recyclable solvent [2], namely 1-(n-Butyl)-3-methylimidazolium chloride, was used to solubilize cellulose, collagen and polyurethane. Some studies reported that ionic liquids could enhance the transdermal absorption of drugs [3,4].

Cellulose is the most abundant semi-crystalline natural polymer, consisting of repeating glucose units bounded by β-1,4-glycosidic bonds [5]. It presents a good hydrophilicity, high sorption capacity and cost-effectiveness, as well as biocompatibility and an ability to maintain moisture, which recommend cellulose for different biomedical or cosmetic applications [6].

Cellulose and its derivatives have found wide applications in various fields. Thus, carboxycellulose has been shown to be effective as a hemostatic, being used in surgical sutures [7]. The incorporation of titanium dioxide-like pigments into the structure of nanocrystalline cellulose has not only increased its resistance to paint degradation, but has also imparted antibacterial properties [8]. Nanocellulose has also paved the way for the design of environmentally friendly, biocompatible materials, that have proven effective as retention agents for heavy metals [9–13]. Cellulose esters have been shown to have good thermoplastic properties and, moreover, have been developed as compatibilizers and reinforcing agents with other polymers [14]. The development of nanostructured cellulose-based structures has expanded the area of use of this biopolymer in wastewater treatment [15,16], the stabilization of carbon nanotubes [17], and the development of new composites for drug transport [18].

Collagen is the most abundant protein in animals, and constitutes the matrix of skin, bones and other tissues. We have considered collagen type I as a component of our biomaterials because it is a triple-helical conformation comprising of three polypeptide chains intertwined in a right-handed manner, and it is one of the main components of the extracellular matrix. It has a fibrillar morphology [19] and exhibits elasticity and mechanical toughness [20]. Some studies demonstrate that the incorporation of substances from the category of flavonoids in the protein matrix of collagen reduces the susceptibility of the latter to oxidative stress, as showed by Lucarini et al. [21]. Cellulose–collagen composites have been shown to have good mechanical properties, which is vital for practical application. Such bio composites have been used successfully as scaffold material in tissue engineering [22,23]. Moreover, the biocompatibility of cellulose and collagen with the human body allowed the design of matrices with an osteogenic effect on mesenchymal stem cells [24].

Polyurethanes have attracted attention for their potential use in medical applications, especially when they are functionalized using different natural compounds [25]. Polyurethanes composites are used as medical implants, such as cardiac pacemakers and vascular grafts, and due to excellent mechanical properties and biocompatibility, they could be used in regenerative medicine. The introduction of microcrystalline cellulose in the base matrix of polyurethane elastomers has resulted in an increase in the mechanical strength properties of the material in question, as well as in the thermal stability [26,27].

These above-mentioned components were chosen due to the importance of the toxicity, safety and environmental compatibility of biomaterials for various applications.

At the same time, the incorporation of different biological agents into biomaterials, and their controlled release, represents a proper way to control different processes such as inflammation, infections or stimulation of tissue regeneration [28]. Different Quercus species have been shown to possess antimicrobial, anti-inflammatory, gastroprotective, hemolytic and antioxidant properties [29]. Since ancient times, these species have been used to treat inflammation diseases, tannins being widely distributed in their compositions. Tannins are plant-based substances which belong to the polyphenols' class (from the polyphenolcarboxylic acid series, or from the phenyl-benzopyran series). Tannins are highly astringent, precipitating substances of a protein nature. At the same time as the coagulation of proteins, there is also an action of retraction of the tissue, thus reducing the action surface, a property that is used to treat wounds.

Lipoic acid is a natural antioxidant compound and an oxidative stress scavenger, and has been used as a drug carrier for pathological conditions characterized by oxidative stress, including cancer and neurodegenerative diseases [30], and also as an anti-inflammatory agent [31]. It is a hydrophobic substance derived from caprylic acid, and contains two sulfur atoms connected by a disulfide bond, which is thus considered to be oxidized [32].

In the light of fact that the skin is the largest organ of the body, in the current study, the addition of lipoic acid and tannin to cellulose–collagen–polyurethane matrix has been studied. The reason for choosing this formulation was the fact that cellulose ensures the mechanical strength of the polymer matrix, polyurethane gives the necessary elasticity for topical application, and collagen gives bioadhesion. Lipoic acid and tannins in oak bark were chosen as bioactive principles due to their antioxidant properties and biocompatibility with the human body.

To our knowledge, until now, no evaluations of cellulose–collagen–polyurethane formulations, comprised of either tannin or lipoic acid, have been reported in the literature. Having in mind that some interactions between the used fillers and matrix could occur, the mucoadhesiveness, the in vitro filler release and the antioxidant activity of the materials were evaluated. The morphology, interactions between components, water sorption capacity and mechanical properties of the materials have also been investigated, by scanning electron microscopy (SEM), Fourier transform infrared spectroscopy (FTIR), Dynamic vapor sorption (DVS) and compression tests.

2. Materials and Methods

2.1. Materials

Cellulose (cotton linters, ~20 micrometers, 240 Da), collagen hydrolysate, a polypeptide made by further hydrolysis of denatured collagen (molecular weight of 96 kDa; due to semantics, and ease of reading, we use the generic name "collagen" for the rest of article), lipoic acid, gallic and ellagic acids were purchased from Sigma-Aldrich (St. Louis, MO, USA) and used without further purification. Oak bark (*Quercus robur* L.) was obtained from a local drugstore.

2.2. Methods

2.2.1. Polyurethane Synthesis

The polyurethanes used in this study were synthesized from polycaprolactone (PCL), methylene diphenyl diisocyanate (MDI) and a mixture of butane diol (BD) and beta-cyclodextrin (β-CD) at a ratio of 9/1 (*w/w*) as chain extender, in dimethylformamide (DMF) solution. Briefly, the PCL was dried in vacuum 1 mm Hg at 80 °C for 3 h. The reaction was carried out under stirring with MDI at the temperature of 80 °C for 1 h and then the DMF solution mixture of BD and β-CD was added and the mass reaction was kept under stirring at 60 °C for 6 h. The polyaddition reaction was stopped with a solution of 5 mL EtOH:DMF 1:1 (*v/v*) at the viscosity of ~7000 cP. Molar ratio between components PCL/MDI/(BD/β-CD) was 3/4/1 [33].

The synthesis route is shown in Scheme 1.

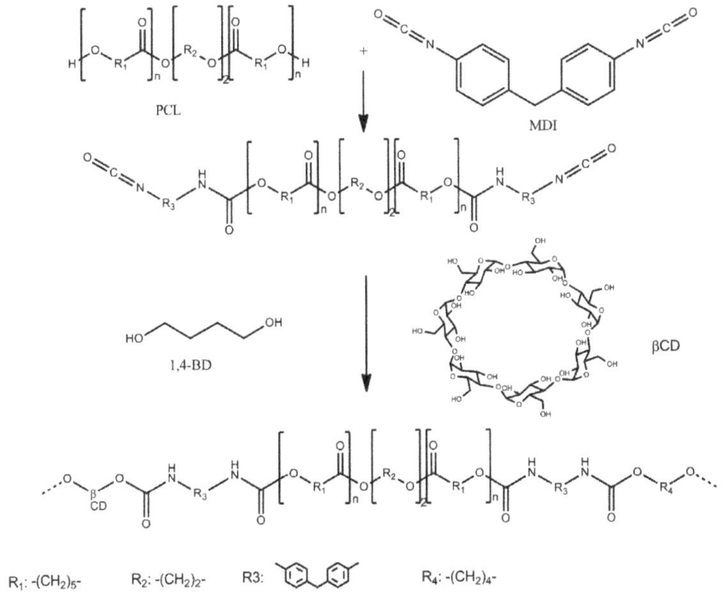

Scheme 1. Polyurethane synthesis.

2.2.2. Tannins Extraction

Tannins extraction was performed according to the method described by Sivakumar et al. [34] with slight modifications (Scheme 2). Oak bark was cut into small sizes in the range of 1 to 2 cm. Then a pre-selected amount of the bark material (100 g) was taken in a clean glass beaker. Extraction was carried out using 300 mL of distilled water. Ultrasound-assisted extraction was carried out twice using a power level of 240 W for 30 min. The aqueous extract was filtered and the tannins were precipitated by saturating the solution with ammonium sulfate. The crude material was subsequently purified by dissolution in acetone and reprecipitation with ethyl ether.

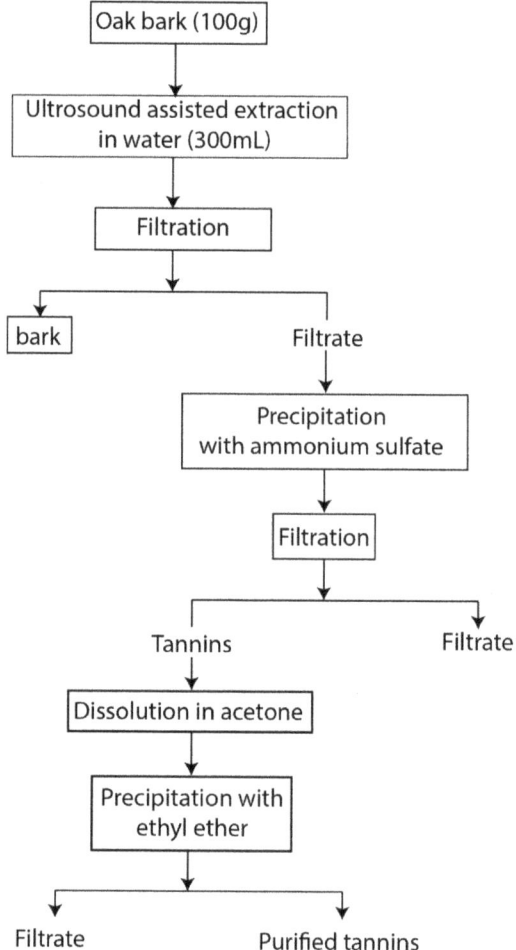

Scheme 2. The workflow for the tannin's extraction.

2.2.3. Preparation of Biomaterials

The reference material (REF) was obtained by dissolution of cellulose (1 g), collagen (0.25 g) and polyurethane (0.25 g) in buthyl-3-methylimidazolium chloride (10 g) under stirring, at a temperature of 100 °C for 8 h. Other biomaterials named LIP and TAN were obtained by addition of 0.15 g of lipoic acid and tannin, respectively, to the cellulose–collagen–polyurethane matrix. After 48 h, the samples were washed with distilled water. The respective amounts of tannin and lipoic acid removed from the

TAN and LIP formulation were evaluated by Ultraviolet–Visible (UV) spectroscopy. It was found that 4.3% of tannin and 2.8% lipoic acid were removed by washing.

2.2.4. Characterization

FTIR Spectroscopy

Fourier Transform Infrared Spectroscopy (FTIR) spectroscopy was used to analyze the possible interaction materials' components. The film samples were measured by a Bruker, Vertex 70 (Billerica, MA, USA) equipped with an attenuated total reflection (ATR) device. All samples were acquired using a diamond crystal with ZnSe focusing element at room temperature. Scanning was performed in a range from 4000 cm^{-1} to 600 cm^{-1} with a spectral resolution of 2 cm^{-1}, with 64 repetitious scans averaged for each spectrum. Prior to measurement, the materials were conditioned at 65% ± 2% relative humidity and 20 ± 2 °C for 48 h.

Scanning Electron Microscopy

Scanning Electron Microscopy (SEM) was used to analyze the cross sections of material using a SEM (FEI QUANTA 200ESEM instrument) with an integrated EDX system: GENESIS XM2i EDAX with an SUTW detector. The samples were analyzed with a low-vacuum secondary electron detector at an accelerating voltage of 25.0 kV, at room temperature and 0.050 Torr internal pressure. The experiment was performed in triplicate.

2.2.5. Bioadhesivity Test

A TA.XT plus® analyzer from Stable Micro Systems (Godalming, UK) was used to evaluate the adhesion force (maximum detachment force) and total work of adhesion as described in a previous paper [35]. The bioadhesion test was performed on a hydrated dialysis tubing membrane (cellulose, Visking DTV14000), in PBS (pH 7.4) at 37 °C, while for the determination of mucoadhesive properties a fresh porcine skin membrane was used. The values given for each sample are the results of five determinations.

2.2.6. Compression Test

The compressive properties of the materials were determined using a Shimadzu Testing Machine EZTest (EZ-LX/EZ-SX Series, Kyoto, Japan) at a compression rate of 1 mm × min^{-1}. This test was performed at 22 °C and was applied on samples, as plates, with 10 mm thickness, 12 mm width and 4 mm height. The setup of the test and the calculations of the elastic modulus were performed in accordance with the procedure already reported for curdlan-based hydrogels [36]. Compressive strength was determined as the compressive stress at 10% strain, while the elastic modulus was calculated as the slope of the initial linear region in the stress-strain curve.

2.2.7. Dynamic Vapor Sorption (DVS)

IGAsorp equipment (Hiden Analytical, Warrington, UK) was used to evaluate the water sorption at atmospheric pressure by passing a humidified stream of gas over the sample. Isothermal studies were performed at humidity between 0% and 95%, in the temperature range from 5 °C to 85 °C, with an accuracy of ±1% for 0%–90% Relative Humidity (RH) and ±2% for 90%–95% RH.

2.2.8. In Vitro Release Studies

The experiments were carried out in a 708-DS Dissolution Apparatus coupled with a Cary 60 UV-VIS spectrophotometer (Agilent Technologies, Santa Clara, CA, USA) at 37 ± 0.5 °C and a rotation speed of 100 rpm, in media with phosphate buffered saline (pH 7.2). The concentration of the released compound was analyzed spectrophotometrically, showing λ_{max} values of 276 nm (TAN) and 287 nm (LIP) at room temperature. The concentrations were calculated based on the calibration curves determined at the same wavelengths. The filler release kinetics were evaluated using the equation proposed by Korsmeyer and Peppas [Equation (1)] [37].

$$\frac{M_t}{M_\infty} = kt^n \tag{1}$$

where M_t/M_∞ represents the fraction of the drug released at time t; M_t and M_∞ are the absolute cumulative amount of drug released at time t and the maximum amount released in the experimental conditions used, at the plateau of the release curves; k is a constant incorporating the characteristics of the macromolecular drug loaded system, and n is the release exponent, which is indicative of the release mechanism.

In the equation above a value of $n = 0.5$ indicates a Fickian diffusion mechanism of the filler from the biomaterial sample, while a value $0.5 < n < 1$ indicates a non-Fickian behavior. When $n = 1$, a case II transport mechanism is involved with zero order kinetics, while $n > 1$ indicates a special case II transport mechanism [38].

2.2.9. (DPPH—2,2 diphenyl-1-picrylhydrazyl) Assay

The 2,2 diphenyl-1-picrylhydrazyl (DPPH) free radical scavenging activity was determined by the method described by Sridhar and Charles [39] with slight modifications. For the DPPH assay, samples with weights varying between 20 and 70 mg were added to the same volume (10 mL) of DPPH methanolic solution (100 µM). Mixtures were shaken and left to incubate for 20 min in the dark at room temperature. A decrease in absorbance was measured at 515 nm against a blank of methanol without DPPH using a Jenway 6405 UV/Vis spectrophotometer. The inhibition percentage of DPPH discoloration was calculated using Equation (2), where $A_{control}$ is the absorbance of control and A_{sample} is the absorbance of the sample.

$$\%inhibition = \left[\frac{A_{control} - A_{sample}}{A_{control}}\right] \times 100 \tag{2}$$

Values are expressed as mean ± SD of triplicate measurements.

2.2.10. HPLC Analysis of Tannins

A Shimadzu Prominence High Performance Liquid Chromatography (HPLC) System equipped with an Alltech Econosil C18 column (4.6 × 250 mm, 5 µm) was used for HPLC analysis. Elution: solvent A (water with 1% AcOH), solvent B (methanol). Gradient: solvent B from 20% to 100% in 60 min. Flow rate 1 mL/min. Temperature 25 °C, injection volume 20 µL. UV detection at 280 nm. The tannins were dissolved in the initial mobile phase at a concentration of 1 mg/mL. Gallic and ellagic acids have been used as standards for identifying the components of tannic extract [40]. The compounds were quantified using the calibration curves method [41,42].

3. Results and Discussion

The design and characterization of novel biomaterials, obtained without chemical modification, is of great relevance for the development of biomaterials with various applications.

3.1. Analysis of Tannins in Oak Bark

The chromatogram in Figure 1 indicates the presence of gallic and ellagic acids, predominant in the aqueous extract of the oak species *Quercus robur* L. [41]. Scientific data stipulate a ratio between ellagic and gallic acid of about 10/1 [42]. In this case, the ellagic and gallic acids were found in amounts of 83.3 and 7.2 mg/g extractum pulvis of oak bark.

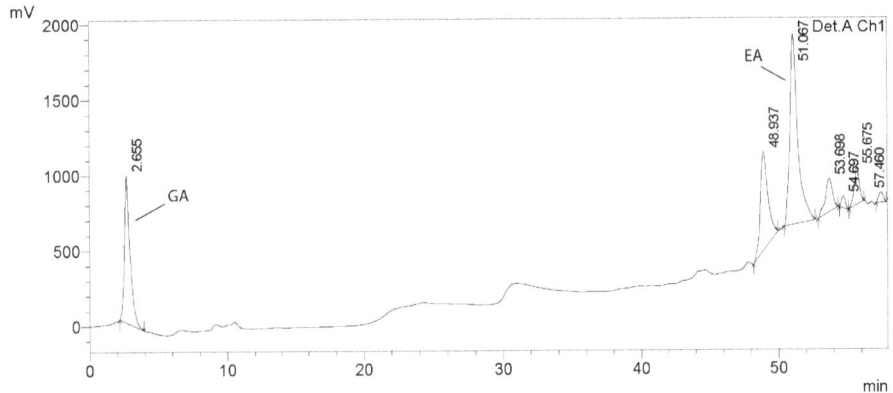

Figure 1. HPLC chromatogram for *Quercus robur* L. extract (GA—gallic acid, EA—ellagic acid).

3.2. FTIR Analysis

FTIR spectroscopy was used to analyze the possible interactions between material components. Figure 2 depicts the FTIR spectra for components of the reference film. The peaks observed in cellulose spectra [Figure 2(1)], in the range of 3660–2900 cm^{-1}, are characteristic for the stretching vibration of O–H and C–H bonds in polysaccharides, while the broad peak at 3334 cm^{-1} is characteristic for the stretching vibration of the hydroxyl group in polysaccharides [43]. The band at 2900 cm^{-1} is attributed to the C–H stretching vibration of all hydrocarbon constituents in polysaccharides. Typical bands assigned to cellulose were observed in the region of 1630–900 cm^{-1}. The peaks located at 1631 cm^{-1} correspond to the vibration of water molecules absorbed in cellulose [44]. The absorption bands at 1427, 1367, 1334, 1029 cm^{-1} and 898 cm^{-1} belong to the stretching and bending vibrations of –CH$_2$ and –CH, –OH and C–O bonds in cellulose [45,46].

Figure 2(2) shows the Fourier transform infrared spectrum recorded for polyurethane. The absorption band at 3323 cm^{-1} corresponds to NH stretching. The sharp peaks at 2858 cm^{-1} and 2925 cm^{-1} are associated with –CH$_2$ stretching, while other modes of –CH$_2$ vibrations are identified by the bands at 1448, 1406, 1334 and 1236 cm^{-1}. In addition, the absorption band at 1631 cm^{-1} is associated with a C=O group in polyurethane. The group of NH bend vibrations is identified by the band at 1631 cm^{-1} [47].

The spectra of collagen [Figure 2(3)] shows the amide A band, associated with N–H stretching, at 3330 cm^{-1}. The amide bands were observed at 1631, 1541 and 1334 cm^{-1}, respectively. Polypeptide backbone C–O stretching vibration was found in the range of 1600–1700 cm^{-1}. C–N stretching vibrations were noted at 1222 cm^{-1} [48].

In Figure 3 are presented FTIR spectra for the obtained materials. Based on the recorded spectra, we can calculate a series of indices that reflect the degree of ordering and the total crystallinity, as well as the strength of the hydrogen bonds, for the studied materials.

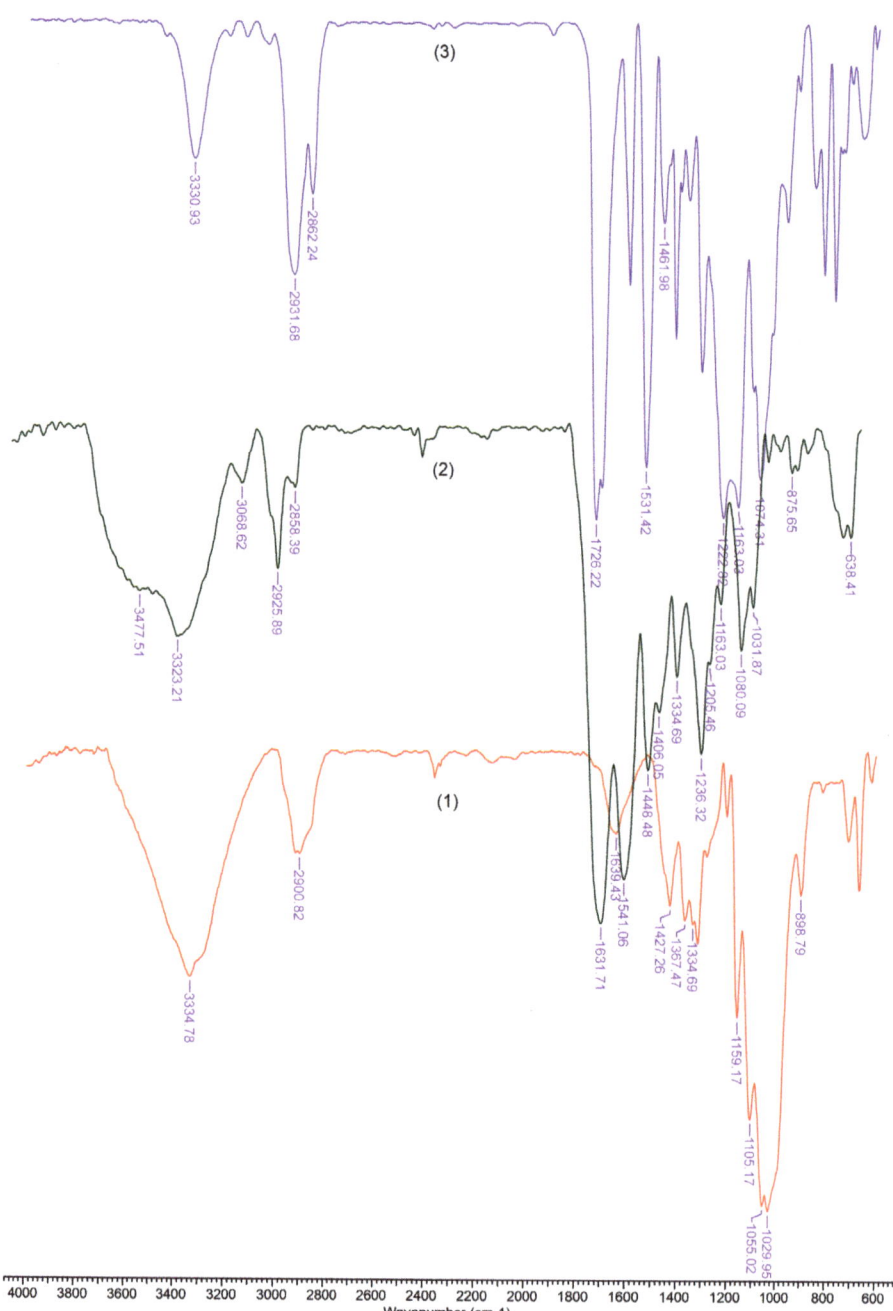

Figure 2. FTIR spectra of reference' components: (**1**)—cellulose; (**2**)—polyurethane; (**3**)—collagen.

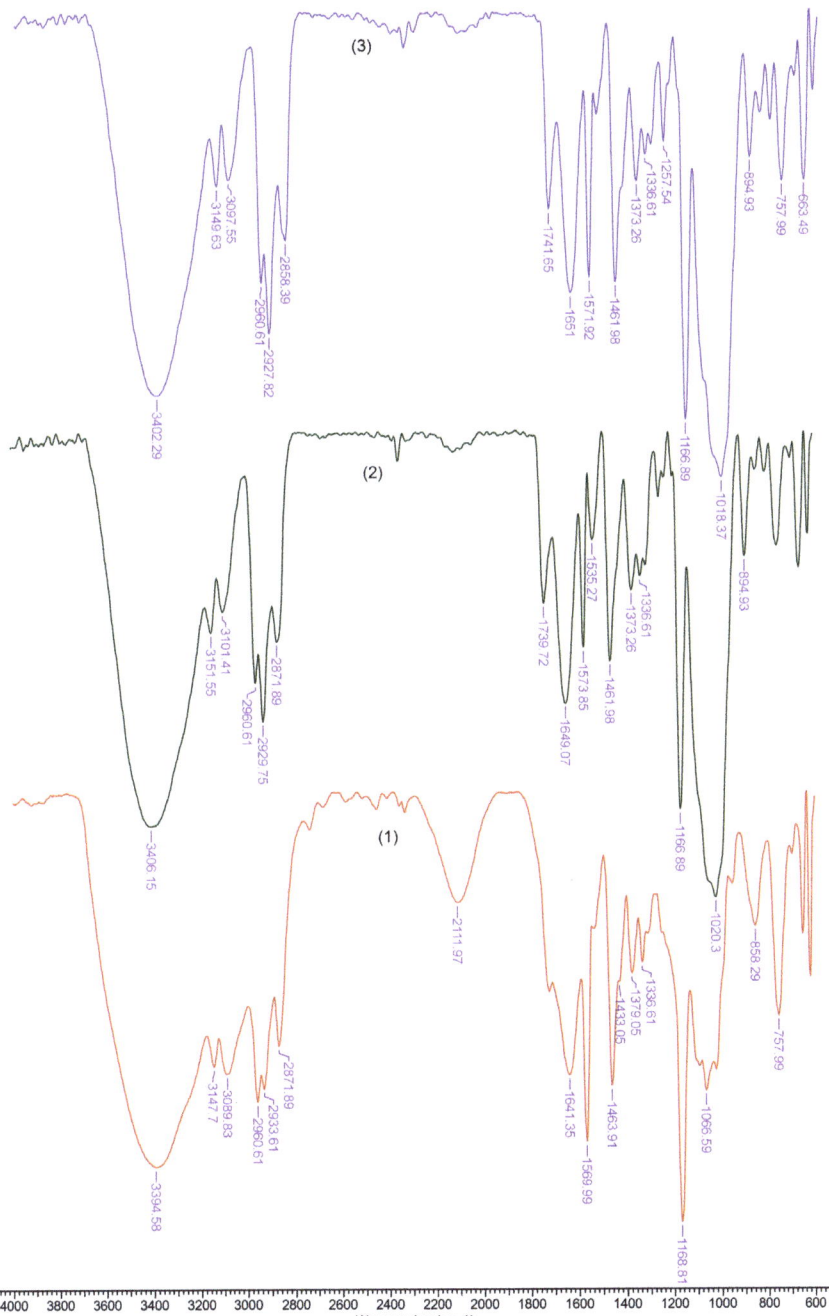

Figure 3. FTIR spectra of the obtained biomaterials: (**1**)—cellulose–collagen–polyurethane; (**2**)—TAN; (**3**)—LIP.

The ratio between the heights of the bands at 1376 and 2902 cm^{-1} was proposed by Colomn and Carrillo [49] as the total crystalline index (TCI). The band at 1437 cm^{-1} is associated with the crystalline structure of cellulose, while the band at 899 cm^{-1} is assigned to the amorphous region in the cellulose. The ratio between the absorbance of the bands at 1437 and 899 cm^{-1} is used as a lateral order index (LOI). Considering the chain mobility and bond distance, the hydrogen bond intensity (HBI) of cellulose is closely related to the crystal system and the degree of intermolecular regularity—that is, crystallinity. The ratio of the absorbance bands at 3336 and 1336 cm^{-1} was used to study the cellulose sample's HBI. The obtained results are displayed in Table 1. The TCI is proportional to the degree of crystallinity of cellulose, and LOI represents the ordered regions perpendicular to the chain direction in the cellulose.

The REF samples exhibited the highest TCI and lowest LOI, which implies the highest crystallinity degree and an increase in ordered regions perpendicular to the chain direction in cellulose. The data from Table 1 show that the LIP material presents the highest LOI and lowest value of TCI. It is possible that a lateral ordered cellulose structure was obtained in the cellulose-collagen-polyuretane matrix by the addition of lipoic acid. At the same time, the HBI value increased as compared to that of the matrix, which means that fewer available hydroxyl groups in the cellulose chain are able to interact by inter- and/or intramolecular hydrogen bonding.

On the contrary, the film comprised of tannin registered the highest value of HBI, suggesting strong interactions between the adjacent cellulose chains, resulting in a high level of cellulose chain packing, due to the numerous phenolic hydroxyl groups attached to the aromatic and heterocyclic rings. This also resulted in greater mechanical properties.

Table 1. Infrared total crystallinity (TCI), hydrogen bond intensity (HBI) and lateral order index (LOI) for the obtained biomaterials.

	TCI (A_{1376}/A_{2902})	LOI (A_{1437}/A_{899})	HBI (A_{3336}/A_{1336})
REF	0.492	1.533	3.459
TAN	0.447	2.132	4.368
LIP	0.406	2.250	4.177

3.3. Mechanical Properties

Compressive strength is an important parameter for the scaffolds used in tissue engineering. Consequently, uniaxial compression tests were applied to materials up to 70% strain, and the obtained stress-strain curves for all samples can be seen in Figure 4A. The tannin- and lipoic acid-containing biomaterials showed a typical linear stress-strain behavior at < 20% initial strain level (Figure 4A), demonstrating that the hydrogel blends changed from a relaxed state to a stressed state to store energy for resisting the compression stress [50]. As the strain level increased, the stress rapidly rose, and a fracture at about 20% strain level was first observed in the reference sample (Figure 4A), suggesting that the energy dissipation inside the network for this sample was not enough to resist the external force applied [51]. However, the reference material was found to be very elastic, with a value of elastic modulus of 4.5 kPa and a compressive strength of 80 kPa (Figure 4B). In general, the addition of filler causes the decrease of elasticity [52], but this influence was only observed for the sample containing lipoic acid (2.95 kPa, Figure 4B).

On the other hand, the addition of filler particles to the cellulose–collagen–polyuretane matrix induced a progressive increase of the compression strength, as well as the strain of the materials. When 10% of lipoic acid was added, the compression strength reached about 105 kPa, while the same amount of tannin increased compression strength up to 139 kPa. This means that the filler used can bear the stress effectively and increase the mechanical strength of the obtained materials.

The mechanical properties of polymer blends usually give an indication of possible interactions between their constituents [53]. The two-fold increase in the compressive strength of the materials containing tannin (TAN) indicate a possible interaction between the matrix and the filler within the blend via hydrogen bonding, as proved above by the highest HBI values (Table 1). The increase in mechanical properties, especially the rigidity of the samples containing tannin, could also be attributed to the decrease of the Brunauer–Emmett–Teller (BET) surface area and water sorption capacity (Table 2), because of the low dissipation energy inside the network. A low BET surface area indicates a reduced porosity of the sample, so the densification region which is normally observed for porous materials [54] was missing in our case, and thus the rigidity of the tannin-containing film increased (Figure 4B). The values of the elastic modulus of the materials prepared in this study are comparable to or even higher than those obtained for other cellulose-containing hydrogels. For example, Kalinoski & Shi [55] added lignin and/or xylan to cellulosic hydrogels, leading to values for the elastic modulus ranging from 20 to 105 kPa.

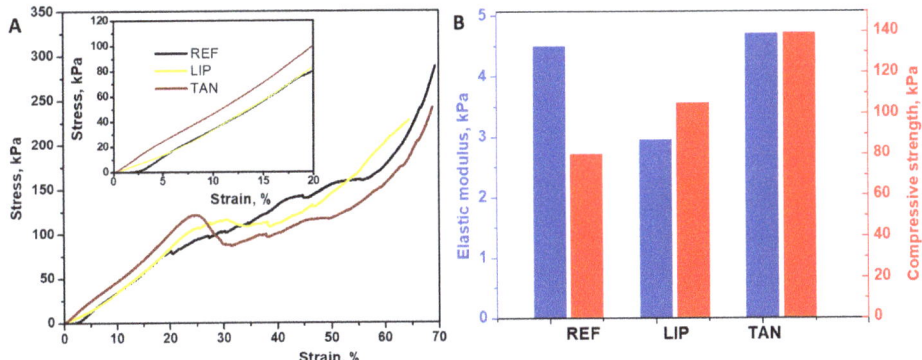

Figure 4. (**A**) Representative compressive stress-strain curves for REF, LIP and TAN at room temperature; (**B**) The values of the elastic modulus and compressive strength for the obtained biomaterials.

Table 2. Dynamic vapors sorption (DVS) parameters for the obtained biomaterials.

Sample	Sorption Capacity % d.b.	BET Data	
		Area $m^2 \times g^{-1}$	Monolayer $g \times g^{-1}$
REF	48.5	331.062	0.094
TAN	17.7	151.100	0.043
LIP	24.8	758.750	0.216

However, the values of the elastic modulus of our materials are weak compared to other cellulose-containing double network materials used for articular cartilage, which exhibited an elastic modulus of 322 kPa [56]. The results in the literature also suggest that by altering the ratios of xylan and lignin to cellulose, one can potentially fine-tune the mechanical properties of cellulosic hydrogels [57]. It is also possible to use chemical cross-linkers when preparing the physically cross-linked materials, which is especially effective in improving the matrix properties. The mechanical properties analysis provide evidence that cellulose–collagen–polyuretahane films with added lipoic acid and tannin become more resistant and less elastic. As our focus is to create physically cross-linked mechanical properties, materials using lipoic acid and tannin without chemical cross-linkers, or using chemical cross-linkers along with physical cross-linking methods, may warrant future study.

3.4. Water Adsorption Isotherms

Adsorption and desorption isotherms express the dependence of equilibrium water content on materials and relative humidity. According to Figure 5, a hysteresis behavior is related to the water uptake capacity values of our materials.

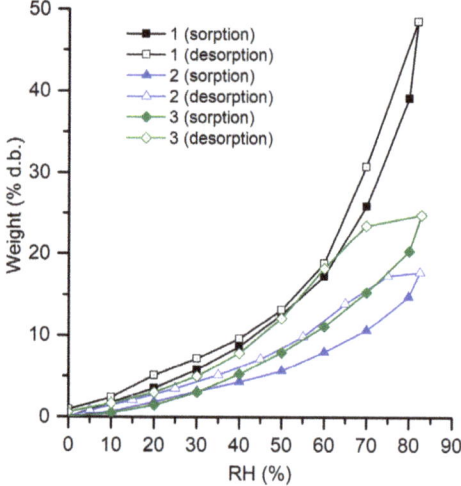

Figure 5. Sorption/desorption isotherms of the studied biomaterials (1—reference sample; 2—TAN sample; 3—LIP sample).

The Brunauer–Emmett–Teller (BET) and Guggenheim–Anderson–de Boer (GAB) equations were used for the modelling of the sorption isotherms, but the BET model better fitted the experimental results as compared to the GAB model. This means that the transport phenomena are associated with the occurrence of monomolecular sorption range (I) and a multi-layer sorption (II). The data from Table 2 evidence a higher amount of water for the REF sample, which could also be correlated with the high elasticity already observed for this sample (Figure 4B). The addition of fillers to the polymeric matrix resulted in a significant decrease of the sorption capacity, the highest diminishing being recorded for the TAN sample. Other authors [58] reported that the adsorption of tannin on the cellulose surface is driven by the hydrophilic domains via H-bonding, and as a result, the hydrophobic domains of the tannin are exposed at the very top of the molecular surface.

3.5. Bio/Mucoadhesivity Properties

The developed materials present a surface layer possessing adhesive properties. Bio/mucoadhesive materials represent a promising tool to achieve site-specific drug delivery [59].

Adhesion of materials to the epithelial tissue is an important property for product safety, efficacy and quality. In Table 3, the results are presented from bioadhesion and mucoadhesion tests on cellulose dialysis membrane and porcine skin, respectively. The TAN addition induced a decrement of 33.56% of bioadhesion force, while lipoic acid presence did not influence this parameter.

Table 3. Adhesive properties of the studied materials.

Sample	Bioadhesion Test		Muchoadhesion Test	
	Adhesion Force (n)	Total Work of Adhesion ($n \times s$)	Adhesion Force (n)	Total Work of Adhesion ($n \times s$)
REF	0.143 ± 0.00205	0.025 ± 0.00286	0.067 ± 0.00339	0.0099 ± 0.00033
TAN	0.095 ± 0.00205	0.020 ± 0.0017	0.087 ± 0.00449	0.0204 ± 0.00041
LIP	0.142 ± 0.00368	0.031 ± 0.00163	0.124 ± 0.0033	0.0317 ± 0.00057

The mucoadhesion test was performed in order to measure the ability of the films to adhere onto the porcine skin. The mucin present in the mucus surface layer of porcine skin is rich in cysteine (>10% of the amino acids) and therefore in the thiol groups which can lead the formation of many disulfide bonds (S–S) [60]. The thiol groups of mucins could interact with the hydroxyl or carboxyl groups of tannins, and these interactions could cause a better adhesion to the cell surface of porcine skin. An increment of 29.85% in mucoadhesiveness of TAN was recorded, and this could mean that the presence of tannin prolongs the material bioavailability, a fact confirmed by the drug release study. The LIP sample exhibited the highest mucoadhesiveness, mucoadhesive force being 1.85 times higher than that of the reference sample, since thiolate compounds are well-known as mucoadhesives [61].

The total work of the adhesion values is in good agreement with those of the adhesive forces.

3.6. Controlled Release of Active Compounds from Biomaterials

The tannin and lipoic acid release profiles from the investigated materials were studied (Figure 6) to evaluate the potential delivery applications.

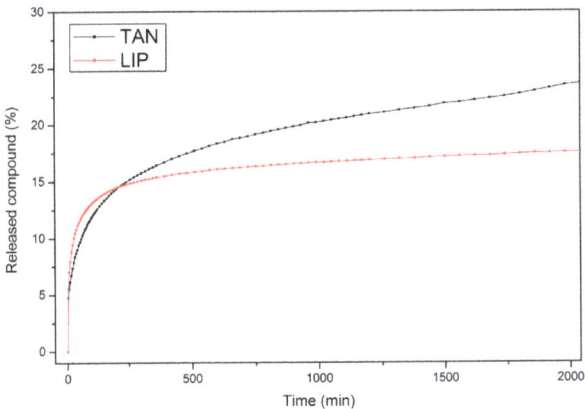

Figure 6. Release profiles of tannin and lipoic acid from the investigated samples.

Comparing results presented in Figure 6 and Table 4, it seems that the highest amount of the active substance was released from the TAN material, which is in a good agreement with the findings regarding adhesive properties (Table 3). For both samples, the correlation coefficient (R^2) was rather high (0.998), and describes the first-order kinetic model. The transport constants (k) and transport exponents (n) were determined from the obtained data. The k value was higher for the LIP sample as compared to the TAN sample. The value of n was less than 0.3 for both samples, and this corresponds to a Fickian diffusion, suggesting that the release mechanisms of both fillers were related to the physical diffusion/filler dissolution interaction of electrostatic forces or hydrogen bonds [62].

Table 4. Kinetic parameters of the fillers released from investigated samples.

Samples	n	R^2_n	k, min^{-n}	R^2_k
TAN	0.294	0.998	0.0308	0.999
LIP	0.235	0.998	0.0500	0.999

n = release exponent, k = release rate constant, R^2_n and R^2_k = correlation coefficient. Corresponding to the slope obtained for determination of n and k.

3.7. Antiradical Activity

It is well known that free radicals are able to induce oxidative stress in biomolecules, thus causing a wide range of degenerative diseases. The free radical scavenging activity of TAN and LIP materials confirms their potential therapeutic value in protecting against oxidative injury. The results from Figure 7 show that the materials are able to scavenge the free radicals, so reducing the oxidative stress.

The material comprised of tannin as the bioactive compound presented the highest ability to scavenge free radicals [63]. This can be explained by the hydrogen-donating phenolic hydroxyl groups attached to the aromatic and heterocyclic rings, which impart good antioxidant activity to the TAN film [59]. It is worth mentioning that free tannin and tannin in the cellulose–collagen–polyurethane matrix exhibited almost the same radical scavenging activity (92.34% and 92.03%). The LIP material exhibited scavenging activity of DPPH radical (83.18%) lower than free lipoic acid (87.05%), probably due to the interactions between the components of the material.

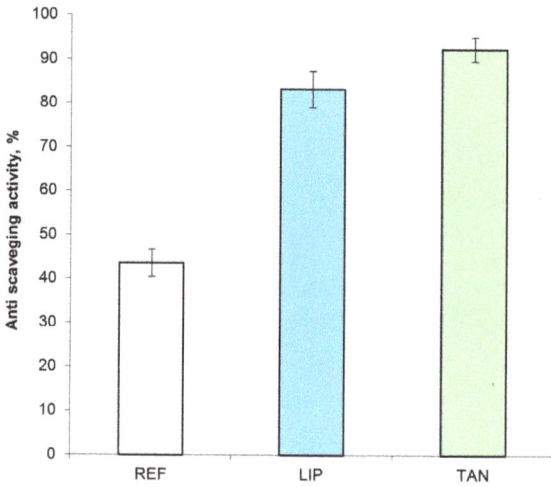

Figure 7. Antiradical activity of the obtained biomaterials.

3.8. Materials' Morphology

Investigating the SEM images (Figure 8) can give different valuable information about film morphology. When filler was added into the matrix, its surface tended to become smoother. The absence of voids, as well as the presence of some discrete micro-domains, suggests the development of hydrogen bond networks between matrix components and fillers, confirmed by the HBI value (Table 1). It also suggests that the compatibility and interfacial strength between the filler particles and cellulose–collagen–polyurethane matrix is improved.

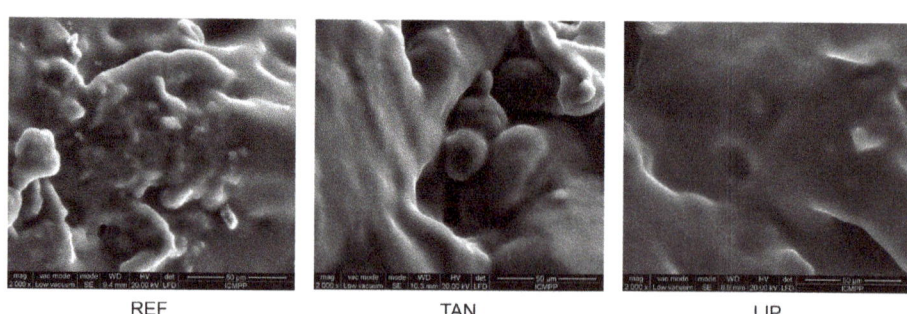

Figure 8. Images of materials.

4. Conclusions

A new biomaterial comprising cellulose, collagen and polyurethane was obtained by dissolution in butyl methylimidazole chloride. Other formulations containing lipoic acid and tannin were developed and analyzed. The addition of filler particles to the cellulose–collagen–polyurethane matrix induced a progressive increase of the compression strength, as well as the strain of the materials, which means that the filler used can bear stress effectively and increase the mechanical strength of the obtained materials. A hysteresis behavior is related to the water uptake capacity values of the materials, while when filler was added into the matrix, its surface tended to become smoother.

Perhaps the most important implication of this study is associated with the hypothesis that the fillers' addition to the polymeric matrix induces improved biological properties, a fact confirmed by the increasing of the mucoadhesiveness, as well as of the anti-scavenging activity. The in vitro release of the used fillers is described by Korsmeyer–Peppas model.

In summary, the obtained results confirm that the prepared materials could be promising carriers for controlled release of TAN and LIP, with potential medical and cosmetic applications. For perspective, we are already considering the incorporation of active principles with anti-cellulite action (for cosmetic applications, obviously) into the polymer matrix, that will prove it has mechanical strength, elasticity and bioadhesiveness.

Author Contributions: Conceptualization and Methodology, I.S.; Writing Original Draft Preparation, I.S. and N.A.; Writing and Review, I.S. and N.A.; Editing, N.A.; Investigation, I.S., N.A., M.V.D., S.V., A.B., B.I.C., L.V., and D.P. All authors have read and agreed to the published version of the manuscript.

Funding: This research received no external funding.

Conflicts of Interest: The authors declare no conflict of interest.

References

1. Stone, S.A.; Gosavi, P.; Athauda, T.J.; Ozer, R.R. In situ citric acid crosslinking of alginate/polyvinylalcohol electrospun nanofibers. *Mater. Lett.* **2013**, *112*, 32–35. [CrossRef]
2. El Seoud, O.A.; Kostag, M.; Jedvert, K.; Malek, N.I. Cellulose in Ionic Liquids and Alkaline Solutions: Advances in the mechanisms of biopolymer dissolution and regeneration. *Polymer* **2019**, *11*, 1917. [CrossRef] [PubMed]
3. Halayqa, M.; Pobudkowska, A.; Domańska, U.; Zawadzki, M. Studying of drug solubility in water and alcohols using drug—Ammonium ionic liquid-compounds. *Eur. J. Pharm. Sci.* **2018**, *111*, 270–277. [CrossRef] [PubMed]
4. Monti, D.; Egiziano, E.; Burgalassi, S.; Chetoni, P.; Chiappe, C.; Sanzone, A.; Tampucci, S. Ionic liquids as potential enhancers for transdermal drug delivery. *Int. J. Pharm.* **2017**, *516*, 45–51. [CrossRef] [PubMed]
5. Zhang, C.; Li, P.; Zhang, Y.; Lu, F.; Li, W.; Kang, H.; Xiang, J.; Huang, Y.; Liu, R. Hierarchical porous structures in cellulose: NMR relaxometry approach. *Polymer* **2016**, *98*, 237–243. [CrossRef]
6. Lee, S.H.; Kim, H.J.; Kim, J.C. Nanocellulose applications for drug delivery: A review. *J. Environ. Sci.* **2019**, *35*, 141–149. [CrossRef]
7. Sharma, P.R.; Joshi, R.; Sharma, S.K.; Hsiao, B.S. A Simple Approach to Prepare Carboxycellulose Nanofibers from Untreated Biomass. *Biomacromolecules* **2017**, *18*, 2333–2342. [CrossRef]
8. Zhan, C.; Li, X.; Sharma, P.R.; He, H.; Sharma, S.K.; Wanga, R.; Hsiao, B.S. A study of TiO_2 nanocrystal growth and environmental remediation capability of TiO_2/CNC nanocomposites. *RSC Adv* **2019**, *9*, 40565–40576. [CrossRef]
9. Chen, H.; Sharma, S.K.; Sharma, P.R.; Yeh, H.; Johnson, K.; Hsiao, B.S. Arsenic (III) Removal by Nanostructured Dialdehyde Cellulose–Cysteine Microscale and Nanoscale Fibers. *ACS Omega* **2019**, *4*, 22008–22020. [CrossRef]
10. Sharma, P.R.; Chattopadhyay, A.; Sharma, S.K.; Geng, L.; Amiralian, N.; Martin, D.; Hsiao, B.S. Nanocellulose from Spinifex as an Effective Adsorbent to Remove Cadmium (II) from Water. *ACS Sustain. Chem. Eng.* **2018**, *6*, 3279–3290. [CrossRef]
11. Yu, H.Y.; Zhang, D.Z.; Lu, F.F.; Yao, J. New Approach for Single-Step Extraction of Carboxylated Cellulose Nanocrystals for Their Use as Adsorbents and Flocculants. *ACS Sustain. Chem. Eng.* **2016**, *4*, 2632–2643. [CrossRef]

12. Sharma, P.R.; Sharma, S.K.; Antoine, R.; Hsiao, B.S. Efficient Removal of Arsenic Using Zinc Oxide Nanocrystal-Decorated Regenerated Microfibrillated Cellulose Scaffolds. *ACS Sustain. Chem. Eng.* **2019**, *7*, 6140–6151. [CrossRef]
13. Sharma, P.R.; Chattopadhyay, A.; Sharma, S.K.; Hsiao, B.S. Efficient Removal of UO_2^{2+} from Water Using Carboxycellulose Nanofibers Prepared by The Nitro-Oxidation Method. *Ind. Eng. Chem. Res.* **2017**, *56*, 13885–13893. [CrossRef]
14. Klemm, D.; Heublein, B.; Fink, H.P.; Bohn, A. Cellulose: Fascinating Biopolymer and Sustainable Raw Material. *Angew. Chem. Int. Ed.* **2005**, *44*, 3358–3393. [CrossRef] [PubMed]
15. Mohammed, N.; Grishkewich, N.; Tam, K.C. Cellulose Nanomaterials: Promising Sustainable Nanomaterialsfor Application inWater/Wastewater Treatment Processes. *Environ. Sci. Nano* **2018**, *5*, 623–658. [CrossRef]
16. Sharma, P.R.; Sharma, S.K.; Lindström, L.; Hsiao, B.S. Nanocellulose-Enabled Membranes for Water Purification: Perspectives. *Adv. Sustain. Syst.* **2020**, 1900114. [CrossRef]
17. Sharma, P.R.; Varma, A.J. Functional nanoparticles from cellulose: Engineering the shape and size of 6-carboxycellulose. *Chem. Commun.* **2013**, *49*, 8818–8820. [CrossRef]
18. Klemm, D.; Cranston, E.D.; Fischer, D.; Gama, M.; Kedzior, S.A.; Kralisch, D.; Kramer, F.; Kondo, T.; Lindström, T.; Nietzsche, S.; et al. Nanocellulose as a natural source for ground breaking applications in materials science: Today's. *Mater. Today* **2018**, *21*, 720–748. [CrossRef]
19. Chang, S.W.; Flynn, B.P.; Ruberti, J.W.; Buehler, M.J. Molecular mechanism of force induced stabilization of collagen against enzymatic breakdown. *Biomaterials* **2012**, *33*, 3852–3859. [CrossRef]
20. Salamanca, E.; Hsu, C.C.; Yao, W.L.; Choy, C.S.; Pan, Y.H.; Teng, N.C.; Chang, W.J. Porcine collagen–bone composite induced osteoblast diferentiation and bone regeneration in vitro and in vivo. *Polymer* **2020**, *12*, 93. [CrossRef]
21. Lucarini, M.; Sciubba, F.; Capitani, D.; Di Cocco, M.E.; D'Evoli, L.; Durazzo, A.; Delfini, M.; Boccia, G.L. Role of catechin on collagen type I stability upon oxidation: A NMR approach. *Nat. Prod. Res.* **2020**, *34*, 53–62. [CrossRef] [PubMed]
22. Chenga, Y.; Lua, J.; Liub, S.; Zhaoc, P.; Luc, G.; Chen, J. The preparation, characterization and evaluation of regeneratedcellulose/collagen composite hydrogel films. *Carbohydr. Polym.* **2014**, *107*, 57–64. [CrossRef] [PubMed]
23. Pei, Y.; Yang, J.; Liu, P.; Xu, M.; Zhang, X.; Zhang, L. Fabrication, properties and bioapplications of cellulose/collagen hydrolysate composite films. *Carbohydr. Polym.* **2013**, *92*, 1752–1760. [CrossRef] [PubMed]
24. Noha, Y.K.; Da Costaa, A.D.S.; Parkd, Y.S.; Due, P.; Kimb, I.K.; Park, K. Fabrication of bacterial cellulose-collagen composite scaffolds and their osteogenic effect on human mesenchymal stem cells. *Carbohydr. Polym.* **2019**, *219*, 210–218. [CrossRef]
25. Brzeska, J.; Tercjak, A.; Sikorska, W.; Kowalczuk, M.; Rutkowska, M. Morphology and physicochemical properties of branched polyurethane/biopolymer blends. *Polymer* **2020**, *12*, 16. [CrossRef]
26. Lei, W.; Fang, C.; Zhou, X.; Li, Y.; Pu, M. Polyurethane elastomer composites reinforced with waste natural cellulosic fibers from office paper in thermal properties. *Carbohydr. Polym.* **2018**, *197*, 385–394. [CrossRef]
27. Stanzione, M.; Oliviero, M.; Cocca, M.; Errico, M.E.; Gentile, G.; Avella, M.; Lavorgna, M.; Buonocore, G.G.; Verdolotti, L. Tuning of polyurethane foam mechanical and thermal properties using ball-milled cellulose. *Carbohydr. Polym.* **2020**, *231*, 115772. [CrossRef]
28. Pereira, R.; Carvalho, A.; Vaz, D.C.; Gil, M.H.; Mendes, A.; Bártolo, P. Development of novel alginate–based hydrogel films for wound healing applications. *Int. J. Biol. Macromol.* **2013**, *52*, 221–230. [CrossRef]
29. Söhretog, D.; Sabuncuog, S.; Harput, U.S. Evaluation of antioxidative, protective effect against H2O2 induced cytotoxicity, and cytotoxic activities of three different Quercus species. *Food Chem. Toxicol.* **2012**, *50*, 141–146. [CrossRef]
30. Lee, B.S.; Yuan, X.; Xu, Q.; McLafferty, F.S.; Petersen, B.A.; Collette, J.C.; Black, K.L.; Yu, J.S. Preparation and characterization of antioxidant nanospheres from multiplea-lipoic acid-containing compounds. *Bioorg. Med. Chem. Lett.* **2009**, *19*, 1678–1681. [CrossRef]
31. Weerakody, R.; Fagan, P.; Kosaraju, S.L. Chitosan microspheres for encapsulation of alpha-lipoic acid. *Int. J. Pharm.* **2008**, *357*, 213–218. [CrossRef] [PubMed]
32. Zhou, Y.; Yu, J.; Feng, X.; Li, W.; Wang, Y.; Jin, H.; Huang, H.; Fan, D. Reduction-responsive core-crosslinked micelles based on a glycol chitosan–lipoic acid conjugate for triggered release of doxorubicin. *RSC Adv.* **2016**, *6*, 31391–31400. [CrossRef]

33. Mândru, M.; Vlad, S.; Ciobanu, C.; Lebrun, L.; Popa, M. Polyurethane-hydroxypropyl cellulose membranes for sustained release of nystatin. *Cellul. Chem. Technol.* **2013**, *47*, 5–12.
34. Sivakumar, V.; Ilanhtiraiyan, S.; Ilayaraja, K.; Ashly, A.; Hariharan, S. Influence of ultrasound on Avaram bark (Cassia auriculata) tannin extraction and tanning. *Chem. Eng. Res. Des.* **2014**, *92*, 1827–1833. [CrossRef]
35. Anghel, N.; Lazar, S.; Ciubotariu, B.-I.; Verestiuc, L.; Spiridon, I. New cellulose-based materials as transdermal transfer systems for bioactive substances. *Cellul. Chem. Technol.* **2019**, *53*, 879–884. [CrossRef]
36. Suflet, D.M.; Pelin, I.M.; Dinu, M.V.; Lupu, M.; Popescu, I. Hydrogels based on monobasic curdlan phosphate for biomedical applications. *Cellul. Chem. Technol.* **2019**, *53*, 897–906. [CrossRef]
37. Korsmeyer, R.W.; Lustig, S.R.; Peppas, N.A. Solute and penetrant diffusion in swellable polymers. I. Mathematical modeling. *J. Polym. Sci. Part B Polym. Phys.* **1986**, *24*, 395–408. [CrossRef]
38. Ritger, P.L.; Peppas, N.A. A simple equation for description of solute release II. Fickian and anomalous release from swellable devices. *J. Control. Release* **1987**, *5*, 37–42. [CrossRef]
39. Sridhar, K.; Charles, A.L. In vitro antioxidant activity of Kyoho grape extracts in DPPH and ABTS assays: Estimation methods for EC50 using advanced statistical programs. *Food Chem.* **2019**, *275*, 41–49. [CrossRef]
40. Mammela, P.; Savolainen, H.; Lindroos, L.; Kangas, J.; Vartiainen, T. Analysis of oak tannins by liquid chromatography-electrospray ionisation mass spectrometry. *J. Chromatogr. A* **2000**, *891*, 75–83. [CrossRef]
41. Arinaa, M.Z.I.; Harisuna, Y. Effect of extraction temperatures on tannin content and antioxidant activity of Quercus infectoria (Manjakani). *Biocatal. Agric. Biotechnol.* **2019**, *19*, 101104. [CrossRef]
42. Elansary, H.O.; Szopa, A.; Kubica, P.; Ekiert, H.; Mattar, M.A.; Al-Yafrasi, M.A.; El-Ansary, D.O.; El-Abedin, T.K.Z.; Yessoufou, K. Polyphenol Profile and Pharmaceutical Potential of *Quercus* spp. Bark Extracts. *Plants* **2019**, *8*, 486. [CrossRef] [PubMed]
43. Rosa, M.F.; Medeiros, E.S.; Malmonge, J.A.; Gregorski, K.S.; Wood, D.F.; Mattoso, L.H.C.; Imam, S.H. Cellulose nanowhiskers from coconut husk fibers: Effect of preparation conditions on their thermal and morphological behavior. *Carbohyd. Polym.* **2010**, *8*, 83–92. [CrossRef]
44. Poletto, M.; Pistor, V.; Zeni, M.; Attera, A.J. Crystalline properties and decomposition kinetics of cellulose fibers in wood pulp obtained by two pulping processes. *Polym. Degrad. Stabil.* **2011**, *96*, 679–685. [CrossRef]
45. Xu, F.; Yu, J.; Tesso, T.; Dowell, F.; Wang, D. Qualitative and quantitative analysis of lignocellulosic biomass using Infrared Techniques: A mini-review. *Appl. Energ.* **2013**, *104*, 801–809. [CrossRef]
46. Fackler, K.; Stevanic, J.S.; Ters, T.; Hinterstoisser, B.; Schwanninger, M.; Salmén, L. FTIR Imaging Spectroscopy to Localise and Characterise Simultaneous and Selective White-Rot Decay within Sprude Woodcell. *Holzforschung* **2011**, *65*, 411–420. [CrossRef]
47. Bonakdar, S.; Emami, S.H.; Shokrgozar, M.A.; Farhadic, A.; Ahmadi, S.A.H.; Amanzadeh, A. Preparation and characterization of polyvinyl alcohol hydrogels crosslinked by biodegradable polyurethane for tissue engineering of cartilage. *Mater. Sci. Eng. C Mater. Biol. Appl.* **2010**, *30*, 636–643. [CrossRef]
48. Riaz, T.; Zeeshan, R.; Zarif, F.; Ilyas, K.; Muhammad, N.; Safi, S.Z.; Rahim, A.; Rizvi, S.A.A.; Rehman, I.U. FTIR analysis of natural and synthetic collagen. *Appl. Spectrosc. Rev.* **2018**, *53*, 703–746. [CrossRef]
49. Colom, X.; Carrillo, F. Crystallinity changes in lyocell and viscose-type fibres by caustic treatment. *Eur. Polym. J.* **2002**, *38*, 2225–2230. [CrossRef]
50. Yue, Y.; Han, J.; Han, G.; French, A.D.; Qi, Y.; Wu, Q. Cellulose nanofibers reinforced sodium alginate-polyvinyl alcohol hydrogels: Core-shell structure formation and property characterization. *Carbohydr. Polym.* **2016**, *147*, 155–164. [CrossRef]
51. Yue, Y.; Wang, X.; Han, J.; Yu, L.; Chen, J.; Wu, Q.; Jiang, J. Effects of nanocellulose on sodium alginate/polyacrylamide hydrogel: Mechanical properties and adsorption-desorption Capacities. *Carbohydr. Polym.* **2019**, *206*, 289–301. [CrossRef] [PubMed]
52. Raschip, I.E.; Fifere, N.; Varganici, C.D.; Dinu, M.V. Development of antioxidant and antimicrobial xanthan-based cryogels with tuned porous morphology and controlled swelling features. *Int. J. Biol. Macromol.* **2020**, *156*, 608–620. [CrossRef] [PubMed]
53. Kar, G.P.; Biswas, S.; Bose, S. Simultaneous enhancement in mechanical strength, electrical conductivity, and electromagnetic shielding properties in PVDF-ABS blends containing PMMA wrapped multiwall carbon nanotubes. *Phys. Chem. Chem. Phys.* **2015**, *17*, 14856–14865. [CrossRef] [PubMed]
54. Felfel, R.M.; Gideon-Adeniyi, M.J.; Hossain, K.M.Z.; Roberts, G.A.F.; Grant, D.M. Structural, mechanical and swelling characteristics of 3D scaffolds from chitosan-agarose blends. *Carbohydr. Polym.* **2019**, *204*, 59–67. [CrossRef]

55. Kalinoski, R.M.; Shi, J. Hydrogels derived from lignocellulosic compounds: Evaluation of the compositional, structural, mechanical and antimicrobial properties. *Ind. Crop. Prod.* **2019**, *128*, 323–330. [CrossRef]
56. Zhu, X.; Chen, T.; Feng, B.; Weng, J.; Duan, K.; Wang, J.; Lu, X. Biomimetic bacterial cellulose-enhanced double-network hydrogel with excellent mechanical properties applied for the osteochondral defect repair. *ACS Biomater. Sci. Eng.* **2018**, *4*, 3534–3544. [CrossRef]
57. Rudzinski, W.E.; Dave, A.M.; Vaishnav, U.H.; Kumbar, S.G.; Kulkarni, A.R.; Aminabhavi, T.M. Hydrogels as controlled release devices in agriculture. *Des. Monomers Polym.* **2002**, *5*, 39–65. [CrossRef]
58. Missio, A.L.; Mattos, B.D.; Ferreira, D.; Magalhaes, W.L.E.; Bertuol, D.A.; Gatto, D.A.; Petutschnigg, A.; Tondi, G. Nanocellulose-tannin films: From trees to sustainable active packaging. *J. Clean Prod.* **2018**, *184*, 143–151. [CrossRef]
59. Zhang, Y.; Chan, J.W.; Moretti, A.; Uhrich, K.E. Designing polymers with sugar-based advantages for bioactive delivery applications. *J. Control. Release* **2015**, *219*, 355–368. [CrossRef]
60. Hauptstein, S.; Bernkop-Schnurch, A. Thiomers and thiomer-based nanoparticles in protein and DNA drug delivery. *Expert Opin. Drug Deliv.* **2012**, *9*, 1069–1081. [CrossRef]
61. Bernkop-Schnürch, A. Thiomers: A new generation of mucoadhesive polymers. *Adv. Drug Deliv. Rev.* **2005**, *57*, 1569–1582. [CrossRef] [PubMed]
62. Mandapalli, P.K.; Venuganti, V.V.K. Layer-by-layer microcapsules for pH-controlled delivery of small molecules. *J. Pharm. Investig.* **2015**, *45*, 131–141. [CrossRef]
63. Ma, M.; Dong, S.; Hussain, M.; Zhou, W. Effects of addition of condensed tannin on the structure and properties of silk fibroin film. *Polym. Int.* **2017**, *66*, 151–159. [CrossRef]

© 2020 by the authors. Licensee MDPI, Basel, Switzerland. This article is an open access article distributed under the terms and conditions of the Creative Commons Attribution (CC BY) license (http://creativecommons.org/licenses/by/4.0/).

MDPI
St. Alban-Anlage 66
4052 Basel
Switzerland
Tel. +41 61 683 77 34
Fax +41 61 302 89 18
www.mdpi.com

Polymers Editorial Office
E-mail: polymers@mdpi.com
www.mdpi.com/journal/polymers